33

W9-CSN-684

Genetic Variation and Disorders
in Peoples of African Origin

The Johns Hopkins Series
in Contemporary Medicine and Public Health

Consulting Editors:

Martin D. Abeloff, M.D.
Samuel H. Boyer IV, M.D.
Gareth M. Green, M.D.
Richard T. Johnson, M.D.
Paul R. McHugh, M.D.
Edmond A. Murphy, M.D.
Edyth H. Schoenrich, M.D., M.P.H.
Jerry L. Spivak, M.D.
Barbara H. Starfield, M.D., M.P.H.

Genetic Variation and Disorders in Peoples of African Origin

JAMES E. BOWMAN, M.D.

*Professor, Departments of Pathology and Medicine, Committee on Genetics,
and Committee on African and African-American Studies,
University of Chicago*

AND

ROBERT F. MURRAY, JR., M.D., M.S.

*Professor of Pediatrics and Medicine,
Department of Pediatrics and Child Health,
and Chairman, Department of Genetics and Human Genetics,
Howard University*

The Johns Hopkins University Press
Baltimore and London

© 1990 The Johns Hopkins University Press
All rights reserved
Printed in the United States of America

The Johns Hopkins University Press
701 West 40th Street
Baltimore Maryland 21211

The Johns Hopkins Press Ltd., London

∞The paper used in this book meets the minimum requirements
of American National Standard for Information Sciences—Permanence
of Paper for Printed Library Materials, ANSI Z39.48-1984.

Library of Congress Cataloging-in-Publication Data

Bowman, James E.
   Genetic variation and disorders in peoples of African origin / James E. Bowman
and Robert F. Murray, Jr.
      p. cm.—(The Johns Hopkins series in contemporary medicine and public
health)
   Includes biographical references.
   ISBN 0-8018-3962-9 (alk. paper)
   1. Genetic disorders.   2. Blacks—Health.   3. Black race.
4. Human genetics—Variation.   5. Variation (Genetics).   I. Title.
II. Series.
   [DNLM: 1. Hereditary Diseases—ethnology.   2. Hereditary Diseases—
genetics.   3. Negroid Race—genetics.   QZ 50 B787g]
RB155.5.B69   1990
616'.042'08996—dc20
DNLM/DLC   90-4014 CIP
for Library of Congress

B
5.5
369
90

To our families

CONTENTS

PREFACE

The initial name for this book was *Genetic Variation in Blacks*, but this title would have been misleading. It was discarded because it excludes the Coloureds of South Africa and some peoples of the Middle East and North, Central, and South America who do not call themselves black but who are descendants of Africans. Within recent memory, blacks in the United States were referred to as colored, and later, as Negro, African-American, and an assortment of unmentionable pejorative names. No matter what African-Americans may be called in the future, they will remain peoples of African origin—at least in terms of some of their ancestors. Many African-Americans are also equally peoples of European, Asian, or Amerindian origin, or of multiple origins; to say otherwise would lend credence to an anachronistic social definition of race and ignores biological reality. But more about this point in the next chapter.

We finally selected the title *Genetic Variation and Disorders in Peoples of African Origin*. (Of course, since *Homo sapiens* probably originated in Africa, we are all peoples of African origin, but such an assertion is pedantic.) Peoples of African origin include not only Africans south of the Sahara but also North and East Africans such as Moroccans, Algerians, Libyans, Egyptians, Ethiopians, and Somalis. The ancestors of these peoples were indigenous Africans—no matter what the external gene contribution. The Egyptians, for example, are now called Middle Easterners. But, the Middle East is an artificial geographical term that was developed by Winston Churchill during World War II to designate a region of strategic importance to the war. Historians and population geneticists should not necessarily be influenced by popular remnants of international hegemony. Egypt is geographically a part of the African continent. We exclude the Europeans of Zimbabwe, South Africa, and Europeans who live in other countries in Africa. Asian Indians, Chinese, and other recent—since the fifteenth century A.D.—immigrants are also excepted.

We will avoid, whenever possible, some widely used words and phrases that we believe are either pejorative, inaccurate, or an anathema to many peoples. Such words include *tribe, miscegenation, barbarians, savages, civilized people, advanced society, underdeveloped country, developing country, New World, Old World*—to name a few. We will explain our objections to some of these words here; some are obvious; and we will explain our reservations about other words later. Even though the word *race* is odious to many—and some scholars deny that there is such an entity—we did not place *race* on our list. Unfortunately, the word *race* is unavoidable—particularly when we must quote the work of others. Most of our references use words such as *black*, *Negroid*, *white*, *Caucasoid*, and *Mongoloid*, all of which admittedly are imprecise. For example, *Caucasoid* includes such genetically diverse peoples as the English and Asian Indians. Neither the ancestors of the English nor the ancestors of the Asiatic Indians came from the Caucasus Mountains of the Soviet Union. On the other hand, no one classification will be generally acceptable. Populations, or ethnic groups, or even races—like their members—are born; they live; they change; they die. *Africans* is also a diffuse categorization, for it includes peoples who are more dissimilar than are Europeans. We prefer terms such as *Europeans*, *English*, *Scots*, *Irish*, *European-Americans*, *African-Americans*, *Nigerians*, *Kenyans*, but most of these groups can also be divided into markedly dissimilar and unrelated peoples, and they can be split ad infinitum. Accordingly, here, we must name peoples as the authors called them—or as they call themselves—with our sincere apologies to all who may be offended.

The word *tribe* is common in writings about certain groups in Asia, Africa, and in the Americas. The Ashanti, the Ga, the Ibo, and the Hausa are not tribes, but they are often referred to as such, even by African scholars. We believe that it is incongruous to refer to peoples whose ancestors were once part of nation empires as tribes, and though many Europeans are descendants of tribes, this word is never used for present-day European groups. (Some white South Africans have recently referred to themselves as a white African tribe, but this is a political maneuver developed by some white South Africans in the hope that such a term will lend credence to their hegemony in South Africa.) Even so, there are situations where *tribe* may be an appropriate term, but it will be used sparingly here.

There is no doubt, however, that *New World* is a notably specious phrase: it ignores geological evidence for the formation of the continents from a common land mass; it perpetuates the myth that the Americas were "discovered" by Europeans; it conceals historical, archaeological, and anthropological evidence that Africans and other peoples arrived in the Americas long before 1492 (see Chapter 1); and it is an affront to Native Americans (Amerindians), who arrived in the Americas at least 30,000 years ago and who developed civilizations that antedated many in Europe.

Why did we dare write a book about genetic variation and disorders in peoples of African origin? Some of our colleagues have taken us to task for this attempt. Did we not know that eugenics and genetics have long been used as scientific weapons against peoples of African origin, and that authorities in education, psychology, sociology, history, political science, law, philosophy, medicine, biology, anthropology, archaeology, religion, and many other fields defame peoples of African origin, and hide under the almost unassailable shield of scholarship? We will indicate such abuses, where appropriate. But the purpose of this book is not to refute every spurious allegation about Africans and their descendants. This would take many volumes, and our task would never end.

Because much of the literature about Africans and their descendants is so debased, some scholars counter in defense and deny that peoples differ. Obviously we dissent. To ignore genetic variation is to disclaim that a three-year-old child can detect differences between West Africans and Swedes. There should be no dispute about the medical significance of genetic variation and disorders. Physicians and other health workers should know that genetic disorders often differ in frequency among populations and that some genetic disorders vary in severity from population to population. Only an ignoramus would include Tay-Sachs disease, or phenylketonuria, or cystic fibrosis high on the list of a differential diagnosis in an African-American child, but in any case of severe anemia, sickle cell disease would be a part of the differential diagnosis in an African, a Greek, or an Italian child. Genetic diversity is important in medical-legal problems, and evolution would be impossible without genetic variation.

Are some genetic variants and disorders more common in peoples of African origin? Yes. If there were no differences, there would be no need for this book. But, even though some genetic variants and disorders are statistically more frequently found in peoples of African origin, they are usually confined to a minority of the population. Other genetic variants and disorders are more common in Europeans, in European Jews, in Asians, or in populations other than peoples of African origin. Many genetic variants and disorders do not differ among populations. And, even though genetic variation is common, we are all more alike than we are unlike.

Readers will find only passing references to I.Q. in this book, for we do not believe that I.Q.—whatever it measures—is racially determined. Those who wish to pursue this subject may walk into almost any library or book store and find shelves of books on racial differences in I.Q. We will not contribute to the logjam.

We hope that this book will reconcile common prejudices that are frequently associated with biological population variation. Rather, population dissimilarity should be treasured, for if all human populations were analogous, many groups of *Homo sapiens* would have been extinct thousands of years ago—perhaps even some of our direct ancestors.

Our editor, Wendy Harris, firmly and gently fostered this book over many years. If either of us ever writes another book, future editors beware, we will judge you by an impossible standard of excellence and understanding.

James E. Bowman is indebted to the Henry J. Kaiser Family Foundation for a Senior Fellowship at the Center for Advanced Study in the Behavioral Sciences, Stanford. The former director, Gardner Lindzey, the librarian, Margaret Amara, and the staff provided an atmosphere of crucial support during the early phases of the manuscript.

We thank our families for their continued support and understanding in developing this book.

Genetic Variation and Disorders
in Peoples of African Origin

1

African Peoples
and Their Migrations

This historical analysis of the migrations of African peoples is limited to a brief review of the movement of African peoples within Africa, since about the first century A.D., and the dispersal of Africans to countries outside of Africa. Williams (1976) challenged the classical European view of African history, which designates three eras: (1) the first period, the fall of the Roman Empire to A.D. 700; (2) the second period, Islamic civilization, A.D. 700 to the coming of the Europeans in the fifteenth century; and (3) the period of European colonization from 1500 to 1960. This concept of African history discounts more than four thousand years of traceable history before the advent of the Roman Empire. Nevertheless, our purpose is not to summarize the history of Africans but to enable the reader to better understand genetic variation in Africans by identifying the peoples with whom they came in contact in their voluntary and involuntary migrations.

We will begin with what is generally regarded as the most important migration within Africa, the movement of blacks from West Africa (the Zande and the Bantu) to the east to the Lake Victoria region, and then eventually south (Figure 1.1). This migration began in about the first century A.D. and continued over approximately the next two centuries. As these peoples advanced, the Pygmies and the Bushmen (San) retreated in their wake. The Pygmies were able to survive surrounded by their more powerful newcomers, but the San were ultimately displaced to their present homelands in the Kalahari Desert of Southern Africa.

THE PEOPLES OF AFRICA

This section on the peoples of Africa is derived mainly from the excellent review of the genetics and history of Sub-Saharan Africa (Ex-

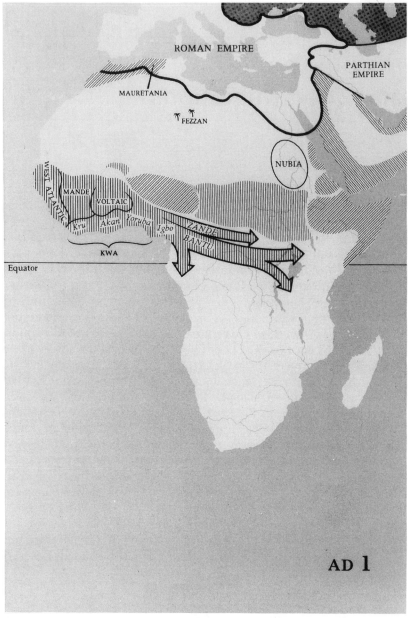

Figure 1.1. Map of A.D. 1
Source: McEvedy 1980, p. 83. Copyright 1980 by © Colin McEvedy

coffier et al., 1987). The authors emphasized that there is no unanimity in the classification of African peoples, and there is no reason why there should be. Formerly, the archaeologists had their heyday. Subsequently, historians, anthropologists, linguists, and now population geneticists and molecular geneticists entered the picture. In this book we use the terms *Africans, black, Sub-Saharan Africans*, to name a few. Nevertheless, these groups can be subdivided ad infinitum or lumped together in various ways. Within all of these groups are populations that differ from one another markedly. In fact, the peoples of Africa or Sub-Saharan Africa are widely dissimilar, more so than Europeans.

Excoffier et al. generally followed the classification of Greenberg (1963), though there are many objections to this division. Even so, this classification is a starting point. Greenberg used historical linguistics as his foundation for the categorization of languages of the African continent, thus linking languages that are supposed to have a common origin.

Africa's four major language families (Niger-Congo, Khoisan, Nilo-Saharan, and Afro-Asiatic) are found in East Africa. Afro-Asiatic, however, is the only family not confined to Africa. In addition to Berber, ancient Egyptian, Cushitic, and Chadic, Afro-Asiatic also includes Semitic, which is widely found in the Middle East and the Arabian peninsula. Excoffier et al. asserted that West Africa is linguistically quite homogeneous; a majority of the languages belong to the Niger-Congo family. In Central and Southern Africa, the Bantu (Benue-Congo)-speaking peoples constitute the largest demographic group of Africa. Even though there is considerable linguistic homogeneity among Bantu speakers, this does not mean that they are genetically, physically, or culturally homogeneous. Less numerous are the Khoisan of Southern Africa. The hunter-gatherer San and the pastoral Khoikhoi can be considered two groups of Khoisan in cultural terms (Excoffier et al., 1987).

THE AFRICAN SLAVE TRADE

The slave trade to the Mediterranean region from Sub-Saharan Africa extended from the early days of the Roman Empire and continued long after the Middle Ages (Davidson, 1980). These slaves were usually highly esteemed and thus did not have the stigma associated with later slavery. Davidson emphasized that East Africa, in contrast to West Africa, had extensive contacts with non-Africans from at least A.D. 10 through slave and other trade with the cities of the Red

Sea, the Persian (Arabian) Gulf, southern Arabia, India, Ceylon, and other countries to the east. By A.D. 800 the Arabs were heavily involved in the slave trade in East Africa. This trade extended to North Africa; the intervening countries in Africa, Europe, the Middle East; and as far east as China. It is evident that there must have been considerable early mixture of African peoples with those of the Middle East. However, no large black minorities are delineated in the Middle East as they are in the Americas (Davidson, 1980). We add, though—as any traveler to the Middle East will testify—peoples of African origin are inextricably mixed with indigenous peoples in this region, but they are not separately identified by external appearance as they are in Europe, the United States, and South Africa.

The Hispanic nations of the Iberian peninsula initiated the slave trade in West Africa in the fifteenth century and were the last to give up this trade (Curtin, 1975). The Dutch, English, Danish, and French dominated the slave trade in the seventeenth and eighteenth centuries, but Brazil and Cuba accounted for the majority of slaves in the nineteenth century. During this period—from the fifteenth to the nineteenth centuries—most of the contact of the Europeans with Africans in Africa was through coastal enclaves that were established with the sanction of local leaders. The majority of the slaves were taken from a 3,000-mile coastal region from Senegal in the north to Angola in the south from lands that were rarely more than several hundred miles inland (Davidson, 1980).

Sources of Slaves from Africa

In discussing the sources of the slaves from Africa, Curtin (1975) described eight coastal regions of shipment of slaves, noting that the interior origins were often unknown (Figure 1.2). The first region is Senegambia, which includes Gambia and Senegal. The second region is usually labeled Sierra Leone, but it extends from the Casamance in the north to Cape Mount in the south, and includes Guinea-Conakry, Guinea-Bissau, a small part of Senegal, and Liberia. The third region includes what the British in the eighteenth century called the Windward Coast. It encompasses the Ivory Coast and Liberia. (Curtin pointed out that this region should not be confused with the nineteenth-century Windward Coast, an area on either side of Sierra Leone.) The fourth region is the Gold Coast, the coastal stretch from Assini on the west to the Volta River on the east; it corresponds generally to the Republic of Ghana. A fifth area is in the Bight of Benin, the land from the Volta to the Benin River (Togo, and the Peoples Republic of Benin). The sixth region is the Bight of Biafra, which is

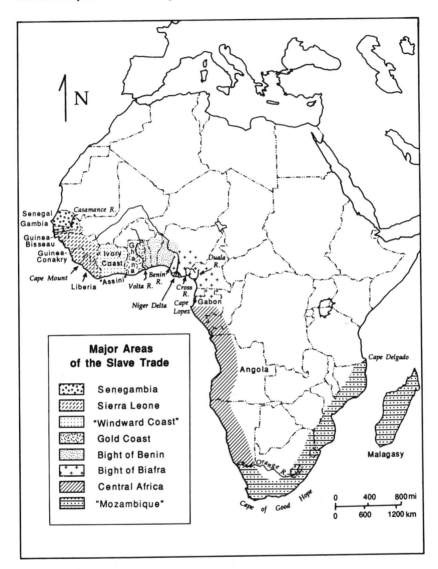

Figure 1.2. Map of Africa illustrating the seven major regions of origin of slaves
Source: Drawn by Brian Urbaszewski

centered on the Niger Delta and the mouths of the Cross and Duala
Rivers to the east. The boundaries were the Benin River on the west,
and Cape Lopez in the south, and it includes Gabon. The seventh
division was called Central Africa. It encompassed a number of trad-
ing regions from Cape Lopez south to the Orange River: Angola now
contains this region. The final region, the eighth, is the coast of south-

eastern Africa from the Cape of Good Hope to Cape Delgado, including the Island of Madagascar (present Malagasy). The word *Mozambique* was used for this region, but few slaves were ever sold from the region of present-day Mozambique.

Regional preferences for African slaves could be significant in any evaluation of genetic variation among blacks in the Americas, or in any other area where slaves were imported. A few examples will be given; for further details, consult Curtin (1975).

Slaves in Europe and the Americas

About 50,000 slaves were imported into Europe during the whole course of the slave trade. Moreover, about 25,000 African slaves were sent to the Atlantic Islands, and about 100,000 were exported to the Portuguese territory of São Tomé, giving a total of 175,000. Many estimates have been made of the number of slaves that were exported to the Americas using the logs of slave ships, documents from slave markets, diaries, and other historical sources. Curtin (1975) concluded that the total number of slaves was from 12 to 15 million between 1451 and 1870, with an error of about 20 percent. Several million died on their way to the ports; and the losses on the Atlantic passage may have been from 10 to 15 percent.

Curtin compared the estimated imports of African slaves to the Americas with censuses of populations of partly or entirely African descent in the year 1950 (Table 1.1). Although the United States and Canada had about one-third of the total African-descended population in North America in 1950, this region had imported less than 5 percent of the slaves. On the other hand, the Caribbean Islands had more than 40 percent of the slave imports, but in the 1950s had only 20 percent of the African-American population. Cuba received 7 percent of the slave trade, but in 1953, the mulatto and black population was 1.5 million, or about 3 percent of the African-American population of the United States. At the end of the slave trade to Saint Dominique in 1791, the number of slaves imported there was more than twice that brought to the United States, but the surviving population was only 480,000, compared with a U.S. slave population of about 4.5 million at the time of the Emancipation Proclamation in 1861.

Curtin analyzed the possible cause of the paradox between the number of imported slaves and present-day population sizes. Although no definite conclusions were drawn, he postulated that if a measure of well-being is the ability to survive and multiply, the slaves in the United States may have been healthier and lived in a more favorable environment. If so, this thesis challenges the general belief

Table 1.1

Relation of Importation of Slaves to Population of African Descent, c. 1950

Region	Slave Imports		Population	
	No.	%	No.	%
United States and Canada	*427*	*4.5*	*14,916*	*31.1*
Middle America	*224*	*2.4*	*342*	*0.7*
Mexico			120	0.3
Central America			222	0.5
Caribbean Islands	*4,040*	*43.0*	*9,594*	*20.0*
South America	*4,700*	*50.0*	*23,106*	*48.2*
Surinam and Guyana			286	0.6
French Guiana			23	<0.05
Brazil			17,529	36.6
Argentina, Uruguay				
Paraguay, Bolivia			97	0.2
Chile			4	<0.05
Peru			110	0.2
Colombia, Panama, Ecuador			3,437	7.2
Venezuela			1,620	3.4
Total	9,391	100.0	47,958	100.0

Source: Curtin, 1975

Note: Totals and percentages are from regions in italics. Numbers should be multiplied by 1,000.

that slavery in Latin America was milder than that found in the Dutch or English colonies. Tropical diseases could have been a factor in differential survival: they were not as prevalent in the United States as in the tropical parts of the Americas. On the other hand, many tropical diseases were rampant in the southern United States until at least the early twentieth century. Malaria, for example, was not eradicated in the United States until just before World War II. But all of this is speculation. For example, the reexportation of slaves to the United States could have been more common than recorded. Immigration from the Caribbean to the United States and Canada has also been extensive.

Many slave buyers characterized African cultures according to stereotypes. For example, various ethnic groups were identified as industrious, faithful, or tending to revolt or run away. Slaves were frequently off-loaded in the Caribbean before entering the United States. Predictably, large markets, such as Jamaica, were not finicky, for if they were too selective, they could not meet the demand. On the other hand, small colonies, such as British Guiana, could afford to be picky. Choices of slaves often followed the prevailing fad. Virginians had no ethnic favorites, but the South Carolina owners did have strong ethnic preferences: they favored slaves from Senegambia: the

Bambara and Malinke were most common there. South Carolinians would also buy slaves from the Gold Coast (Ghana), but they rejected slaves from the Bight of Biafra, who were viewed as too independent, uncontrollable. Accordingly, the origins of slaves from South Carolina differed considerably from the other areas of the British trade.

Exports—transfer of slaves from one market to another—determined the ethnic makeup in other regions of the slaveholding South. Since South Carolina was the source of slaves for North Carolina and Georgia, the ethnic origins of slaves in these regions reflected the South Carolinian predilections. Slaves for other mainland plantations usually came from the Chesapeake Bay ports.

Africans in Canada had various origins. Hill (1981), has written a history of blacks in Canada: the first black resident was a six-year-old child who arrived in 1628 from Madagascar to what was then New France. Even though the law of France forbade slavery, special dispensation was given to Canada on a limited basis by Louis XIV in 1689 to facilitate agriculture. Full permission for slavery was granted by Louis XIV in 1709. In 1713, the French territory of Acadia was granted to the British by the Treaty of Utrecht. Subsequently, settlers from New England moved to the area, which was renamed Nova Scotia. Hill surmised that black slaves were introduced to this area at that time, because slaves helped to build Halifax in 1749. The slaves were usually skilled craftspeople, and they were resold to the American colonies when they were no longer needed. The British continued slavery when they conquered New France in 1760. When the American Revolutionary War broke out, the British offered to free slaves who would join their troops, even though they had supported slavery in New France. Naturally, numerous slaves moved north, and many were taken to Canada at the end of the war in British ships, although many slaves who moved north with their owners remained slaves.

Slavery was abolished de facto in Lower Canada in 1803 in a decision of the chief justice of Lower Canada that slavery was inconsistent with British law. Even though this decision did not legally do away with slavery, it was abolished in practice, because Canadian slaveholders no longer had the protection of the courts. Upper Canada soon followed. Slavery gradually diminished to such an extent that few slaves remained in Upper Canada when the British abolished slavery throughout the empire in 1833. This edict stimulated additional migrations of blacks from the United States, and the Underground Railroad was to play a major role in the flight to freedom. No one knows how many blacks escaped to Canada through the Underground Railroad, but Hill gave a conservative figure of about 30,000 between 1800 and 1860. Many blacks died or disappeared along the way; others "passed for whites" after they reached Canada.

Slaves were not the only source of blacks in Canada; there were many free black settlers from the United States. Many blacks migrated to Canada, even though the climate, particularly in Nova Scotia, was severe. Accordingly, numerous blacks took a British governmental offer of transport to Sierra Leone where a colony of free blacks was to be established. Others resettled in other provinces of Canada.

It is difficult to determine the number of blacks in Canada today, because racial classifications are not used in census forms. Apparently, some black groups are now attempting to introduce racial categories on census forms to facilitate the allocation of funds for economic and social programs that would benefit blacks. In a similar move, some Latin American and Caribbean populations—who formerly did not wish to be classified as black—now call themselves black. Thus, even though racial classification is rejected by many scientific scholars, political and economic considerations often necessitate the retention of racial classification by descendants of Africans in the United States and in other countries in the Americas.

During the slaveholding era, many slaves were exported directly from Africa to the Spanish Americas; others were reexported from British colonies. Close to 1,552,000 Africans were sent to the region, and nearly 3,646,800 slaves were exported to Brazil. The French slave trade could not meet the demands of the French islands in The Western Hemisphere. The shortage was made up illegally by buying slaves from other sources, such as the markets in North America, and those of the British, Danish, and Dutch slave traders. It is estimated that between 690,000 and 720,000 slaves were imported into Saint Dominique alone. About 500,000 slaves were shipped to Dutch America, including present-day Guyana. Only 28,000 slaves were sent to the Virgin Islands (Curtin, 1975).

Harris (1974) points out that the peoples of Latin America are frequently composed of mixtures of white, black, and Amerindian populations, in about equal proportions, except in Mexico, where the black contribution may be lower. Some countries do not follow this broad generalization, as we shall see.

Variable mixtures of Africans and Europeans are found in the semitropical lowlands of Central and South America. Amerindians are found mainly in the highlands of Central and South America, and here there are also populations with European and Amerindian ancestors. Farther south—in nontropical Latin America—Europeans eliminated the indigenous Amerindians, and Africans were not imported. Thus, the populations of Argentina are almost completely of European origin.

Before the Europeans came, the lowland Amerindians were agriculturalists who practiced slash-and-burn agriculture in scattered small groups. These groups migrated periodically. Highland Amerindians, however, had dense, fixed populations and developed elaborate architectural monuments in Mexico, Ecuador, Peru, and Bolivia. The highland Amerindians never conquered the lowland Amerindians, who could easily scatter into the forests. Later, Europeans were also unable to conquer and enslave lowland Amerindians, for the Amerindians disappeared into the jungle, just as they had for centuries against highland peoples. The Europeans invaded the Americas and began to develop the land. Africans were imported to fill the labor void left by the elusive lowland Amerindians.

On the other hand, the highland Amerindians were stable and had already been organized by the Incas. Once the Incas had been either eliminated or converted, the indigenous Amerindians worked the land as before, which meant that the Europeans did not import Africans into the highlands. This accounts, in part, for the scarcity of Africans in the highlands of South America today.

Disease also played a role in the early substitution of highland Amerindian labor for African labor. Amerindians were decimated by the diseases of Europeans—particularly measles, smallpox, respiratory infections, and malaria. It is believed that malaria may have been introduced into the Americas by Columbus's men on their second voyage in 1493. But, as we shall see, malaria could have been introduced before Columbus by Africans, other Europeans, or Middle Easterners—or all of the foregoing. Still considerable evidence suggests that malaria is not ancient in the Americas.

In the first place, the Amerindians feared malaria, and they had no remedies for it (Effertz, 1909). As a general rule, if peoples do not have remedies for diseases that are common in their region, it suggests that the disease has been recently introduced. It matters not whether or not a prescribed treatment is effective, for even in modern societies many disorders are treated by ineffective medication or by nonsensical surgery. Second, the Amerindians were decimated by malaria (Effertz, 1909)—further evidence of the lack of long-standing contact with the disease. Third, malaria is found in many animals other than humans in Africa and in Asia (Dunn, 1965), such as anthropoid apes, cercopithecoid monkeys, Malagasy lemurs, rodents, porcupines, artiodactyls, bats, and flying lemurs. On the other hand, malaria is found only in humans and in the cebid monkey in the Americas. Two types of malaria are found in this monkey: *Plasmodium brasilienum* and *Plasmodium simium*. *Plasmodium brasilienum* is morphologically and clinically identical to *Plasmodium malariae* (a human malaria), and *Plasmodium simium* is morphologically and clinically identical to *Plasmodium vivax* (a human malaria). Unless parallel and convergent evolution has occurred twice—which is highly improbable—malaria in the cebid monkey is of human origin. Even though some evidence indicates that malaria in the Americas is not as ancient as that in other parts of the world, it may have been present long before the arrival of Columbus.

Pre-Columbian Africans in Latin America and the Caribbean

Von Wuthenau (1975), Sertima (1976, 1983), and Bradley (1981) maintained that Africans and Middle Easterners arrived in the Americas long before Columbus and his men. It is commonly known that other Europeans (the Vikings, for example) arrived in the Americas long before Columbus, but the historical, archaeological, and anthropological evidence should eventually supplant the mythology that these were the only people who antedated Columbus.

Von Wuthenau (1975) asserted: "In all parts of Mexico from Campeche in the east to the south coast of Guerrero, and from Chiapas, next to the Guatemalan border, to the Panuco River in the Huasteca region (north of Veracruz), archeological pieces representing Negro or Negroid people have been found, especially in Archaic or pre-Classic sites. This also holds true for large sections of Mesoamerica and far into South America—Panama, Columbia, Ecuador, and Peru" (p. 58).

Sertima noted that on Columbus's second voyage to Española (Haiti and the Dominican Republic), the Indians of Española "said that there had come to Española black people who have the tops of their spears made of a metal which they call gua-nin." Columbus had forwarded "samples to the Sovereigns to have them assayed, when it was found that of 32 parts, 18 were gold, 6 of silver and 8 of copper" (p. 11). Sertima traced the origin of the word *guanin* to the Mande languages of West Africa.

These spears stimulated Columbus to find another route to the Americas from the Guinea Coast (in Africa), which he later did in 1498. This voyage ended in the Americas on the island of what is now Trinidad. Columbus also came within sight of the mainland of South America, but for some reason he did not land. Some of Columbus's men did reach the shore and brought back symmetrically woven cotton cloth worked in colors like those found in Guinea and Sierra Leone.

Sertima related that in 1513 Balboa and his men explored the

Figure 1.3. Precolumbian art from Mexico. Olmec head, San Lorenzo V, approximately 1100 B.C.; height 1.86 meters
Source: Von Wuthenau, 1975. Copyright © 1975 by Alexander Von Wuthenau. Reprinted by permission of Crown Publishers, Inc.

isthmus of Darien (a narrow neck of land between the two Americas). They came upon an Indian settlement in which there were a number of African captives. The Spaniards believed that pirates from Ethiopia had established themselves after the wreck of their ships. (In those days, "Ethiopia" was used as a general term for Africa.) Africans were also encountered in an island off Cartagena, Colombia, in the early sixteenth century. Many other contacts were made with Africans by the early Spanish explorers.

Figure 1.4. Precolumbian art from Mexico. Boy with African features holding up altar with trophy heads; height 30 centimeters

Source: Von Wuthenau, 1975. Copyright © 1975 by Alexander Von Wuthenau. Reprinted by permission of Crown Publishers, Inc.

Archeological evidence supports these contacts. In 1862 a huge granite head of an African was found in the Canton of Tuxtla near a place where the most ancient of pre-Columbian statues were discovered. Modern methods of dating show that some of the African stone heads found among Olmec artifacts and in other parts of Mexico and Central America date from as early as 800 to 700 B.C. (See Figures 1.3 to 1.5.) Sertima and other scholars have concluded that "there cannot now be any question but that there were visitors to the New World from the Old in historic or even in prehistoric time before 1492" (Sertima, 1976, p. 25).

African heads were found in the Mexican state of Tabasco that date from about 800 B.C. Huge African heads and Egyptian bas-reliefs were also discovered at San Lorenzo in Vera Cruz. In collecting this evidence of early Africans in the Americas, Sertima also described artifacts, oral traditions, and botanical, linguistic, and cultural data.

Further evidence that foreigners arrived in the Americas long before the advent of Columbus is provided by the enigmatic legend and the representations of Quetzalcoatl in Mexico. Carrasco (1984) recounts the legend: "Quetzalcoatl arrives in Tlapallan and disappears

Figure 1.5. Precolumbian art from Mexico. African stone head from Veracruz, Mexico. Classic. American Museum of Natural History, New York, NY

Source: Von Wuthenau, 1975. Copyright © 1975 by Alexander Von Wuthenau. Reprinted by permission of Crown Publishers, Inc.

over the water telling his people not to mourn too much for he will return and that a bearded people will eventually rule the land" (p. 30). The beards are important: Amerindians do not have beards unless there has been admixture with non-Amerindians. When one of us (J.E.B.) was puzzled by the bearded figures of Chichen Itza in Mexico in 1964, he was told of the legend and shown many pre-Columbian figures in private collections which resembled African figurines. It has been suggested that the unchallenged landing of Cortez was facilitated by the legend of the promised return of the god Quetzalcoatl and his prediction about bearded people.

Nevertheless, inexplicably, both Sertima and von Wuthenau assert that Professor de Garay's studies among the Lacandon Mayans of Chiapas, Mexico, showed sickle hemoglobin, thus inferring a possible African influence in this group. But this claim is erroneous. One of the authors (J.E.B.) was a member of de Garay's several expeditions to the Lacandones and performed the blood genetic studies in Chicago. Sickle hemoglobin and G-6-PD deficiency were both absent, and Rh blood group analyses were not consistent with African ancestry. AK^2, was absent (see Chapter 5), as it is in West Africans, but this allele is also rare to absent in Southeast Asians (Bowman et al., 1967). Malaria (and perhaps other diseases of Europe, Africa, and Asia) may have been introduced in the Americas many hundreds of years before Columbus. Future studies of African peoples in Latin America should take these archaeological, historical, and anthropological findings into consideration.

PROBLEMS OF RACIAL CLASSIFICATIONS

Studies of migrant Africans and their descendants are complicated by inconsistencies in the classifications of peoples of African origin. Today in the United States children of black-white matings are called black or African-American; several years ago they were called colored or Negro. One black ancestor—at least as far back as a great-grandparent—labels a person black, but even this convention varies from state to state. This custom follows the rule of hypodescent (Harris, 1974) and arises from the determination of the controlling group to maintain social, political, and economic subjugation of the subordinate population. But these social and economic considerations were not important to the early settlers in the American colonies (Bennett, 1984). Whites were identified as Englishmen or Christian. Apparently, the word *white* appeared late in the seventeenth century as a result of slavery and the emigration of blacks. Bennett remarked that the first blacks were called Blackamoors, Moors, Negers, and Negars. *Negro*, a Spanish and Portuguese word for black, did not come into fashion until the latter part of the seventeenth century.

Myrdal (1962) described in poignant detail the vagaries of racial classification of blacks in the United States with its hypodescent preponderance. This section relies on his classic work, with the exception that the term *black* is substituted for *Negro*. Myrdal pointed out that the imported slaves by no means represented pure "races." Many Africans already had an admixture of Caucasoid genes from Mediterranean peoples, and more genes from whites were added during the slave trade. The Portuguese who settled on the Guinea Coast in the early days of the slave trade in the fifteenth century rarely took Portuguese women with them: they were noted for their intermingling with native populations. A large proportion of slaves brought to the mainland Americas were slaves reexported from the West Indies. Many of these slaves did not come directly from Africa, but had already been reexported from Spain and Portugal. Myrdal maintained that the importation of slaves into these regions was in practice by the beginning of the sixteenth century, and by 1539, the importation figures reached 10,000 to 12,000 a year. Some of the offspring of slaves remained in the Iberian peninsula. The slaves who were brought to the West Indies were further mixed with whites and Indians. The slave importation became increasingly large from the West Indies, and after the U.S. Emancipation Proclamation the immigration of blacks was largely from the West Indies.

Extensive intermingling of various African groups occurred. The slaveholders purposefully broke up tribal groups to decrease their unified resistance against slavery. (Of course, intermingling between African groups originated in Africa long before slavery, because of migration, commerce, wars, and an extensive internal slave trade.)

Black and white indentured servants and slaves commingled as well. Amerindians and blacks were often held as slaves together. As the number of black slaves increased, many local Amerindians gradually disappeared into the black population. Myrdal asserted that the relationship of black and white indentured servants had much the same social basis as black and Amerindian relationships. At that time they had a common bond that promoted interracial sex relations.

Most of the intermixture between whites and blacks in the United States was between white men and black women, and as a mulatto generation came into existence, the mulatto women were preferred as sex objects. Along with the stratification of a lower slave status for blacks, various states passed laws against interracial marriage and sex relations, but this did not diminish illegal sex.

Myrdal pointed out that interracial sex was not limited to Southern rural plantations but was prevalent in the Southern cities and in the North, where there were many free blacks. The Civil War was a particularly fruitful period of intermixture. The Northern Army left uncounted Yankee genes in the Southern black population, and in the white population, for that matter.

Myrdal analyzed "passing for white" and considered the practice to be the result of intermixture (we do not use the term *miscegenation*, because today it is considered to be pejorative by some). It is likely that passing has been going on since blacks arrived in the United States. Passing may occur only during certain activities, for example employment or recreation. Passing may be temporary, permanent, voluntary, or involuntary. It may happen with or without the knowledge of the passer. Although many attempts have been made to ascertain the scope of passing, the extent is unknown.

Why should passing by blacks be a source of concern to both blacks and whites? If an individual has both black and white ancestors, why must he or she be considered black? This is a hangover from the subordinate position of blacks. Page (1987) described the case of "a woman without a race." A secretary sued Suffolk, Virginia, officials in 1983, alleging that she was discriminated against because she was black. She lost her case. The same secretary sued the same officials in 1987 and claimed that she was discriminated against because she was white. The woman has green eyes, reddish hair, a very fair complexion. Her immediate family always thought of themselves as white; they knew of no black ancestors. On the other hand, the woman's lawyer surmised that someone in the family was probably black, because both the secretary and her parents were listed as black on their birth certificates. The secretary had been assigned to all-black schools in preintegration days and had lived among blacks as a black person. The secretary believed that she was being passed over for promotion in her city job and so she sued as a black woman. The judge ruled against the secretary because after looking at her and her parents, he stated that he could find no reason why she could call herself black. The secretary obtained another job in Norfolk where all of her supervisors were black. Once more she believed that she was being confined to a dead-end job. This time she sued as a white women. The attorney for the city of Norfolk, however, claimed that the woman could not have it both ways, and cited her earlier arguments. In a newspaper editorial, Page speculated that the secretary probably felt that she was a woman without a race, because in most countries she would be considered white—merely because she looked white. And even in South Africa she would be labeled Coloured or even white. Page argued that the judge's assertion that the secretary is white because she looks white makes practical sense; however, the judge contradicted the age-old American dogma that one drop of black blood makes a person black. Further, Page remarked that other minorities would not encounter the secretary's problem. An Asian or Native American would not be considered Asian or Native American just because that person happened to have one Chinese or Navajo ancestor three or four generations ago. Even the Census Bureau has finally decided that people are Hispanic if they call themselves Hispanic on census forms. Page quoted Alvin Poussaint, a black professor of psychiatry at Harvard, who has commented that black life is determined by white standards, and that black Americans lost something crucial when they lost the ability to define themselves. Page concluded that someday people should be able to call themselves whatever they want.

Williams (1976) quotes a Sumerian legend: "What became of the Black People of Sumer?" the traveler asked the old man, "for ancient records show that the people of Sumer were Black. What happened to them?" "Ah," the old man sighed. "They lost their history, so they died." Williams remarks that scholars in the United States who would classify anyone with one black ancestor as black would classify a North African black with any amount of white blood as white. Thus black Pharaohs became white.

Dominant and subordinate groups are not so visible in Brazil. Accordingly, the rule of hypodescent does not pertain: instead, racial classification is extremely complex. In some classifications, there are more than forty categories. The divisions are so intricate that siblings are often assigned different racial groups. Racial identification criteria include physical appearance, education, wealth, and social class— to name a few. There is an old aphorism in Brazil: "money whitens" or "money bleaches." An individual may be born in one category and die in another. The Brazilian model also prevails in some other countries of South America, Central America, and in some Caribbean countries.

Why does racial classification differ in Brazil and the United States? Harris (1974) estimated that in 1715 in the United States the total white population was 375,000, and there were 60,000 slaves. In Maryland, North Carolina, and South Carolina the ratio was almost 3 to 1 in favor of whites. On the other hand, in Brazil, out of a population of 300,000, 100,000 were Europeans—a ratio opposite to that in the American colonies. In 1819—a century later—out of an estimated 3,618,000 Brazilians, 834,000 (less than 20 percent) were white. In 1820 in the United States, the total population was 9,638,453, and of this number, 7,866,797 (more than 80 percent) were white. Blacks never constituted more than 38 percent of the population of the South in the United States. This proportion declined to less than 25 percent in 1940 in the South and less than 10 percent for the entire country in that year. In Brazil the minority white population was confronted with the options of accommodation with a majority population or apartheid. Accommodation was chosen. Children of Europeans and blacks and children of Europeans and Amerindians were classified into groups intermediate between white on the one hand and black and Amerindian on the other. These groups formed buffers between the ruling Europeans and the Africans and Amerindians who were not mixed.

In the United States, subjugation of the black minority was facilitated because an intermediate (mulatto) group was not recognized. Children born of relations between the dominant white group and the subordinate black group were placed in the black group, and thus remained subordinate. Today, in the United States, blacks perpetuate this racial classification because it is economically and politically salutary. A multigroup classification—like that in Brazil—would attenuate the potential political power of African-Americans.

A brief analysis of the early colonial period in Brazil adds to our understanding of racial divisions in Brazil. This summary is borrowed from Burns (1980). The rulers of Portugal usually sent out on their conquests soldiers, adventurers, and petty criminals condemned to exile. Portuguese women rarely accompanied the men during the first century of colonial Brazil. Consequently, there appeared almost a "new race," the mameluco or cabolco, a blend of European and Amerindian. The Crown was anxious to bring the Amerindians within the empire as christianized subjects and thus was loath to enslave them unless they resisted the Portuguese. Hence, some Indians were enslaved under the subterfuge that they had been taken in a "just war." However, the growth of the sugar industry created a shortage of labor. Enter the African.

Blacks had been imported into Portugal as early as 1433, and the Portuguese were well acquainted with West Africa. The nagging reservations about using Amerindians as slaves were not applied to the Africans, and Africans were well adapted to the labor needed in the Americas. Accordingly, forced migrations of millions of Africans began about 1538 and continued vigorously until at least 1850. The African origins were varied: Guinea, Dahomey (present-day Peoples Republic of Benin), Nigeria, the Gold Coast (Ghana), Cape Verde, São Tomé, Angola, the Congo, and Mozambique, among others.

Burns (1980) suggested three main African contributors to Brazil. First, Sudanese groups, of which the Yoruba and Dahoman predominate. These peoples originated from African regions that later became Liberia, Nigeria, Ghana, and the Peoples Republic of Benin. The Yoruba groups are scattered throughout Brazil, but their descendants appear to be concentrated in Bahia and Maranhao. Muslim Guinea-Sudanese groups comprised the second classification. These black males, of whom the Hausa were the best known, were also found mainly in Bahia. The third contributors, Bantu groups from Angola, the Congo, and Mozambique, were found principally in Rio de Janeiro and Minas Gerais. The Portuguese men took full advantage of the black women and, as a consequence, a mulatto population rapidly appeared. The Amerindians declined in number, but the European and African influx continued.

Fields (1982) quoted a (probably apocryphal) conversation between the late Papa Doc Duvalier of Haiti and a U.S. journalist about the population of Haiti. The interchange points up the vagaries of racial classification. The American asked the president about the proportion of whites in the population of Haiti. Duvalier's answer—to the astonishment of the American—was "98 percent." According to Fields, the reporter thought that Duvalier had misinterpreted his question. He repeated his question and obtained the same answer. The American finally asked Duvalier how he defined white, and Duvalier retorted by asking the American how he defined black. The answer was that anyone with black blood was classified as black in the United States. Duvalier nodded and replied that anyone with any white blood in Haiti was classified as white.

Even though apartheid is prevalent in South Africa, the minority Europeans there have been more flexible in their racial classification than are whites in the United States, for, as in Brazil, whites are a minority in South Africa. The South African categorizations that we will use in this book are white, Coloured, black, San, and Khoikhoi. The classification of the peoples in South Africa differs among the various authorities in the field. Jenkins and Tobias (1977) summarized a 1971 conference on "the Peoples of Southern Africa." More than forty participants from the Republic of South Africa attended. It was pointed out that before this conference a confusing array of names were used for indigenous Africans; namely, Africans, blacks, black and brown South Africans, Negroes, Bantu, Bantus, Bushmen, San, Masarwa, Khoisan, Khoikhoi, and Hottentots. (No mention was made as to whether peoples other than Europeans were invited, or whether any non-Europeans were consulted.) It was decided that peoples who had been called Bushmen should be called San; peoples who had been named Hottentot should be termed Khoikhoi. Apparently, the words *San* and *Khoikhoi* were selected because the respective peoples called themselves such. Oddly, however, it was decided that populations who were linguistically Bantu should be called Negro or South African Negro, even though West, East, and South Africans never refer to themselves as Negro. In fact, they resent this name. Accordingly, we will refer to these peoples as they wish to be known: South African black.

An interesting example of a political and economic justification for racial classification is the categorization of Chinese and Japanese in South Africa. Japanese were formerly classified as Coloured, but when trade with Japan became important to South Africa, the Japanese were reclassified: white. As of 1990, the Chinese were still classified as Coloured.

This analysis has reviewed some of the political, social, and economic bases for classifications of some African peoples in the United States, Latin America, the Caribbean, and South Africa. None of these justifications corresponds to biological or any other reality. Indeed, there *is* no biological reality to classification of humans, for we are all inextricably mixed. This assertion is dramatically substantiated by Finkelstein (1949):

> The usual genealogical table, which mentions only two remote ancestors and many descendants, conceals the interesting fact that an inverted table might be arranged, indicating from how many different individuals we are descended. Theoretically the number of our ancestors in any generation is 2^n where n represents the number of generations back. Thus the number of ancestors of any one of us for 50 generations would be 2^{50}, or 1,125,899,906,842,624. This is a larger number of people than have lived on the face of the globe since its inception: it is obvious that there was considerable interbreeding among the ancestors of modern man, and that probably the whole world is kin. (P. xxvii)

It is evident that categorizations of human populations are capricious. But once the rationales for most classifications are known, and even though they may not be biologically sound, at least they are understandable. Even so, these categorizations lead to the following conundrum: some blacks from the United States could fly to Brazil and be called white; they could then fly to South Africa and be classified as black, Coloured, or white; they could next proceed to North Africa and be considered white or Arab. In this book, we will search for genes of African origin—wherever they may be found.

Anthropometry and Skeletal Variation

A perennial problem in anthropometry is the standardization of measurements. Unfortunately, there was no consensus at a special conference on standardization of anthropometric techniques and terminology that was held in March 1967 at the Aerospace Medical Research Laboratories, Wright Patterson Air Force Base, Ohio (Hertzberg, 1968). A set of ten dimensions was introduced: height, weight, sitting height, acromial height, waist height, gluteal furrow height, dactylion height, chest circumference, waist circumference, hip circumference. Nineteen other dimensions were added to these; we need not bore readers with them here. Hertzberg, for one, objected to the additional measurements because he believed that although they were useful, they were impractical. The techniques recommended by the group may be found in the references given by Hertzberg. Other data were also recommended on each subject (such as age, race, economic status, place of origin) so the anthropometric values could be interpreted.

There was a long discussion on measuring stature. Stature can be measured in approximately four ways: (1) standing erect, but not stretched; (2) standing against a wall, with back flattened and stretched to maximal height; (3) standing away from the wall, with back stretched to maximal height; and (4) recumbent (Hertzberg, 1968, p. 6). Reports on stature should always record the technique of measurement because, as was pointed out, metrical differences between (1) and (2), for example, can average as much as 1 centimeter. Indeed, accurate stretch measurements also depend on a subject's motivation. The body length is greater when lying down than standing, because the longitudinal compression of gravity is absent. Another factor is that body length is usually greater in the morning than at night. Unfortunately, anthropometric data often omit this kind of information. The measurement techniques of some of the various dimensions are shown in Table 2.1.

Table 2.1
Standard Anthropometric Measurements (Recommended Terminology)

Measurement	Technique
Heights	Vertical distances measured from the floor as the subject stands, or sits, or from a horizontal surface on which the subject sits
Breadths	Horizontal, lateral, or transverse dimensions
Depths	Horizontal, anterioposterior, or sagittal dimensions
Lengths	Dimensions along the long axis of a body segment, frequently implying no direction
Reaches	Dimensions along the long axis of the arm. These dimensions may be viewed as a special form of Length, and usually are measurements in the horizontal plane in front of the subject, unless another direction is indicated.
Circumferences	Surface dimensions in a plane around a body segment or area. Although the plane is frequently perpendicular to the long axis of the segment, the segment may occupy any position in space. The plane of the tape may be horizontal, vertical, or at any angle in between.
Curvatures	Surface dimensions between two points on the undulant surface of the body. This terminology replaced "Arcs," which had geometrical implications not necessarily found in anthropometry.
Prominences	Dimensions by which one point on the body protrudes from another: ear prominence and nose prominence were given as examples

Source: Hertzberg (1968), pp. 7–8.

The conference left many unresolved concerns. The minimal list of dimensions for biological description was not determined. Questions remained about the standard location of chest circumference. Location poses no difficulty in males if the level of the nipples is specified. On the other hand, the level of the nipples varies considerably in females, even at a given age. Accordingly, the chest circumference was standardized at the level of the fourth sternocostal articulation, but the preferred location remained open. Even well-established dimensions (such as waist circumference, lower thigh circumference, ankle circumference, hand circumference, hand length) were not resolved. Since new dimension sites continue to be developed, the standardization was recognized as an ongoing process.

Jamison and Zegura (1974) analyzed measurement error from two or more investigators to ascertain anthropometric variance. During the initial research, the two investigators divided the measurement battery so that each took half of the measurements. Two years later one of the investigators repeated the measurements previously taken by the other. Replicate anthropometric measurements were examined on twenty male and twenty-two female Eskimos. For the hand measurements and all of the vertical facial dimensions one of the authors obtained smaller means than the other. The discrepancies in vertical facial dimensions arose because one of the investigators consistently located the nasion lower than the other. The same investigator obtained larger results for vault dimensions, facial breadths, nose breadth, and ear height and breadth. The authors cautioned that it is assumed that errors should be random rather than systematic and therefore the means of each variable will be unaffected even if several investigators independently take measurements. However, this study demonstrated that errors in anthropometry can be systematic. Analysis of variance indicated that twelve of the sixteen variables could be comparably measured by two investigators. The variables with readily defined endpoints yielded the highest correlation, namely, head length, head breadth, minimal frontal breadth, bizygomatic breadth, bigonial breadth, symphyseal height, nose breadth, ear breadth, and hand length (crease).

STATURE

Parent-child correlations for stature and midparent child regression for stature (the average of the parents) were investigated in a cross-sectional sample of 806 Philadelphia children six to twelve years of age (Malina, Mueller, and Holman, 1976). There were 422 black and 384 white children in the sample. The children's statures were measured, but the parental heights were reported by questionnaires. Parent-child correlations for the white sample were consistently higher than those for blacks. These differences were statistically significant for father-daughter ($P < 0.01$), father-child ($P < 0.01$), and all midparent relationships ($P < 0.02$). The heritabilities of stature as given by the midparent child regression were 37 percent for blacks and 49 percent for whites. Absence of assortative mating among blacks and diminished parent-child environmental similarity were likely factors to explain these findings, but greater stepparentage among black fathers could not be excluded.

Assortative mating for stature, weight, and the ponderal and Quetelet's indices were examined in a large sample of Philadelphia blacks and whites (Mueller and Malina, 1977). The husband-wife correlation for stature was significant and positive in whites ($r = 0.34$, d.f. $= 382$), but negligible in blacks ($r = 0.06$, d.f. $= 420$). When the couples were divided into statural mating combinations on the basis of short, medium, and tall, the white spouses' stature showed an approximately linear relationship to one another. The distribution of the black spouses' statures was not completely independent, even though the husband-wife correlations were close to zero. Elements of both positive and negative assortative mating were found among blacks, which resulted in an excess of mating types in which the husband was shorter than the wife—except at the heterogeneous extremes of the bivariate array.

BIRTH WEIGHT

Naylor and Myrianthopoulos (1967) investigated the relation of ethnic and selected socioeconomic factors to human birth weight. The data of the Collaborative Study on Cerebral Palsy, Mental Retardation, and other Neurological and Sensory Disorders of Infancy and Childhood were used. Approximately 60,000 pregnant women were evaluated from the first months of pregnancy through labor and delivery up to the seventh year of the child's life. The study population was about 49 percent black, 42 percent white, 8 percent Puerto Rican, and 1 percent a variety of other ethnic groups. Multiple regression and covariance analyses of 20,000 births suggested that the association of socioeconomic variables to birth weight differed among the three ethnic categories. Family income appeared to be more important in Puerto Ricans than among blacks and whites, and education of the Puerto Rican mother had an opposite effect to that of the whites. The mother's education was not significant in blacks. Foreign birth of the mother, rural birth of the mother, presence of the husband in the household, and low density (number of persons per room) all had positive effects on birth weight in the three ethnic categories. Attempts to account for racial differences in terms of regression on socioeconomic factors failed to remove the possibility that white babies were about 130 grams heavier than black babies. Whether Puerto Rican birth weights approximated white or black birth weights depended on how income was treated. Interestingly, when income was uncorrected for family size, birth weight increased with income, but when income was on a per capita basis, birth weight decreased with rising income. When raw family income was used for adjustment, Puerto Rican birth weights approximated those of black newborns, but Puerto Rican birth weights were similar to those of white newborns when the per capita adjustment was used.

GROWTH

Schutte (1980) investigated growth differences between lower- and middle-income black male adolescents by measuring height, weight, lean body mass (by way of total body water analysis), and total body fat on 203 black U.S. males aged ten to nineteen years. In this sample, 105 boys were from lower income and 98 were from middle-income families. The lower-income boys were significantly shorter and lighter and had significantly smaller lean body weight compared with their middle-income counterparts. Differences in body fat were not observed. Growth differences between income levels were pronounced after eleven years of age, which suggested that the the smaller size of the lower-income group is a result of a delay in the adolescent growth spurt. Schutte postulated that the growth of the lower-income group may reach the level of the middle-income group during late adolescence, but this has not been proved.

BODY SIZE AND FORM

Spurgeon, Meredith, and Meredith (1978) investigated the body size and form of children of predominantly black ancestry living in West and Central Africa, North and South America, and the West Indies. Stature, sitting height, hip width, arm, calf circumferences, and body weight were measured in black children of Richland County, South Carolina. Lower limb height and three indices of body shape were derived from these measurements. The sample size exceeded 200 for each of five age-sex groups, and represented boys and girls aged six years, boys and girls aged nine years, and boys aged eleven years. The data were then compared with those of other populations. The black children in the study were taller than black children studied since 1960 in Angola, Chad, Ghana, Liberia, Nigeria, Senegal, Uganda, Anguilla, Barbados, Cuba, Guyana, Jamaica, Nevis, St. Kitts, St. Vincent, and Surinam. The children of well-to-do black families in Accra and Ibadan were no taller or heavier than black children of Richland County when measured without regard to socioeconomic status. In hip width, the averages of Richland County black children were larger than those for children of the Hutu and Yoruba, and in arm girth they were larger than children of the Hutu and Tutsi. Almost identical hip/lower limb indices were found in the black populations in Africa, Cuba, and the United States.

Spurgeon and Meredith (1979) compared blacks in Richland County, South Carolina, with whites in the same county at age fifteen years. The two ethnic groups, measured during 1974–77, had similar values for arm girth, calf girth, and body weight. When the black youths were compared with white age peers, they were shorter in sitting height, longer in lower limb height, narrower in hip width, longer in lower limb height relative to sitting height, and narrower in hip width relative to lower limb height.

RELATIONSHIP BETWEEN LEG LENGTH AND SITTING HEIGHT

Meredith (1979) disputed the common claim that African-Americans have somewhat longer legs in relation to sitting height than Europeans as a result of interbreeding between Africans and Europeans. Statistics from twenty-two published studies recorded arithmetic means on one or more human age groups for distance from vertex to soles (total height) and distance from vertex to the subischial plane (sitting height). The total height minus sitting height established the mean lower limb height. The skelic index was provided by the following formula: lower limb height times 100 divided by sitting height. The skelic index values at ages eight, nine, thirteen, and fifteen years of Bantu children of the Nyakyusa of Tanzania, youths of the Digo of Kenya, and Sara children of the Republic of Chad all showed higher indices than those for the corresponding age-sex groups of black U.S. children. Skelic indices also were evaluated for adult Africans and African-Americans. The values for African-Americans were all below 100, but ten of those for African men were 100 or higher. With one exception, the mean lower limb height of African-American women did not exceed 95 percent of sitting height. Among eleven groups of African women, the mean lower limb height was 95 percent of mean sitting height.

BODY FAT

Densitometric and anthropometric studies were analyzed on a biracial sample of 242 subjects aged six to sixteen years in Bogalusa, Louisiana (Harsha, Frerichs, and Berenson, 1978). The black children had less body fat than white children, and boys had less fat than girls. The median densities were 1.060, 1.049, 1.044, and 1.035 grams per milliliter for black boys, white boys, black girls, and white girls, respectively. The authors concluded that each race and sex group required different standards to estimate body composition parameters.

Skinfold measurements were taken on six standard body sites in a stratified random sample of 278 children aged seven to fifteen years in a biracial community in southern Louisiana (Harsha, Voors, and Berenson, 1980). White children had thicker skinfolds than black children for the same body weight with the exception that the subscapular skinfold was relatively thicker in blacks. The investigators postulated that the racial difference in fat distribution may indicate a genetic adaptive trait under circumstances demanding both a caloric reserve and facilitation of convective heat loss in tropical climates and that peoples whose ancestors are derived from northerly populations have thicker skin-fat folds.

SKELETAL VARIATION

Skeletal variation is difficult to evaluate. First, the literature is replete with spurious beliefs. Studies of racial variation of the skull and associated brain size are particularly biased, but we will not discuss much of this literature, for Gould (1978, 1981), among others, has done a masterful job. We will, however, review one of the most important papers on the subject—that of Tobias (1970). Second, it is difficult, if not impossible, to study genetics without family studies. Bone specimens of families are rarely available; thus deductions frequently must be made from single specimens. Third, population variation in skeletal material may be reported under the assumption that the variation is genetic—a frequently erroneous conclusion. Even though family studies of skeletal material are often impossible, X-rays have been used in some investigations. Today, however, radiological surveys of children would probably be rejected by committees that oversee ethics in clinical investigations of human subjects. Nevertheless, with these and other reservations, nonmetrical variation is common, and often has considerable medical consequences in addition to anthropological significance.

Nonmetric and Metric Traits

Corruccini (1974) examined the importance of nonmetric traits in population studies by studying 76 discrete variants in 321 human skulls from the Terry Collection at the Smithsonian Institution. He observed that discontinuous variant frequency distributions and discrete traits in isolation are not of paramount value in genetic studies, but may be valuable in analyzing the population genetics of extinct groups in conjunction with other types of data.

Carpenter (1979) made a comparative study of metric and nonmetric traits in 317 crania from the same collection. Twelve metric variables and fifteen nonmetric variables were scored. The metric variables were found to be significant sex and race discriminators, but the nonmetric variables were not. The nonmetric variables were found to be better age discriminators than the metric variables. Carpenter concluded that nonmetric traits by themselves have very little discriminatory value and should be used only to supplement other osteological measurements and observations.

On the other hand, nonmetrical variation in estimating human population admixture may not be useful in making quantitative conclusions about genetic relationships (Wijsman and Neves, 1986). To test the utility of discrete cranial traits for estimating genetic distances among populations, estimates of admixture were obtained for gene frequency data and nonmetric data in São Paulo mulattos (M). The gene frequency data served as a control that the three populations are related. Estimates of admixture were obtained by using São Paulo whites (W) and blacks (B) as parental populations, and by estimating the parameter of admixture (m). While the gene frequency data indicated distances among the three populations which are compatible with the linear mode of admixture, the nonmetric data showed significant deviations from the model. This finding indicated that the frequencies of the nonmetric traits in these experiments are not a linear function of genetic distance. Such a conclusion invalidates the use of nonmetric traits in making quantitative conclusions about genetic relationships. The authors suggested that there is a need for the study of other skeletal characters for the estimation of genetic distances, as well as approaches for such investigations through the analysis of hybrid individuals.

Skeletal Maturation

Singer and Kimura (1981) reviewed skeletal maturation of African children and children of African ancestry and studied a group of African children in South Africa. These authors related that black children in Africa and children of African ancestry have a genetic potential for more rapid skeletal development in early life than white children, but later in childhood the skeletal maturity of black children falls behind that of whites. Singer and Kimura analyzed growth of body height and weight and skeletal maturation of Hottentot (Khoikhoi), Rehoboth Baster, and Cape Coloured children. For body height and weight, comparisons were made with U.S. black and white children, Hutu boys, Bantu-speaking Baganda girls in Uganda, and Cape Coloured children. The African-Americans and the whites showed the greatest and the Khoikhoi showed the smallest values for height and weight among these populations; the Rehoboth Basters, Cape Coloureds, and East Africans had values approximately intermediate between them at all ages. The Rehoboth Basters and Cape Coloureds attained almost the same height as the U.S. blacks and whites after the age of eighteen years.

These authors then reviewed other growth indices and pointed out that carpal centers are present more often among African-American neonates than among white newborns. African children are also advanced skeletally when compared to white children who live under better economic circumstances, but black children lose their lead in ossification at about eighteen years. Singer and Kimura's study (1981) showed that on the British standard, the skeletal ages of the Khoikhoi and their related populations were lower at all the recorded ages. Further, the Khoikhoi were always the most skeletally delayed of the three populations in the study.

Comparative Skeletal Mass

Cohn et al. (1977) examined the skeletal mass and radial bone mineral content in black and white women and reviewed the prevalence of osteoporosis in blacks. These authors indicated that osteoporosis is rare in blacks and that there is a low prevalence of osteoporosis among peoples of African origin and that black women are less susceptible to fracture and osteoporosis than are white women. In these investigations, the difference in total bone mass in black and white populations was derived from measurements of bone mass by radiography or biopsy. Cohn and colleagues studied total skeletal mass in the form of total body calcium measured directly by total body neutron activation analysis. Although there was no significant difference in stature, black women had a greater skeletal mass and bone mineral content of the radius than did age-matched white females. When the total body calcium values were corrected for body size, the total body calcium was still higher for the black women but not as high for the total body calcium values. It was suggested that the larger muscle mass in black women is partly a determinant of their increased skeletal mass and partly responsible for their resistance to osteoporosis and fracture of the skeleton.

Skull Thickness, Brain Volume, and Achievement

Craniometry has long been the domain of anthropologists, other scientists, and charlatans. Many who study the human skull have attempted to show that the volume and thickness of skulls of blacks differ from those of whites, with the assumption that skulls with low volumes have small brains and small brains indicate low intelligence. Today, I.Q. testing—with the same abuses—has supplanted craniometry for the measurement of intelligence (Gould, 1981)—but remnants of craniometry remain.

Estimation of the thickness of the skull in living subjects is performed at selected points along the radiogram, but there are no consistent sites of measurement. Adeloye et al. (1975) studied skull thickness in a hospital population of 300 black and 200 white males and females in the United States. Measurements were made at four arbitrarily selected reference points on the lateral skull film 3 centimeters superior and inferior to the lambdoidal suture, respectively. The thickness of the skull increased during the first two decades of life in all groups. A peak was reached in the fourth to sixth decades, after which the figures fluctuated. Females of both races had significantly thicker parietal and occipital bones than did males. The frontal bone was thicker in the white male than in the black male; and the parietooccipital bone was thicker in the black male than in the white male. Adeloye and co-workers examined the old belief in a racial relationship between thickness of the skull and brain weight, the idea that "the best filled skulls have thinner walls." No significant differences were noted between the races.

It has long been believed that blacks have smaller brains than whites. The literature is replete with attempts to prove this point. Since most of the anthropological literature on the subject is written by Europeans, European South Africans, or European-Americans, it would indeed be odd if the results were skewed in favor of blacks. Be that as it may, the paper that finally punctured the myth of "white brain superiority" was written by a white South African, based on a lecture given at the University of Cape Town, South Africa (Tobias, 1970).

The size of a brain may be estimated by weighing it or by determining its volume, but if brain size is estimated by volume of cranial capacity, there is a major drawback: the cranium accommodates much more than brain tissue. The nonbrain tissue proportion of the cranial cavity has been estimated to range from 10 to 30 percent (Tobias, 1970). Further, the brain shrinks with age.

Tobias (1970) discussed bias in reports that claimed that the brain weight of Africans is lower than that of Europeans. For example, Swan (1964) reported that the main cranial volume of Europeans is from 100 to 175 cubic centimeters greater than that of Africans, and that the mean brain weight is about 8–10 percent greater. Tobias pointed out that the mean brain weights of blacks ranged from 1,276 to 1,355 grams in three series of studies, and that the mean brain weights of whites ranged from 1,301 to 1,355 grams in eleven series of investigations. Tobias emphasized that these are ranges of averages, and that the ranges of individual values are wider, with considerable overlap. The black populations that were sampled had averages above and below modern human means, when uncorrected for general body size and other related variables. This is the key to the controversy. Tobias postulated that one would expect people with bigger bodies to have bigger brains and formulated the following syllogisms:

1. Taller people tend to have bigger brains; heavier people tend to be taller; therefore, heavier people appear to have bigger brains.
2. Heavier people, adjusted for height, do not have bigger brains; taller people have bigger brains, irrespective of weight; therefore, brain weight is positively correlated with body height, but not with body weight.

Other variables must also be considered. Brain size varies with sex, age at death, nutritional state in early life, nonnutritional environment in early life, source of sample, occupational group, cause of death, lapse of time after death, temperature after death, anatomical level of severance of the brain at autopsy, presence or absence of cerebrospinal fluid, and presence or absence of meninges and blood vessels. Tobias (1970) concluded that unless corrections have been made for differences in body height in comparisons of mean brain sizes of different populations, valid statements cannot be made on interracial differences. Further, since no studies have made such corrections, all of the reported comparisons are invalid.

As we have indicated, it has been assumed that relatively small brain sizes are incompatible with high intelligence. Yet we know that some humans with normal intelligence have brain sizes three times that of the average of other humans. According to Tobias (1970), brain size is one of the most variable parameters in modern humans. Only body weight and height exceed brain size in the range differences between the highest and lowest. Moreover, no evidence exists for differences in behavior between individuals with large and with small brains. Tobias also could not substantiate that the cerebral cortex of blacks is thinner than that of whites. Tobias concluded:

> The exploding of the myth about brain size and grey matter differences leads to the realisation that, in essence, the truth does not gain accep-

tance by mere repetition of a set of facts. Hypotheses are not confirmed by statement and restatement of the hypotheses. Science requires rather the patient testing of the facts, the repeating of early studies by more modern, better controlled and better standardized methods, the constant re-examination and critical re-appraisal of premises and assumptions, the elasticity of mind which permits, nay demands, old hypotheses to be modified when new facts emerge which cannot be adequately explained by them, resistance to the tendency to develop a vested interest in a particular viewpoint, avoidance of ascribing motives to scientists of opposite viewpoint, in favor of the unbiased examination of the evidence they advance, the eschewing of premature hypotheses based on over-tenuous evidence. It is to these stern and rigorous demands of scientific method . . . that my address is dedicated. (P. 22)

Gould (1978) reviewed the ranking of races based upon cranial capacity and came to the same conclusions as Tobias. Gould pointed out that nineteenth-century intellectuals discoursed endlessly on the subject of racial differences. Gould reexamined the raw data of Samuel George Morton, who had the world's largest collections of skulls—which exceeded 1,000—and found that Morton's summary tables (which placed whites above Amerindians, and blacks on the bottom) were based on "apparently unconscious finagling." When the data were reinterpreted, all races had approximately equal capacities.

Tear Duct Size

Post (1969) discovered that measurements of the external opening of the tear duct (nasolachrimal canal) of skulls of African-American males and white males showed that the openings were larger in African-Americans. In both groups, the openings of the apertures increased with age. A sample of white female skulls confirmed the age distribution: the African-American female skulls had smaller openings than the white female skulls at ages younger than fifty, but older skulls revealed no age differences. Post believed that the racial differences might represent ecological adaptation. He pointed out that dacrocystitis is associated with stenosis of the tear duct and that its rarity among blacks, contrasted with high frequencies among whites, may be explained in part by underlying skull morphology. He further postulated that since chronic dacrocystitis may lead to and augment infection in several other nasal regions, it is likely that tear duct size may have been subject to natural selection before the era of modern therapy.

Facial Features

The association between nasal shape, prognathism, and the shape of the maxillary and dental arch was examined with samples of 167 skulls from Europeans excavated in the north of Holland dated between A.D. 754 and 1580 and 165 skulls from the Republic of Mali, dated from the eleventh to the eighteenth centuries (Glanville, 1969a). Two aspects of the problem of explaining population differences in nasal shape and prognathism were examined. First, if selection has acted directly on the shape of the nose, what is the effect of variation in nasal height and breadth on the degree of prognathism and on other measurements of the skull? Second, if it is assumed that selection has modified primarily the length of the cranial base, the degree of prognathism, and the shape of the maxillary dental arch, what would be the effect of variation in these features on the dimensions of the nose?

A significant interrelationship was noted among the dimensions of the nose, the length of the cranial base, the degree of prognathism, and intercanine distance in both population samples. Variation in one of these parameters was associated with variation in one or more of the others. Granville postulated that interpopulation differences in the size and shape of the external nose are accompanied by differences in prognathism and the shape of the maxillary dental arch and palate. The study, however, did not indicate which, if any, of these features has been a primary source of variation. Granville observed that the morphology of the nose in Africans and Europeans differs in several respects in addition to height and breadth. Prognathism is accompanied by an increasingly broad and short nose. High correlations exist between nasal breadth and the distance between the upper canine teeth. It was suggested that both nasal shape and the maxillary arch prognathism may be subjected to selection by environmental stress.

Cleft Lip and Cleft Palate

A review of genetic studies of close relatives of propositi with cleft lip with or without cleft palate (CL[P]) (Woolf, 1971) postulated that (1) relatives have a significantly increased incidence of CL(P) but not of isolated cleft palate (CP); (2) the close relatives of individuals with CP have a significantly increased incidence of CP but not of CL(P); and (3) when concordance is present in parent-child combinations, the type of anomaly that appears in the parent is usually the type that is found in the child. Woolf further asserted that since the lip develops during the fifth to eighth weeks of gestation and the palatal region about the ninth week, a mechanism that alters the first developmental process may, in turn, alter the second, but the second may be altered independently of the first.

Woolf's review included families of 496 individuals with CL(P). The results were in agreement with reports that CL(P) is more common in males than in females; CLCP is about twice as frequent as CL, and the cleft more often involves the left side in unilateral malformations.

Woolf then tested the thesis that CL(P) is inherited in a polygenic manner. If CL(P) is polygenically inherited, the frequency of CL(P) should be similar in the children of propositi and in the propositi's parents' siblings. Woolf (1964) found that the frequency of affected children of propositi was 4.58 percent. The frequency of affected siblings of parents of the propositi was 4 percent in the 1971 study. The polygenic hypothesis was supported and recessive inheritance was rejected because the frequency of affected parents was less than half that observed for siblings and children. Woolf stated that natural and social selection against infants with CL(P) probably resulted in sample bias.

Polygenic inheritance was delineated from dominant inheritance. In polygenic inheritance, there should be a decrease in the frequency of affected individuals as one proceeds from first-degree, to second-degree, to third-degree relatives; however, in dominant inheritance, the frequency of affected individuals decreases by one-half in each successive group of relatives. The following frequencies were found: first-degree relatives (siblings), 3 percent; second-degree relatives (aunts and uncles), 0.65 percent; third-degree relatives (first cousins), 0.36 percent. Since the population incidence of CL(P) is estimated to be 1.2 per 1,000, the frequencies in these groups of relatives are about thirty-three, five, and three times, respectively, greater than that found in the general population. These results were compatible with polygenic inheritance.

The relationship between CL(P) and CP was then analyzed (Woolf, 1971), with the thesis that if CL(P) and CP are genetically distinct, then the frequency of CP in the relatives of people with CL(P) and the frequency of CL(P) in the relatives of people with CP should occur as would be expected in the general population. Woolf concluded that the overall frequency of CP in the relatives of those affected by CL(P) was similar to that observed in the general population; however, an increased frequency of CP was found in close relatives (first, second, and third degree) of the people who had CL(P).

Krogman (1967) reviewed the literature on the prevalence of CL(P) in populations. The total frequency per 1,000 births is higher in Europeans and European-Americans (0.80–1.66, average, 1.27) than in African-Americans (0.55). The frequency of CL(P) is higher in Japanese than in Europeans.

Teeth and Jaw Features

Goodman (1965) reviewed genetic factors in dentofacial development and suggested that the development of the facial bones and the anatomy, number, and size of the teeth are each determined by interrelated polygenic systems—in other words, genes that appear to affect one system apparently may affect another. These polygenic systems result in continuous variability, which may be viewed as normal variation, but the extremes of normal variation are looked upon as abnormal. Major genes give rise to discontinuous variation.

Evolution of Teeth

Many alterations occurred in the face, jaws, and teeth during the course of evolution from reptiles to mammals (Krogman, 1967). First, the dentition was altered from many sets of teeth (polyphyodent) to two sets of teeth (deciduous and permanent). Second, dentition evolved from all teeth alike (homodent) to different types of teeth (heterodent); namely, incisors (I), canine (C), premolars (P), and molars (M). Third, the timing and sequence of appearance of teeth and other important changes in bony structure evolved. Finally, isomerism—the reduction in number of bone or tooth elements by direct loss or by fusion—developed from mammal-like reptiles' 5-1-4-7/4-1-4-7, to mammals' 3-1-3-4/3-1-3-4, to primates' 2-1-2-3/2-1-2-3, from sixty-six to forty-four to thirty-two teeth in about 200 million years. In this nomenclature, the numbers above the line refer to the maxillary teeth; the numbers below the line refer to the mandibular teeth. From left to right in the primates these are the central and lateral incisors (2), canine (1), first and second premolars (2), and first, second, and third molars (3).

Eruption of the Permanent Teeth

Garn and Lewis (1963) reviewed the shift of the sequence of eruption of the permanent teeth from insectivores through the primates. In the insectivores, the molar teeth appear before the deciduous anterior teeth begin replacement. On the other hand, in Europeans, for example, the deciduous anterior teeth are largely replaced by their permanent incisors before the second and third molar teeth extend the effective size of the dental arch. In the insectivores, the eruption pattern is M1M2M3. . . . The development of the anterior teeth is scattered among the molars in humans; for example, M1 . . . M2M3, or M1 . . . M2 . . . M3. There is considerable variability in the sequence of tooth eruption in humans. The permanent sequence of eruption is not simply M1I1I2P1P2M2M3; it may be (M1I1)I2 (P1CP2M2)M3. (The parentheses enclose sequences within which eruption is variable.) In fact, there are so many other sequences that all of the permanent teeth could be enclosed in parentheses.

De Melo, Freitas, and Salzano (1975) studied the eruption of permanent teeth in 302 white and 904 black children six, nine, and twelve years of age. Differences between the two groups were not large; however, the black subjects were generally more precocious at the beginning of the process, particularly the girls, in both the upper and lower incisors. The dissimilarities disappeared at nine and twelve years of age.

Size of Teeth

Moss, Chase, and Howes (1967) made a comparative odontometric study of the permanent postcanine dentition of U.S. whites and blacks. The mesiodistal and buccolingual dimensions of the crowns were measured along with the total tooth height and crown height, and the individual root lengths and widths. Blacks' tooth crown and root dimensions were significantly different from those of the white population in that they tended to be larger, but only sporadically. Comparative data indicated that the size of the maxillary and mandibular teeth of blacks as well as the shape of the mandibular, but not the maxillary teeth, of U.S. blacks was roughly intermediate between those of U.S. blacks and South African blacks. The authors suggested that the thesis that a "hybrid" population may possess intermediate values of crown index and of crown module is supported by their analysis of Khoikhoi-South African black hybrids with their presumptive parental stocks. Nevertheless, part of this assumption could be questionable: the ancestors of U.S. blacks were mainly from West Africa, not South Africa; thus South African black measurements may bear no relationship to U.S. black ancestry.

Since little is known about the tooth status of antebellum slave groups, Harris and Rathbun (1989) studied a nineteenth-century series of skeletons of American black slaves from a cemetery near Charleston, South Carolina, for tooth crown diameters (odontometrics). Unexpectedly, small crown sizes were found in a sample of thirty-six skulls. Explanations for the small diameters of the crowns of the plantation slaves were admittedly speculative. The authors postulated that the finding involved kin-based divergences or the natural intragroup differences in African slave sources. Undernutrition was excluded as an explanation because the sample skeletons were of average stature and other bodily dimensions. The authors compared their samples to a contemporary (and "well-off," as they stated) sample of mid-South black college students from a previous longitudinal growth study and a study of West Indian black slave series. The U.S. students attended Meharry Medical College, a predominantly black medical school. The students from this medical school were somewhat improperly described by Harris and Rathbun: they were *relatively* well-off, but not all were from the South and many were not well-off.

A characteristic set of differences from white samples derived from a study of Garn et al. (1968) included retention of large maxillary lateral incisors, and disproportionately large premolars and molars in the African slave and in the African-American student sources.

Absence of Teeth

Congenital (or hereditary) absence of teeth ranges from edentia to single or multiple defects. Children with low birth weight also have missing teeth (Keene, 1965; Krogman, 1967). Absence of teeth—such as the left third molar tooth—is associated with late formation of the remaining posterior teeth in affected children and in their siblings (Garn and Lewis, 1963a; Garn, Lewis, and Vicinus 1963b). The formation of M2, for example, is delayed when the third molar is missing. In such instances, there would be an excess of children who are P2M2, and the rarer (in Europeans) M2P2 sequence would be common when the third molar is present. Mandibular M3 is absent more often than maxillary M3, followed by maxillary I2. Since absent M3 is the only absent tooth anomaly to vary in distribution in human populations, we will confine the major part of this discussion to absent M3.

Third Molar Teeth

Before we discuss absence of third molar teeth, a racial comparison of the ages of eruption of third permanent molars is indicated. Chagula (1960) analyzed the age of eruption of third molars in East African males by oral examination of living males, skeletons, and radiographs on jaws with absent third molars, and then compared the ages of eruption with that of other populations. Chagula quoted the earlier work of Suk (1919) that the third molars erupted earlier in Zulus than in Europeans.

The oral studies comprised 900 East African males (Bantu, the largest group; Nilotes; and Nilo-Hamites) ranging in age from six to twenty-five years. Skeletal examinations included 188 adult skulls. Union of the two bones at the occipitosphenoid synchondrosis was the criterion for adulthood. Radiographic studies were confined to those jaws in which no third molar was visible on examination of the 188 adult jaws. In this series only about 1.6 percent of the mandibular third molars were congenitally absent. In some African boys, the third molars began erupting about age thirteen; by the age of twenty, all third molars had erupted in 75 percent of the Africans. Chagula (1960) observed that the previous studies of Suk (1919) showed that in U.S. whites the third molar was usually unerupted at age eighteen. Further, the eruption of third molars in Filipino and East Indians is earlier than that of Europeans. Chagula also indicated that the presence or absence of third molars contributes very little to the dental and skeletal evidence for determining age in Africans. One can only infer that if the third molars of an East African male are erupted he is at least fourteen years old. Data for females are not available.

Chagula (1960) then speculated on the possible reasons for the early eruption of third molars in East Africans. Early loss of deciduous teeth was rejected because the third permanent molars have no temporary (deciduous) precursors. Chagula could not exclude the possibility that gene complexes that control the eruption of permanent teeth begin to function earlier in Africans.

Garn, Lewis, and Vicinus (1963b) stated that absent M3 cannot be treated as an independent anomaly: absent M3 is related to agenesis of other teeth, differences in sequence tooth polymorphism, and delayed timing and movement of the third molar teeth in the siblings of affected individuals. Monogenic and polygenic hypotheses have been proposed to account for this anomaly.

Krogman (1967) reviewed the literature of missing M3 and found it to be most frequent in Mongoloids (Chinese, 17–32 percent; Eskimos, 25–28 percent; Amerindians, 13 percent). It was less frequent in Europeans, 7–25 percent (for mandibular M3); and least frequent in peoples of African origin. The prevalence of missing M3 in this group is about 11 percent in African-Americans and 0 in Africans. Both maxillary and mandibular M3 are absent more often in females. Right and left differences are equivocal.

Third molar agenesis was studied in a Brazilian population of 490 males by Crispin et al. (1972). Previously, admixture estimates of nearby populations showed the following: 60 percent white, 10 percent Amerindian, and 30 percent black. The frequency of missing third molars was about 8 percent per quadrant. Absence of the four third molars was found in 2 percent of the group. No differences were observed when the study population was separated into white or black by physical appearance. These findings suggest that those who were classified as black also had a significant proportion of European genes.

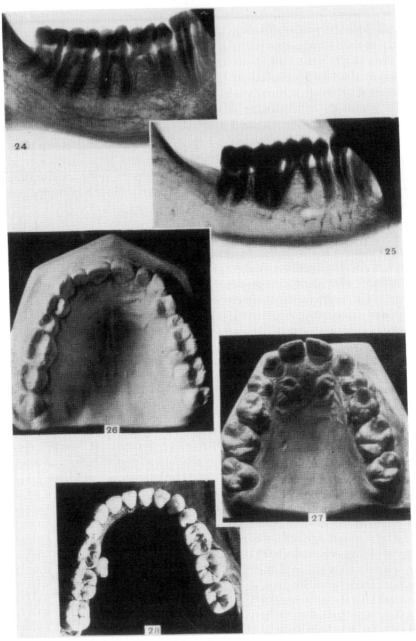

Figure 2.1. Persistence of milk molar. The photographs with numbers 24 and 25 show that peristence of the milk molar is associated with absence of the second premolar. Number 26 shows that the second incisor, the second premolar, and the third molar are bilaterally

Bolk (1916) examined more than 30,000 skulls from inhabitants of Amsterdam who died during the nineteenth century. Persistent milk molar teeth occurred about once in 400 skulls (0.25 percent). About 14 percent of the skulls had absence of the third molars. Bolk also investigated the frequency of the occurrence of both variations in one individual. Among thirty skulls with a persisting milk molar, there were fourteen—nearly one-half—in which the third molar was absent. Bolk postulated that humans are on their way to having absent third molar teeth, but at the same time a process of compensation is beginning at the anterior end of the molar region. Consequently, the second molar is substituting for the second premolar. Accordingly, the future set of teeth in humans will differ from its present formula by the loss of the second incisor, the loss of the second premolar (for which will be substituted the second milk molar), and the third molar. Specimens in which the second premolar persists and the third molar is absent in the lower jaw are common. These variations are also clinically significant.

Bolk (1916) described the clinical importance of detailed knowledge of variations in teeth. Figure 2.1 shows the lower jaw of a young woman. In the right half of the jaw, the second molar persists and there are only two molars. Radiologic studies of the jaw showed absence of the second premolar as well as the third molar on this side. The second milk molar persisted on the left side, but the third molar was absent on the right side. Thus Bolk maintained that this half was on the same evolutionary level as the right side. But there was a difference. The second premolar was absent on the right side, but it was developed on the left. The second premolar, however, had not pushed out the second milk molar, because it erupted on the inner side of this tooth. Bolk postulated that the left half of the jaw may be considered as an intermediate form between the normal condition and the final state that has already been obtained by the right side. There was no second premolar on the right side. The dental lamina had lost the ability to produce the germ from which this tooth could

---

diminished—an intermediate form of the normal dentition of today and the future dentition. Number 27 represents future dentition. The second premolar is suppressed; this tooth is substituted by the second milk molar; and the third molar is absent. Number 28 shows persistence of the second milk molar and only two molars on the right half of the jaw. On the left side, the second milk molar persisted and the third molar was lacking, as on the right side. The second premolar, however, was developed, had not pushed out the second molar, but had erupted on the inner side of the tooth

Source: Bolk, 1916. Reprinted with the permission of Alan R. Liss Inc., publisher and copyrights holder.

arise. The dental lamina still produced this germ on the left, but it was not strong. Accordingly, the progressive variation is complete on the right side of the jaw, but not on the left.

The association between persistent milk molar teeth and absence of the third molar teeth has important surgical implications. If a dentist wishes to bring the teeth into a normal state, the second molar on the left side should be extracted to bring the second premolar into its normal position. If, by chance, the second milk molar on the left side is extracted when there is no premolar in the jaw, there will be a gap in the arch, because the premolar is absent. Accordingly, dentists should never extract a second milk molar before they have determined that there is a premolar substitute.

Second Incisor of the Upper Jaw

Bolk (1916) noted that the second incisor is often absent in the upper jaw. He found the second incisor missing in several generations and in members of a family. Bolk claimed that this anomaly was found in about 2 percent of Germans, but figures were not given. Another anomaly consisted of a supernumerary conical incisor next to the median plane behind the second incisor. Bolk stated an earlier report noted that this supernumerary tooth occurs more frequently in "Indians" (Amerindians?) than in whites.

Paramolar Structures of the Upper Dentition

In 1916, Bolk described an additional cusp that occurs on the buccal surfaces of upper and lower permanent molars in humans. Bolk believed that paramolar cusps are the continuation of the milk dentition. During evolution of the mammalian teeth, paramolar cusps became rudimentary and finally were eliminated from the row of functional teeth. Bolk did not find a single human specimen with a paramolar structure in the lower jaw in more than 30,000 specimens. These cusps, however, have subsequently been found on the premolars, the lower first molars, and the upper and lower deciduous molars (Kustaloglu, 1962). This investigator reviewed the presence of these structures in upper deciduous and permanent molars in various populations. Anomalous cusps were not found among whites, blacks, Filipinos, and Hawaiians. Southwestern Amerindians showed a higher occurrence of these cusps in both deciduous and permanent molars, when compared with other populations.

The Distomolar

The distomolar is a fourth molar tooth (Bolk, 1916). In Bolk's extensive studies of skulls, distomolars were not found in the mandible. The tooth is usually rudimentary and is believed to be a variation that results from a dental germ in excess of the normal number. It is produced by an unusually prolonged dental lamina. A well-developed fourth molar is very rare. In Bolk's collection of more than 30,000 skulls from inhabitants of Amsterdam who died during the nineteenth century, none had a regularly formed fourth molar with four or five cusps, but several were rudimentary. Bolk gave no figures, but stated that the distomolar among Africans is more frequent and may reach a full degree of development. (Recall that third molar teeth are rarely absent in Africans.)

Shovel-Shaped Incisors

Lingual marginal ridging on the maxillary central incisors (UI1) consists of enamel ridges on the mesial and distal margins of the lingual surface of the incisors. Marginal ridging creates an apparent depression of the medial lingual area (E. F. Harris, 1980). The condition was first described by Hrdlička (1920), who labeled the anomaly "shovel-shaped incisors."

The highest frequencies of this variation are found in Mongolian populations, where the prevalence is often 100 percent. There are also sex differences. Harris found that UI1 lingual shoveling is statistically more common in females than in males in Europeans, Asians, Amerindians, and markedly so in Pacific Islanders. Sex differences were not found in African-Americans. All of the samples in Harris's analysis did not, however, show sex differences, for only twelve out of the thirty-eight samples had higher frequencies of this variant in females. Hanihara (1963) studied this anomaly in Japanese, Japanese-whites, whites, Japanese-blacks, and blacks. He graded the shovel-shaped incisors into 0 (no shoveling); 1 (semishovel-shaped); and 3 (shovel-shaped). These are illustrated in Figure 2.2. As expected, the shovel-shaped anomaly was frequent and well developed in Japanese and infrequent or not well developed in whites and in blacks. Japanese-white and Japanese-black children showed intermediate forms. Table 2.2 shows the distribution of UI1 lingual shoveling in various populations.

Rotation of Maxillary Central Incisors

Mesial rotation of the maxillary central incisors results in a winglike appearance of the central upper incisors because of the turned position of the teeth in their sockets. The distal part of the crown is rotated labially, and the mesial part is in a lingual position (Escobar et al., 1976). The reverse position of mesolabial rotation (counterwinged teeth) also occurs. Bilateral rotation of upper central incisors is shown

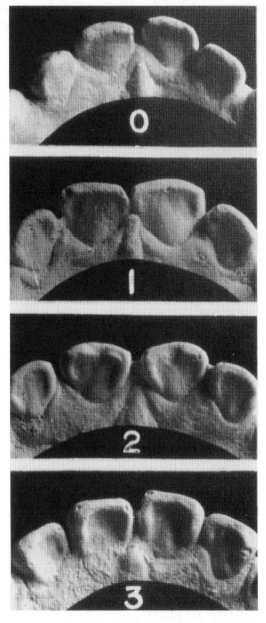

Figure 2.2. Shovel shape of the deciduous upper central incisors. 0, no shovel shape; 1, semi-shovel shape; 3, strong shovel shape

Source: Hanihara, 1963. Reprinted with permission of Pergamon Press PLC. Copyright © 1963

Table 2.2
Distribution of Shovel-Shaped Incisors in Various Racial Groups

			Frequency	
Population	No. of Samples	Sample Size	Males %	Females %
Europeans	6	4,475	22	25
African-Americans	2	878	43	42
Asians	5	4,656	89	92
Polynesians	7	2,964	48	56
Melanesians	7	1,626	58	73
Amerindians	11	4,538	73	75

Source: Combined from thirty-eight different samples by E. F. Harris (1980).

in Figure 2.3. Diagrams of various forms of rotated central incisors
are shown in Figure 2.4.

The frequency of rotated central incisors ranges from 22 to 38
percent among Amerindians, to 10 percent for Japanese, to 3 percent
for Chicago whites (Dahlberg, 1963). We found no surveys for this
anomaly in African peoples, but we have seen this anomaly in Africans and in African-Americans.

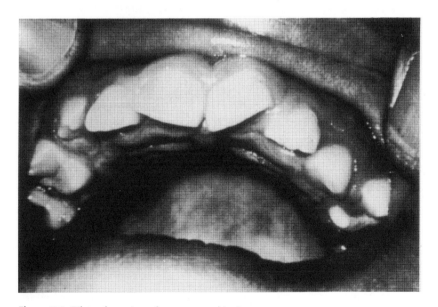

Figure 2.3. Bilateral rotation of upper central incisors
Source: Escobar et al., 1976. Reprinted with the permission of Alan R. Liss, Inc., publisher and
copyrights holder

A	B	C	D	E
Bilateral Winging	Unilateral Winging	Straight	Unilateral Counterwinging	Bilateral Counterwinging

Figure 2.4. Classification of rotated upper central incisors. The labial side of the teeth is up, and the lingual side is down
Source: Dahlberg, 1963. Reprinted with permission of Pergamon Press PLC. Copyright © 1963

Canine Distal Accessory Ridge

The canine distal accessory ridge is a nonmetrical tooth crown characteristic that has been reported only by Scott (1977). The canine distal accessory ridge is located on the lingual surface of the tooth between the medial lingual ridge and the distal marginal ridge. It is not centered between these two ridges, but lies in close contact with the distal marginal ridge. Scott emphasized that weak expressions of the variant frequently approximate the distal occlusal border—a cautionary note for those who wish to study this trait. The distal accessory ridge can be either present or absent. When present, it ranges from a small swelling to a prominent ridge as large as or even larger than the distal marginal ridge.

More than 1,000 dental casts (full mouth impressions) of living

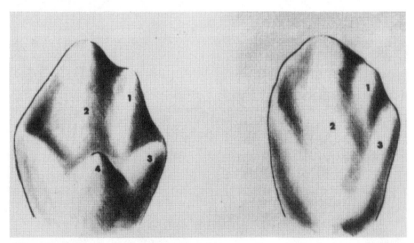

Figure 2.5. Crown features of the upper and lower canines: 1, distal accessory ridge; 2, medial lingual ridge; 3, distal marginal ridge; 4, canine tubercle (quasi-continuous variant found only on upper canine in modern human groups)
Source: Scott, 1977. Reprinted with the permission of Alan R. Liss, Inc., publisher and copyrights holder

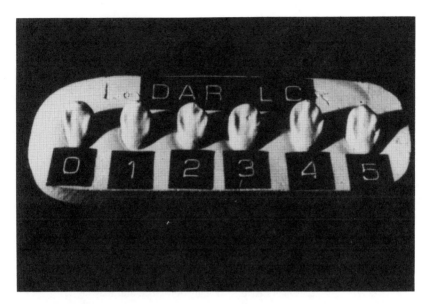

Figure 2.6. Ranked classification for the variable expression of the distal accessory ridge on the lower canine
Source: Scott, 1977. Reprinted with the permission of Alan R. Liss Inc., publisher and copyrights holder

Southwest U.S. Amerindians, U.S. whites, and Asian Indians were studied. Crown features of the upper and lower canines (Scott, 1977) are shown in Figure 2.5, and Scott's ranked classification for the distal accessory ridge is shown in Figure 2.6. The classification has six grades—from absence to varying degrees of prominence. Size is the primary criterion. The ridges exhibit only minor variation in form.

Carabelli's Trait

Pits and extra cusps are found in deciduous and permanent upper molars. The classification of this anomaly, called Carabelli's trait, is varied, but the one that is described by Hanihara (1963) is used in Table 2.3. The pits and cusps are shown in Figure 2.7.

In Table 2.4, types 2 and 3, 4 and 5, and 6 and 7 were combined for ease of classification. Carabelli's cusp has a very high frequency in whites and is of relatively low frequency in Japanese and in black children. The mixed racial groups show almost the same pattern of distribution and are closer to that of the white group. Carabelli's pit showed almost no racial differences in the five groups, which suggests a different type of inheritance for this anomaly.

Hassanali (1982) studied the incidence of Carabelli's trait in Ken-

Table 2.3
Classification of Carabelli's Pit and Cusp

Type	Description
0	No cusp or pit is present.
1	A shallow groove on the mesial side of the lingual surface suggests a trace of a pit.
2	There is a shallow groove or depression without any change in the curvature of the lingual surface.
3	The depression or pit is somewhat deeper than in Type 2 but no bulge is observed on the lingual surface.
4	The appearance is similar to Type 3, but there is a slight eminence on the lingual surface of the protocone.
5	The eminence becomes stronger than in Type 4; however, the cusp extends smoothly to the rest of the lingual surface without interruption.
6	Carabelli's cusp is completely encircled by a groove, so it appears to form a fifth cusp.
7	Carabelli's cusp is strongly developed and may be larger than the hypocone.

Source: After Hanrihara (1963).

yan Africans and Asians and reviewed the literature. A four-grade classification of Kraus (1959) was used: (0) no evidence; (1) pit; (2) groove with ridge; (3) small tubercle; and (4) notable tubercle or cusp. Table 2.5 shows the frequencies obtained from Hassanali's study and his review of frequencies from the literature.

Mylohyoid Bridge

The mylohyoid bridge consists of anomalous bone formation that transforms a part of the mylohyoid groove on the medial aspect of the mandible into a canal (Lundy, 1980). This anomaly is believed to be inherited, but the nature of the inheritance is unknown. Two basic types of bridges have been reported. The first, and most common, pattern I, extends from a point a few millimeters anteroinferior to the mandibular foramen to the anterior margin of the insertion of the medial pterygoid muscle (Figure 2.8). Pattern II may be perforated (Figure 2.9), or discontinuous (Figure 2.10), or it may extend anteroinferior to the corpus (Figure 2.11).

The mylohyoid bridge was studied in the Khoisan of South Africa, and the incidence of this anomaly was reviewed in various populations (Lundy, 1980). Curiously, even though the mylohyoid bridge had been viewed as a Mongolian marker, the Khoisan have one of the highest frequencies of this anomaly. Lundy analyzed twelve popula-

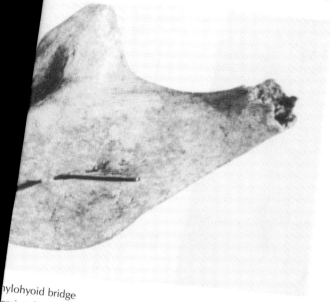

nylohyoid bridge
eprinted with the permission of Alan R. Liss Inc., publisher and copyrights

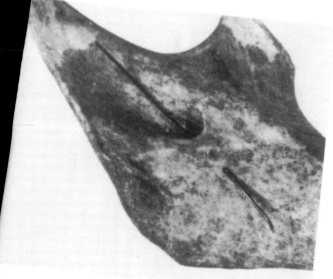

II mylohyoid bridge
0. reprinted with the permission of Alan R. Liss Inc., publisher and copyrights

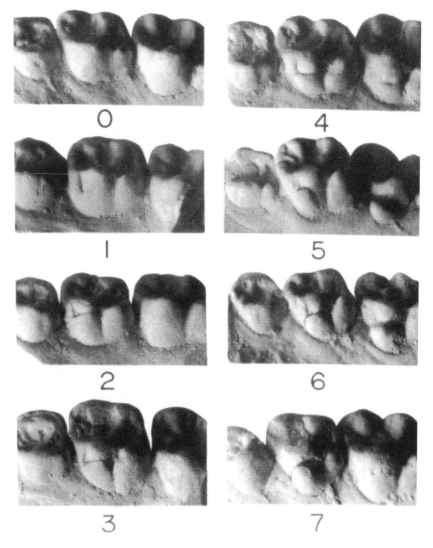

Figure 2.7. Carabelli's cusp on the deciduous upper second molars. Three types of Carabelli's pit and four types of the cusp are shown. (0) Absent Carabelli's cusp. Carabelli's pit: (1) A shallow groove on the mesial side of the lingual surface indicates a trace of a pit. (2) A shallow depression or groove is present without change in the curvature of the lingual surface. (3) The depression or pit is slightly deeper than in number 2, but there is no bulge on the lingual surface. Carabelli's cusp: (4) Similar to type number 3, but there is a slight eminence on the lingual surface of the protocone. (5) The eminence is stronger than in number 4, but the cusp extends smoothly to the rest of the lingual surface without interruption. (6) Carabelli's cusp is encircled by a groove so that it appears to form a fifth cusp. (7) Carabelli's cusp is strongly developed and may be even larger than the hypocone

Source: Hanihara, 1963. Reprinted with permission of Pergamon Press, PLC. Copyright © 1963

Table 2.4
Frequency of Carabelli's Pit and Cusp in Deciduous Upper Second Molars (%)

Population	No.	0	1	2 + 3	4 + 5	6 + 7
Japanese	185	32.4	30.8	24.9	6.5	5.4
Japanese-white	71	14.1	19.7	42.3	18.3	5.6
White	56	5.4	19.6	39.3	14.3	21.4
Japanese-black	41	17.1	19.5	43.9	17.1	2.4
Black	51	19.6	29.4	39.2	2.0	9.8

Source: Hanihara (1963).

tions that are considered Mongoloid and nine that are not Mongoloid. When the incidences were grouped into two sets of data and compared, an F value of 1.38 was obtained. The tabulated F values of 3.31 ($P < 0.05$) and 5.74 ($P < 0.01$) and the high correlation show that there are no significant differences in the two sets of data. The population studies are shown in Table 2.6.

Torus Palatinus

The torus palatinus is a bony ridge in the midline of the hard palate and varies in form, location, and size (Lasker, 1950). Surveys of skulls and of living individuals showed a high frequency in Eskimos, Ainus, Icelanders, Laplanders, Northern Europeans, and Northern Asiatic peoples. This variation was believed by Lasker to be inherited because in a previous study he had found only one instance of a discordant pair in 134 pairs of monozygous twins. Some pedigrees show an autosomal dominant pattern; however, a higher frequency in females suggests that the inheritance might be one of X-linked autosomal dominance.

Torus Mandibularis

Interestingly, the prevalence of the torus mandibularis (a hyperostotic anomaly of the mandible) is high in Eskimos, Laplanders, Ostiaks, Icelanders, and the San; sporadic in Europeans and Amerindians; and unknown in the mandibles of Africans, Australians, and Melanesians (Drennan, 1937). In a sample of twenty-eight San mandibles, there were nine examples of the anomaly, an incidence of 32 percent. The same author also noted a prevalence of 32 percent of the torus palatinus in the same collection, even though a previous investigator did not find this variation in twenty-seven San skulls. The torus mandibularis and torus palatinus were associated in 67 percent of the specimens. Drennan also examined ten skulls from the Khoikhoi and found that only one specimen had a slight tubercular torus mandibularis associated with a slight torus palatinus.

Table 2.5
Carabelli's Trait Frequency from Various Studies (%)

Study	No.	Absence 0	Pit and Groove 1 + 2	Tubercle and Cusp 3 + 4	Total Trait (%)
Hassanali (1982) Africans (Kenyan)	298 casts	33.0	44.5	22.5	67.0
	248 skulls	31.5	46.3	22.2	68.5
	242 casts	26.9	57.5	15.7	73.1
Scott (1980) Bantu	812 teeth	31.7	49.5	18.8	68.3
Kieser (1968) South Africans (coloureds)	163 skulls	54.1	39.2		
Shapiro (1949) South Africans (Bantu)					
Barnes (1969)					

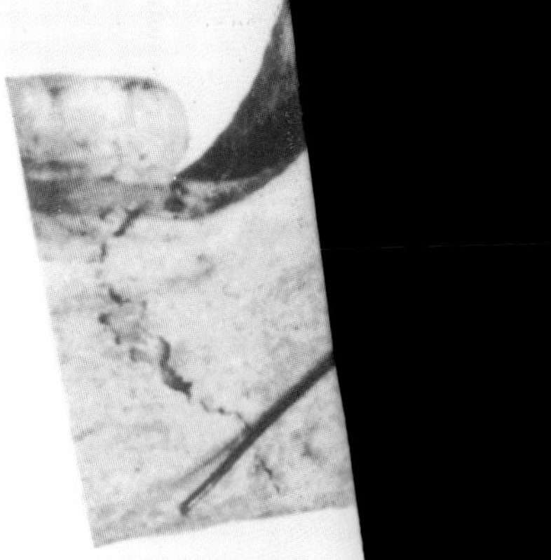

Figure 2.8. Pattern I n
Source: Lundy, 1980. R
holder

Figure 2.9. Pattern
Source: Lundy, 198
holder

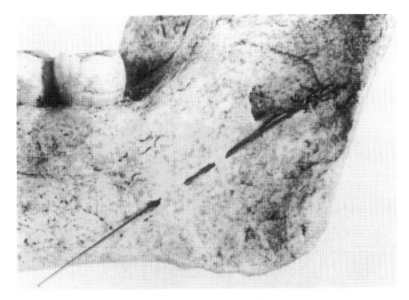

Figure 2.10. Discontinuous form of pattern II mylohyoid bridge
Source: Lundy, 1980. Reprinted with the permission of Alan R. Liss Inc., publisher and copyrights holder

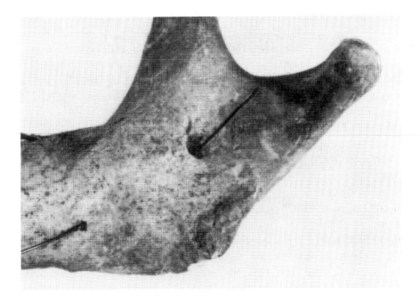

Figure 2.11. Extended form of pattern II mylohyoid bridge
Source: Lundy, 1980. reprinted with the permission of Alan R. Liss Inc., publisher and copyrights holder

Table 2.6
Distribution of Mylohyoid Bridge

Population	N	%
Africa, Khoisan	146	32.2
South African black (Bantu)	544	12.3
U.S. black	234	15.4
U.S. white	180	16.1
Europe, French	844	0.5
Pre-Columbian (Peruvian)	244	17.6
Eskimo, Greenland	288	8.0
Eskimo, Alaska	529	5.5
Aleut	267	30.0
Amerind, Nadene (Alaska and Canada)	126	26.2
Amerind, Plains	580	33.8
Amerind, Pueblos	578	13.0
Amerind, Pacific Northwest	282	19.1
Asian Indian	350	4.9
Thai	273	5.1
Japanese	208	2.9
Ainu	104	6.7
Australia, Aborigines	605	6.1
Hawaiian	865	5.3

Source: Modified from Lundy (1980).

Drennan (1937) recalled that the early Dutch settlers at the Cape in South Africa frequently referred to the local people as "Chinese Hottentots" because of the light yellow skin color and the slanting eyelids. He suggested, as have many others, a connection between the San and Mongoloid peoples.

HAND ANOMALIES

Polydactyly (extra fingers and toes) is discussed in Chapter 11. The classification of brachydactyly (shortening of the fingers and toes) is shown in Table 2.7. In types A-1, C, and E, one or more of the metatarsal bones are absent and all are associated with short stature. All of these anomalies are classified under dominant inheritance by McKusick (1989). Hertzog (1967) classified type A-3 (brachymesophalangy V) as recessive. The brachydactylies are rare and are usually of no anthropological significance; brachydactyly type D and brachymesophalangy V are exceptions.

Table 2.7
Classification of Brachydactyly

Type	Description
A=1	The abnormality is confined to the middle phalanges; the shortened phalanges may be fused to the distal phalanges.
A=2	The middle phalanx of the index finger is affected.
A=3	The middle phalanx of digit V is affected (bracho-mesophalangy V).
A=4	The middle phalanges of digits III and V are short, and there is radial clinodactyly (radial curvature at the joint between the proximal and middle phalanges) of finger IV.
A=5	The middle phalanges are missing.
B	The middle and terminal phalanges are short or absent; apical dystrophy (amputation-like defect).
C	Defect in middle and proximal phalanges of digits II, III, and V (brachomesophalangy II, III, V).
D	Short, broad distal phalanx in thumb.
E	One or more metacarpals or metatarsals are absent.

Source: Modified from Bell (1951); McKusick (1989); Fitch (1979).

Brachydactyly Type D

Brachydactyly D (stub thumbs) is transmitted as an autosomal dominant trait and occurs in about 1.6 percent of all Jews in Israel irrespective of their communal origin (Goodman, Feinstein, and Hertz, 1984). Interestingly, even though a complete survey was not done in the original study in Israel, Goodman and co-workers found that four of twenty-nine (13.8 percent) of unrelated Jews with a short thumb also had a short fourth toe. These authors also mentioned an earlier (1957) study that surveyed U.S. whites and blacks: 0.41 percent and 0.10 percent, respectively, had brachydactyly D and one in ninety-five had a short fourth toe. Goodman and co-workers (1984) also studied a large Sephardi Jewish family in which brachydactyly D was segregating, and in which a wide range of expressivity was noted, varying from a single thumb affected to both thumbs and fourth toes. These authors indicated that this variable expressivity is also compatible with dominant inheritance.

Brachymesophalangy V

The term *brachymesophalangy V* (BMP-V) was first used by Pol (1921) to describe shortening of the middle phalanx of the fifth digit. Garn et al. (1967) stated that the anomaly involves reduction in length and an increase in the breadth of the affected phalanx.

Hertzog (1967) studied brachymesophalangy V in 200 Philadelphia black children and in 96 Chinese children. Twelve people with shortening of the middle phalanx of the fifth digit were found in the Chinese population. None were found in the black population. Hertzog also reviewed the prevalence in other groups and found that BMP-V is much more frequent in Mongolian populations than in Europeans or in Africans. Singer, Kimura, and Gajisin (1980) studied the anomaly in Nama-speaking Hottentots (Khoikhoi), Rehoboth Basters (in Namibia), and Cape Coloured children. They found no children with the abnormality. These workers reviewed the criteria for BMP-V and summarized the frequency in various populations. A modified version of their study is found in Table 2.8. It is evident that the prevalence of this anomaly (in high-frequency populations) is almost without exception much higher in females than in males. (The differences are so striking that statistical confirmation is not necessary.) The sex differences suggest that factors other than simple dominant inheritance may be important in the manifestation of this defect in Mongolian populations. Unfortunately, most studies of BMP-V do not measure the presence or absence of a cone-shaped epiphysis, as did the following investigation. (A cone-shaped epiphysis is one in which a portion of the epiphysis projects distally toward the adjacent epiphysis.)

Garn et al. (1972), in an analysis of juveniles with BMP-V, indicated that brachymesophalangia-V alone without cones is separately inherited without apparent sex bias, but BMP-V with the cone epiphysis of mid-5 and the cone-epiphysis of mid-5 alone are both inherited as a complex and with a marked excess of females over males. Photographs of BMP-V with cone-shaped epiphysis and a cone-shaped epiphysis alone are shown in Figure 2.12. The subjects in this study were 8,854 juvenile participants in the Ten State Nutrition Survey of 1968–70 between the ages of three and one-half and twelve. Of the total number of children, 185 had brachymesophalangia-V and 51 percent of those affected also had a cone-shaped epiphysis. Separately, 163 males and females had a cone-epiphysis of mid-5 and 58 percent of those with cones also had BMP-V. Thus an association was found between BMP-V and cone-epiphysis, with a greater proportion of those with cones showing BMP-V than vice versa.

Table 2.8
Frequency of Brachymesophalangia V in Various Populations

	Males		Females		Both	
Population	No.	%	No.	%	No.	%
Philadelphia blacks					200	1.0
Khoikhoi					33	0.0
Cape Coloured	114	0.0	96	0.0		
Ecuadorian blacks					94	2.1
Ecuadorian Mestizos					182	3.1
U.S. whites					1,000	1.0
Ohio whites					647	0.6
English whites					800	1.0
Australian whites	60	3.3	60	3.3		
Iranians					226	0.9
Costa Ricans					1,841	1.3
Nicaraguans					1,722	1.9
Guatemalans					1,744	2.0
Pima Indians	532	13.9	551	24.9		
Blackfeet Indians					199	1.0
Amerindians	363	6.33	385	9.09		
Peru Indians					40	15.0
Peru Indians					238	5.0
Polynesians			164	1.2		
Australian aborigines	308	13.6	299	20.7		
Micronesians	48	6.3	56	14.3		
Philadelphia Chinese	51	5.9	45	20.0		
Hong-Kong Chinese			247	4.9		
Japanese	8,191	12.5	7,770	19.7		
American Japanese[a]	516	13.4	474	28.7		
Ainu	51	2.0–3.9	70	4.3–15.7		

Source: Modified from Singer, Kimura, and Gajisin (1980).
[a]Pooled from twelve different studies.

Garn et al. (1972) examined these associations through the siblings of those with BMP-V and ascertained whether the siblings of index cases resembled them. Brachymesophalangia-V alone segregated out separately from cones alone and the combination of brachymesophalangia and cones. The authors surmised that simple dominance may be excluded even though examples of parent to child transmission have been reported. They suggested that either recessive inheritance or more complicated modes of transmission may be postulated, especially for the variant including cones. As we mentioned previously, it is unlikely that a radiographic study could be repeated in the United States. Thus the mode of inheritance of these variants may be unknown until nonradiographic methods for skeletal variation in live humans are developed.

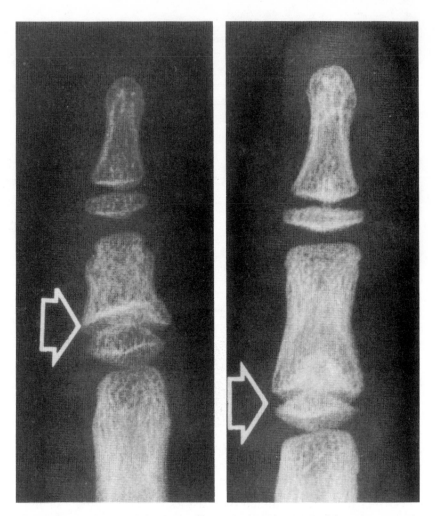

Figure 2.12. Brachymesophalyngia-V with a cone-shaped epipyysis (left) and a cone epi-physis without brachymesophalangia (right). Brachymesophala-V alone is inherited as an entity, while the combination of BMP-V with a cone-shaped mid-5 is influenced separately with a major influence of sex

Source: Garn et al., 1972. Reprinted with the permission of Alan R. Liss Inc., publisher and copyrights holder

Some of the highest frequencies of BMP-V are also found in Australian aborigines, but other so-called Mongolian markers have not been found in this group. One exception to high frequencies of BMP-V in Mongolian populations is the very low frequency of this anomaly in Blackfeet Indians, but this population, unlike other Amerindians, also has a very high frequency of the A blood group allele (Mourant, 1976).

Buschang and Malina (1980) showed that frequencies of BMP-V are based upon indices that are not always comparable, and that most population surveys for this anomaly did not record other anomalies. This is unfortunate, because BMP-V is associated with several other conditions, such as clinodactyly and cone-shaped epiphysis. (Clino-dactyly is an incurved little finger as a result of shortening of the middle phalanx of the finger on the radial side.)

Brachymesophalangia V was more frequent in 212 individuals with Down syndrome (21 percent) than in 14,197 Europeans (1.4 percent) (Garn, Gall, and Nagy, 1972). Notably, none of the twenty-eight Down syndrome children had a cone-epiphysis on mid-5, although 47 per-cent would be expected.

Buschang and Malina (1980) studied BMP-V from samples of chil-dren of different backgrounds. They used two indices. Index 1 was based on the ratio of the width to the length of the fifth mid-phalanx. It consistently produced higher frequencies than index 2. Index 2 was based on the ratio of the length of the fifth to the length of the fourth midphalanx. Index 1 was more selective of BMP-V alone, and index 2 primarily selected BMP-V with clinodactyly with or without a cone-shaped epiphysis. Index 2—which defines shape—selected both the shortest and widest midphalanges, but index 2 detected differences in length. The authors suggested that index 2 provided a more suitable criterion of BMP-V in comparative studies, because of possible popu-lation differences in the shape of the phalanges. The two indices that were used to identify BMP-V showed the highest frequencies among children of European ancestry (Pennsylvania white; Canadian West German) and lower frequencies in Pennsylvania black children.

The Carpal Bones

The human carpus consists of eight bones in two rows—proximal and distal (Figure 2.13). The proximal row contains—from the radial to the ulnar side—the scaphoid, lunate, triquetral, and pisiform bones. The proximal row has the trapezium, trapezoid, capitate, and hamate bones. The carpal angle is defined as the angle resulting from the intersection of two lines, one tangent to the proximal edges of the lunate and scaphoid and one tangent to the proximal edges of the lunate and triquetrium (Harper, Poznanski, and Garn, 1974). These authors studied hand films from the Ten State Nutrition Survey be-tween 1968 and 1970. The carpal angle was greater in blacks than in

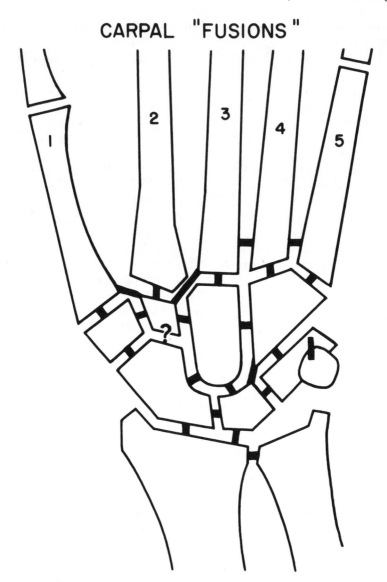

Figure 2.13. Scheme of the carpus to show the locations of recorded fusions
Source: O'Rahilly, 1957. Reprinted with the permission of Lippincott/Harper & Row,
Philadelphia, PA

whites, increases with age during childhood, and at most ages is greater in males than in females. The angle was also greater on the left than on the right.

Harper and co-workers (1974) also indicated that the size and shape of the bones of the carpus (in particular, the scaphoid, lunate, and triquetrium, which determine the carpal angle) can reflect malformations. The carpal angle is increased in arthrogryposis, diastrophic dwarfism, epiphyseal dysplasias, frontometaphyseal dysplasia, otopalatodigital syndrome, Pfeiffer syndrome, spondyloepiphyseal dysplasia, and trisomy-21. The angle is decreased in Madelung deformity, dyschondrosteosis, Turner syndrome, Morquio syndrome, and Hurler syndrome.

Carpal Fusions

Abnormalities of the carpal bones occur in a variety of syndromes (Poznanski and Holt, 1971). This discussion will be restricted to carpal fusion anomalies. O'Rahilly (1953) suggested that the presence of a single cartilaginous primordium—where there are usually two—may favor the subsequent fusion of the two ossification centers. He indicated that the situation is similar to the formation of the scapholunate that is found in a number of mammals. Carpal fusions may also be associated with generalized skeletal disorders.

The carpal anomalies discussed in this section involve fusions in the same row; however, the carpal fusions associated with other abnormalities usually involve fusions of the proximal and distal row of bones (Poznanski and Holt, 1971). Levine (1972) agreed with O'Rahilly (1957) and Cockshoot (1959) that the term *fusion*—as applied to isolated anatomical variations—is a misnomer. O'Rahilly (1957) stated that the first step in development of carpal fusion is either failure of separation of adjacent cartilages in the embryo or partial fusion of cartilaginous anlagen at an early stage in embryonic development. Ossification begins from two centers in a single cartilaginous primordium. The existence of a single cartilage—where there are usually two—inevitably leads to the subsequent fusion of the two ossification centers. Levine looked upon the variants not as a fusion but as a nonseparation of cartilaginous carpals.

Levine also believed that there is a genetic basis for certain types of carpal synostoses. Cockshoot (1959) found capitate-hamate fusion in identical twins. Carpal fusions are also frequently associated with hereditary skeletal disorders, such as symphalangism, the Ellis-van Creveld syndrome, and the hand-foot-uterus syndrome. Carpal fusions vary in incidence among populations.

Table 2.9
Frequency of Os Lunato-Triquetrium in Various Populations

Population	Males No.	Affected	Frequency (%)	Females No.	Affected	Frequency (%)	Sex Not Recorded No.	Affected	Frequency (%)	Reference
African origin										
Yoruba	514	43	8.4	409	31	7.6				Cookshoot (1959)
Ghanaian	500	32	6.4	100	5	5.0				Smitham (1948)
Baganda							80	3	3.8	Dean and Jones (1959)
Baganda							236	5	2.1	Davis (1959)[a]
Wadigo							360	24	6.7	Mackay (1952)
							1,360	14	1.03	
South African										
Black	201	7	3.5	193	11	5.7				Levine (1972)
Coloured	133	1	0.8	132	3	2.3				Levine (1972)
African-American	3,207	33	1.0	4,336	86	2.0				Garn et al. (1971)
European origin										
Scandinavian							2,100	3	0.1	Lonnerblad (1935)
German	5,356	4	0.07	6,307	8	0.12				Arens (1950)
U.S.							5,000	4	0.08	Garn et al. (1971)
U.S.							743	1	0.13	O'Rahilly (1953)
South African	226	0	0.00	219	0	0.00				Levine (1972)
Asian Indian	120	0	0.00	120	0	0.00				Levine (1972)
Amerindian	307	0	0.00	120	0	0.00				Garn et al. (1971)
Mexican-American	636	0	0.00	953	1	0.10				Garn et al. (1971)
Puerto Rican	173	0	0.00	129	1	0.70				Garn et al. (1971)
Japanese							1,400	1	0.07	Wetherington (1960)

[a] Quoted by Cockshoot (1959).

Triquetral-Lunate Fusion

The frequency of the os lunato-triquetrium (or triquetral-lunate fusion) in various populations is summarized in Table 2.9. This abnormality is common in populations of African origin, but is rare in Europeans, Asian Indians, Amerindians, Mexican-Americans, and Puerto Ricans. The disorder is more frequent in females than in males (Garn et al., 1971). The ratio of females to males was 2 : 1 (80 percent in females and 0.38 percent in males). Family studies did not correspond to recessive or dominant inheritance. An X-linkage was excluded for the following reasons: (1) there was male-to-male transmission; (2) the proportion of affected sisters among the siblings of male propositi did not exceed the proportion of affected brothers of female propositi; and (3) both equaled the proportion of affected sisters of female propositi. Garn and colleagues concluded that the mode of inheritance was likely to be multifactorial with a major sex influence.

The frequency of unilateral versus bilateral lunato-triquetral fusion is rarely available, because only one wrist is usually surveyed. In studies where bilateral X-rays were performed, bilateral fusion was twice as common as unilateral fusion (Levine, 1972; Cockshoot, 1959; Dean and Jones, 1959; Smitham, 1948).

Capitate-Hamate Fusion

Fusion of the capitate and hamate bones is less common than fusion of the lunate and triquetral. Levine (1972) found two examples of this anomaly in 585 blacks in South Africa, an incidence of 0.34 percent. The abnormality was not found in whites, Coloureds, or Asian Indians in that country. Cockshoot (1959) found capitate-hamate fusion in 0.5 percent of 923 Yoruba (from Nigeria). In a group of 680 Ghanaians, Smitham (1948) found the anomaly in 5, a prevalence of 0.73 percent. Both lunato-triquetrium and capitate-hamate fusion were found in one subject in the latter series.

THE ARMS

Glanville (1967) studied the mechanical interaction of the humerus and ulna during articulation in skeletons from African and European populations. The humerus is viewed as articulating on a stationary ulna, and the trochlea of the humerus may be imagined as a cylinder that revolves between the arms of the ulnar process. When the arm is extended at the elbow, the olecranon process of the ulna fits in the olecranon fossa of the humerus and when flexed the coronoid process occupies the coronoid fossa. The humeral septum is between the fossae in which a perforation is sometimes present. The perforation has been associated with a greater range of movement at the elbow. The prevalence of perforation was 47 percent in the African sample and 6 percent in the European. Glanville indicated that although both genetic and environmental factors have been proposed to explain the perforation, the cause is unknown.

JOINT HYPERMOBILITY

Santos and Azevedo (1981) reviewed racial distribution of joint hypermobility and studied this variation in a sample of 3,000 schoolchildren ages six to seventeen years in Bahia, Brazil. A review of the literature indicated that joint hypermobility is probably a continuously distributed variable that is influenced by age, sex, ethnic group, and biological and environmental factors. Apparently the effect of race on joint mobility has been investigated in only a limited number of populations. Icelanders have been found to have a greater range of mobility of wrist and hip than Swedes, but there were no differences between populations for metacarpalphalangeal and shoulder joints. Eskimos have a greater prevalence of generalized joint laxity than Amerindians. Indians of Cape Town are more loose-jointed than white South Africans, and whites in general have a smaller range of movement than do blacks.

Unlike previous studies of joint hypermobility, which concentrated on differences between races, Santos and Azevedo (1981) assessed the proportional effect of black admixture in a sample of mixed Brazilian children. Race was classified by means of skin color, facial features, and hair type according to a five-point scale of black admixture: white (W), light mulatto (LM), medium mulatto (MM), dark mulatto (DM), and black (B). The overall frequency of generalized joint hypermobility was 2.3 percent. The darker the children, the lower the frequency of affected children, but this racial effect was also associated with age. The highest frequency of generalized joint hypermobility was found in the youngest and less black mixed children, but among the oldest and darker children there was no joint hypermobility.

VARIATION IN NUMBER OF VERTEBRAE

Variation in vertebral number differs among populations and is believed to be inherited (Kuhne, 1932, 1936). Bornstein and Peterson (1966) found that race and sex influenced the expression of variation in vertebral number and that the total (male and female) variation in presacral vertebrae is characteristic of a population. Kaufman (1977) described a method for studying presacral plus sacral vertebrae and analyzed the total number of precoccygeal vertebrae (PCV). The number of presacral and sacral vertebrae is recorded by articulating each vertebra serially. Attention was paid to the junctional sections between spinal regions. An atlas that is unilaterally assimilated to the occipital is counted as a vertebra. The sacrococcygeal junction is carefully defined. If the first coccygeal vertebra is ankylosed to the sacrum, it is not included in the count. The rare occurrence of a cervical rib on an otherwise typical C7 was sufficient to reclassify the vertebra as thoracic. A first lumbar vertebra that had a lumbar rib or ribs was classified as lumbar.

Bornstein and Peterson (1966), however, counted vertebra that had cervical or lumbar ribs as thoracic, but since Kaufman's method (1977) did not alter the total vertebral number, comparisons among various investigations are valid. A last lumbar vertebra that showed sacralization was counted as sacral if the transverse process was enlarged and developed to form part of the sacroiliac joint and thus bound a sacral foramen. If the first coccygeal vertebra was ankylosed to the sacrum, it was not included in the precoccygeal count. The modal PCV count is 29.

The prevalence of vertebral variants in Europeans and Asians differs from that in blacks. Kaufman (1977) studied PCV variation in South African blacks, Bushmen (San), and U.S. blacks, and reviewed other investigations in blacks, Europeans, Japanese, and Eskimos. Sexual dimorphism was found in all populations. There was a higher frequency of an increase in total PCV counts (PCV counts of 30 and 31) in males than in females. On the other hand, a decrease in total precoccygeal number (28) was found more often in females than in males—but this decreased number was infrequent. The population studies are summarized in a modified form in Table 2.10. South African blacks have high frequencies of an increased number of PCV (30 and 31 group) as compared with other populations. The next highest frequencies are found in U.S. blacks. Europeans, Japanese, and Eskimos have lower frequencies.

Table 2.11 shows the results of a study of nonmetrical and metrical variation in presacral vertebrae of one hundred whites and one hun-

Table 2.10
Total Precoccygeal Vertebrae (PCV) in Various Populations

| | 28 PCV | | | | 29 PCV | | | | 30 and 31 PCV | | | | Total | |
| | Male | | Female | | Male | | Female | | Male | | Female | | | |
Population	No.	%	No.	%	No.	%	No.	%	No.	%	No.	%	M	F
South African blacks	4	1.6	8	5.1	144	58	112	71.3	100	40.3	37	23.6	248	157
San	0	0.0	0	0.0	9	56.2	10	83.3	7	43.8	2	16.7	16	12
U.S. blacks	2	0.8	6	2.4	161	64.4	207	82.8	87	34.8	37	14.8	250	250
European	3	1.5	6	3.6	157	78.8	141	86.0	39	19.6	17	10.3	199	164
Japanese	1	0.9	2	3.4	91	74.6	46	78.0	30	24.5	11	18.6	122	59
Eskimo	2	2.1	0	0.0	74	77.9	75	89.3	19	20.0	9	10.7	95	84

Source: Modified from Kaufman (1977).

Table 2.11
Variation in Presacral Vertebrae

More Frequent in Blacks	More Frequent in Whites
25 presacral vertebrae	Vertebra 7: elongated
Divided superior articular surface of the atlas	Transverse process with irregular transverse foramen
Vertebrae 23 and 24	Bifurcated cervical spinous process
Absence of accessory processes	

Source: Lanier, 1939.

dred blacks. The frequency of bifurcation of cervical spines in several populations was also reviewed by Lanier (1939) and is summarized in Table 2.12.

THE PELVIS

Iscan (1983) attempted to assess race from the pelves of 400 skeletal remains from whites and blacks of both sexes from the Terry collection. There were one hundred in each group. Three measurements were selected: the biiliac breadth was measured from an osteometric board as the maximal distance between the iliac crests; the anterior-posterior height was measured from the sacral promontory to the pubic crests of both sides and the average of the two recorded; and

Table 2.12
Frequency of Bifurcation of Cervical Spines in Various Populations (%)

Population	Cervical Vertebrae						Sample Size
	2	3	4	5	6	7	
Europeans	100	91	91	95	71	0	22
European Americans (males)	97	89	94	08	55	0	100
Japanese	99	95	91	86	34	0	100
Javanese	97	80	90	79	73	2	approx. 60
African-American (males)	81	32	43	62	41	1	100
Fuegians	35	35	90	65	20	10	20
Bantu	24	13	26	42	21	0	65–70
San	18	0	9	17	0	0	11–12

Source: Modified after Lanier (1939).

the transverse breadth was ascertained as the maximal distance between the arcuate lines of the pelvic inlet. In all variables, whites of both sexes exceeded blacks of the corresponding sex. Overall, females appeared to show more predictive results than males. Iscan cautioned that the nutritional history of the samples was questionable and that the socioeconomic status was low: modern samples should be tested.

ANTERIOR FEMORAL CURVATURE

At least three studies confirm that there are differences in anterior femoral curvature between U.S. blacks, U.S. whites, and North American Indians (Stewart, 1962; Walensky, 1965; Gilbert, 1976). Femurs of blacks had the least curvature, and femurs of Amerindians were the most curved. Femurs of coastal Peruvian Indians and Ecuadorian Indians were less bowed than those of samples from North American Indians (Gilbert, 1976). Walensky (1965) postulated that the relative straightness of femurs of U.S. blacks occurred because femurs from this group are generally longer, denser, and of greater diameter than those of other populations. Accordingly, such femurs are more resistant to the bending forces that are imposed by body weight and muscular action. Walensky also suggested that postural habits may also account for the greater curvature of the femurs in North American Indians. On the other hand, femurs from South American Indians were close to those of U.S. whites, and the greater curvature of shorter bones is difficult to explain. Gilbert (1976) summarized some hypotheses that may account for the population differences, such as postural habits, environmental and nutritional factors, and genetic differences. Genetic differences were favored. See Figure 2.14 for an illustration of some differences in curvature.

Fractures of the hip in elderly women are more common than fractures of the hip in men. Walensky and O'Brien (1968) reviewed several articles on hip fracture (Green and Gray, 1939; Gyepes et al. 1962) in which a higher incidence of hip fractures was found in whites than in blacks. Walensky and O'Brien attempted to determine whether anatomical differences might account for the population differences. The study material consisted of one hundred femurs from African-American females and one hundred femurs from white females. Measurements of oblique length, angle of inclination, length of femoral neck, neck shaft angle, and anterior-posterior diameters of the femoral neck and diaphysis were made (see Figure 2.15). Differences in neck length, neck (transverse), neck (anterior-posterior), shaft (transverse), and shaft (anterior-posterior) were not significant.

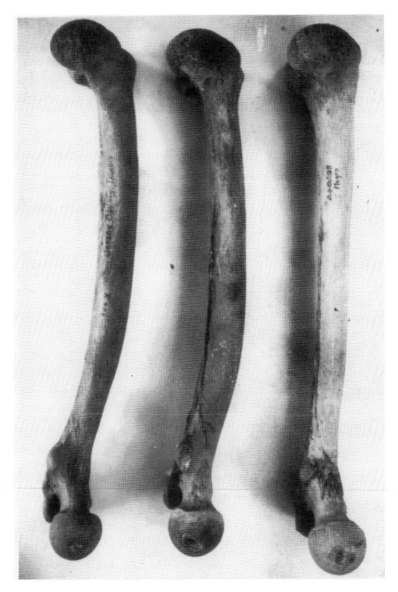

Figure 2.14. Racial differences in the amount of femoral curvature
Source: Walensky, 1965. Reprinted with the permission of Alan R. Liss Inc., publisher
and copyrights holder

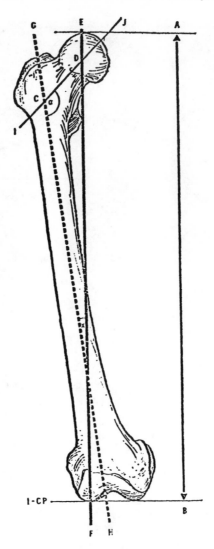

Figure 2.15. Femoral measurements: A, neck shaft angle; AB, oblique length; CD, neck length; EF, load axis; GH, anatomic axis; I-CP, infracondylar plane; IJ, neck axis; X, angle of inclination
Source: Walensky and O'Brien, 1968. Reprinted with the permission of Lippincott/Harper & Row, Philadelphia, PA

Table 2.13
Racial Differences in Femoral Measurements

Measurement and Racial Group	Right				Left			
	Mean	S.D.	t	P	Mean	S.D.	t	P
Physiological length								
White	416.8 ± 2.9	20.4	3.1	<0.01	418.3 ± 2.8	19.5	3.1	0.01
Black	431.2 ± 3.4	24.1						
Neck-shaft angle (°)								
White	121.4 ± 0.97	6.8	2.8	<0.01	124.7 ± 0.92	6.5	1.8	0.01
Black	125.1 ± 0.35	2.5			126.7 ± 0.68	4.8		
Angle of inclination (°)								
White	12.8 ± 0.35	2.5	4.5	<0.001	11.3 ± 0.35	2.5		N.S.
Black	10.5 ± 0.38	2.7			10.8 ± 0.47			

Source: Modified from Walensky and O'Brien (1968).

Femurs from black females were longer, heavier, more dense than those of white females and possessed a larger neck-shaft angle and smaller angle of inclination. Data on measurements that showed significant differences are given in Table 2.13. Walensky and O'Brien postulated that femurs of black women are better equipped to withstand the stresses of weight bearing, especially in advanced age, and that the femurs of white women are more susceptible to fracture.

3

Normal and Abnormal Pigmentation

Pigmentation is apparently more complex in humans than in other mammals, probably because there may be more modifier genes that produce multiple phenotypes (Brues, 1974). Quevedo et al. (1974) distinguished constitutive skin color—the genetically determined level of melanin pigmentation that is developed in the absence of exposure to solar radiation or other environmental influences—and facultative skin color (suntan), the increase in melanin pigmentation above the constitutive level. Facultative color change is reversible, or nearly so. (Hyperpigmentation of skin is also induced by diseases, such as Addison disease, or by endocrine change in pregnancy but these forms of pigmentation will not be discussed here.)

DIFFERENCES IN MELANOSOMES

Szabo et al. (1969) studied melanosome complexes in the Malpighian cells of the epidermis in Europeans, Amerindians, Japanese, Chinese, and blacks by electron microscopy and correlated the structure and distribution of melanosomes with skin color. Melanosomes in the keratocytes of Europeans (Figure 3.1, top) and Mongoloids (Figure 3.1, middle) are grouped and surrounded by a membrane. The melanosome complexes are in basal cells up to the stratum granulosum and are dispersed inside fully keratinized cells in the stratum corneum. The groupings of melanosomes in Mongoloids are usually more compact than those of Europeans, because there is less ground substance between the melanosomes of Mongoloids. The distribution of melanosomes in the basal keratocytes of blacks is different. The melanosomes are longer and wider than those of Europeans and Mongoloids, and they are dispersed (Figure 3.1, bottom). Presumably, the dispersed melanosomes in the basal keratocytes of blacks gives the

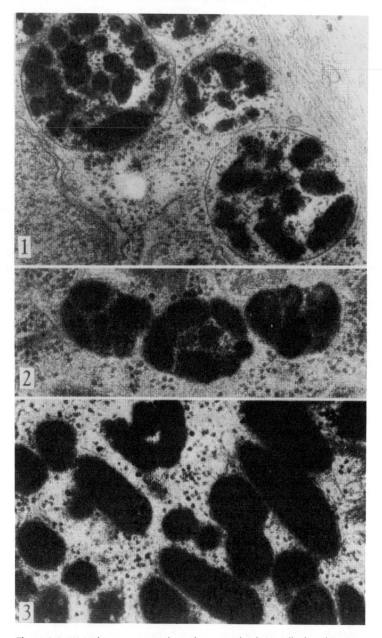

Figure 3.1. (1) Melanosome complexes from a Malpighian cell of a white European. The complexes are surrounded by a membrane and contain small particles besides the stage IV melanosomes (× 50,000). (2) Melanosome complexes from a Malpighian cell of an Oriental (Chinese). There is less ground substance between the melanosomes than in number 1 (× 50,000). (3) Melanosomes in a Malpighian cell of a black. Almost all melanosomes are individually dispersed and are much larger than those of a white or an Oriental (× 50,000)

Source: Szabo et al., 1969. Reprinted by permission from Macmillan Magazine Ltd., G. Szabo and T. B. Fitzpatrick. Copyright © 1969

skin a more uniform and a more dense color than that of Europeans. This, in turn, allows black skin to absorb, attenuate, and scatter solar rays.

SKIN PIGMENTATION MEASUREMENT

The measurement of skin color is compounded by the lack of an objective technique. Older studies matched skin color from a tile and a spinning color top (Lees and Byard, 1978). Modern investigations of skin color using reflectance spectrometry are more objective. Unfortunately, the two major portable units, the British-made E. E. L. (Evans Electroselenium, Ltd.) and the Photovolt Model 610, sample the visible spectrum at different wavelengths. Most of the investigations of skin color in Europe and in Africa are performed on the E.E.L. instrument, but measurements of skin reflectance in the Americas are performed on the Photovolt.

Lees and Byard (1978) attempted to standardize these machines by using multiple regression equations from reflectance spectrometry on Black Caribs and on Creoles from Belize. The authors concluded that although their studies offered an improvement over earlier regression estimates, different coefficients are needed for light-skinned populations. This is because reflectance curves differ in light- and dark-skinned peoples. They also suggested that data are needed from many populations to improve interinstrument comparability. But no one has proposed the obvious: investigators should standardize their work on one machine, which could eliminate many years of work. This suggestion is reinforced by Lees et al. (1979). Although an improved regression formula was developed, these authors admitted that further work is needed on light-skinned populations.

AGE FACTORS

Cross-sectional studies of age variation in skin color in samples of Sikhs who lived in Britain (Kahlon, 1976) demonstrated that pigmentation increases until the preadolescent stage in early life and then diminishes until adulthood. The factors responsible for the variation of pigmentation with age were not identified, but Kahlon suggested that age-related variations in melanocytes, variation in hormones, and delay in melanogenic activity are the primary factors.

A study of skin reflectance in ninety-nine white and thirty-eight black infants twenty-five to forty-four weeks old assessed the validity of changes in color in the skin which are used by pediatricians to determine gestation in the neonatal period (Post et al., 1976). No significant sex differences were found. Black and white infants had similar reflectance values until thirty-two weeks, after which whites became lighter skinned and blacks became darker skinned. These changes resulted from alterations in the production of melanin, and changes in distribution of hemoglobin, carotene, and bilirubin. There were also alterations in skin thickness and turbidity. The foreheads of both black and white infants who had never been exposed to ultraviolet light were measured within hours after birth: on average, they were 10–12 percent darker than the medial aspect of the upper arm. These findings indicated that the capacity to tan cannot be ascertained by a comparison of skin that is normally protected from the sun with skin that is not. This conclusion is most important, because differences in reflectance between the forehead and the upper arm as an indication of tanning capacity have been accepted for many years.

SKIN COLOR GENETICS

Although the genetics of skin color has been characterized as a model of quantitative inheritance, the mode of inheritance of skin color is still unknown. Stern (1953) studied the distribution of skin color in U.S. blacks by a spinning color top technique. He reported that models of four, five, and six gene pairs showed the least deviation from the data. Harrison and Owen (1964) used different methods and proposed that the number of units of pigmentation is probably three or four. Stern (1970) revised his estimates and postulated that three and four gene pairs fit more closely, except at the extremes of color range.

Brues (1974) suggested that genetic determinants for pigmentation probably include the following: polygenes for skin color only; one locus for depigmentation of the eye that does not affect skin or hair; one pleiotrophic gene for reduction of pigmentation in all regions; and one or more loci with multiple alleles that produce blondness or rufosity of the hair in symmetrical patterns over the body. Post and Rao (1977) studied the reflectance of skin in a sample of 154 black and 191 white same-sex twin pairs to determine the effects of genetic and environmental factors. Measurements were made with a Photovolt reflection meter on the forehead, inner upper arm, and flexor surface of the forearm. The measurements were reduced to one index—skin color, which was analyzed by the path analysis system of Rao et al. (1974). The major variance components—racial, residual genetic, and common environmental factors—were estimated, respectively, as 67, 5, and 22 percent.

Byard (1981) reviewed the quantitative genetics of skin color and emphasized that a strong genetic component has always been assumed. She pointed out that since controlled genetic crosses cannot be used, as in other animals or plants, human geneticists have had to take advantage of naturally occurring situations that approximate genetic experiments in other animals. Skin pigmentation has been viewed as the most amenable for study human quantitative traits, because variation in skin color among some groups is much larger than variation within some groups. Byard postulated that a major stumbling block in our understanding of the genetics of skin color is the lack of a satisfactory evolutionary explanation for variation in pigmentation among groups. On the other hand, since there are few—if any—completely satisfactory evolutionary explanations for other human genetic traits, this lack of knowledge does not necessarily explain our ignorance of the genetics of skin color.

NATURAL SELECTION AND VITAMIN D

Neer (1974) suggested that the availability of radioactively labeled vitamin D of high specific activity has led to many new discoveries in the field of vitamin D metabolism. Vitamin D is almost inert biologically until it is hydroxylated. Initially, hydroxylation occurs in the liver. This results in 25-OH vitamin D, a compound with weak activity. The compound is further hydroxylated in the kidneys to 1,25 (OH)2 vitamin D—a potent stimulant of intestinal calcium and phosphorus absorption, bone mineral mobilization, and bone growth.

When vitamin D is reduced or not present, because of lack of exposure of the skin to ultraviolet light, absence of vitamin D in food, or impaired vitamin D hydroxylation by the liver or by the kidneys, the following changes can be observed: there is decreased absorption of calcium and mobilization of bone calcium; a drop in serum calcium and inorganic phosphate; defective mineralization of new bone; and mechanical weakness of bone. The bones of vitamin D-deprived children bend under weight and become deformed. This disorder, known as rickets, is characterized by bow legs, knock knees, and scoliosis. Puberty, pregnancy, and lactation predispose a woman to osteomalacia, or adult rickets (Loomis, 1967). Adult rickets may produce a contracted pelvis, which may interfere with normal birth. This deformity, in turn, leads to brain damage of infants and to infant morbidity and mortality. Accordingly, the selective effects of adult rickets could have considerable evolutionary significance.

An excess of vitamin D also leads to disease. Vitamin D intoxication leads to decreased or absent hydroxylation of vitamin D by the liver and the kidneys. Intestinal calcium absorption and mobilization and serum calcium levels rise. This leads to skeletal mobilization of minerals, causing hypercalcemia, bone loss, and pathological calcium deposits in muscle, skin, heart, blood vessels, and kidneys (Neer, 1974).

SUNLIGHT EXPOSURE

In 1967 Loomis resurrected an evolutionary hypothesis of skin pigmentation that had been developed by Murray (1934). The theory postulates that on exposure to sunlight, individuals with black skin make less vitamin D than do individuals with white skin. Accordingly, people who have black skin are at a disadvantage in northern climates where exposure to the ultraviolet rays of the sun is less intense than it is in the tropics. Loomis proposed that the rate of synthesis of vitamin D in the stratum granulosum of the skin is regulated by pigmentation and keratinization of the overlying stratum corneum. This allows only regulated amounts of solar ultraviolet radiation to penetrate the outer layer of the skin and reach the region where vitamin D is synthesized. Different types of skin—white (depigmented and dekeratinized), yellow (mainly keratinized), and black (mainly pigmented)—are adaptations of the stratum corneum. The stratum corneum maximizes ultraviolet penetration in northern latitudes and minimizes it in southern latitudes. Loomis further stated that the reversible summer pigmentation and keratinization that are activated by ultraviolet radiation (suntan) maintain physiologically constant rates of vitamin D synthesis despite the seasonal variation in solar ultraviolet radiation in the northern latitudes.

Neer (1974) listed several experiments and conditions that would substantiate the theory that skin pigmentation significantly reduces vitamin D synthesis:

1. Skin pigment should be proved to decrease vitamin D production by skin irradiated in vitro or in vivo.
2. Dark-skinned people should have lower vitamin D levels than white-skinned people exposed to equally weak physiological ultraviolet irradiation.
3. Dark-skinned people should have more clinical vitamin D deficiency than white-skinned people exposed to equal natural ultraviolet light at high temperature latitudes.

4. The effects of skin pigment on blood vitamin D levels and intestinal calcium absorption should be eliminated if absorption by the skin is bypassed with oral vitamin D supplementation.

Neer pointed out that of these four conditions, (1) and (2) have not been adequately investigated; (3) has been claimed without epidemiological support; and (4) cannot be evaluated until (2) and (3) are established.

To explain the lack of native light-skinned people in the tropics, Neer maintained that white-skinned people should have more clinical vitamin D intoxication than black-skinned people at tropical latitudes, but this has not been observed. Further, tanning of light skin could complicate the picture, for this would decrease the likelihood of vitamin D intoxication. On the other hand, some individuals with very light skin do not tan, particularly albinos. Perhaps vitamin D intoxication should be considered in individuals with albinism in the tropics.

To determine the effect of increased pigmentation of the skin on the cutaneous production of vitamin D_3, Clemens et al. (1982) studied vitamin D concentrations in two lightly pigmented white and three deeply pigmented black volunteers after exposure to a single standard dose of ultraviolet radiation (UVR). Exposure of the white subjects to one minimal erythemal dose of UVR increased their serum vitamin D concentrations by up to 60-fold twenty-four to forty-eight hours after exposure. This dose did not significantly change the serum vitamin D concentrations in the black subjects. Reexposure of one black subject to a dose of ultraviolet radiation six times larger than the standard dose increased the circulating vitamin D to concentrations similar to the levels recorded in the white subjects after exposure to the lower dose. The authors concluded that increased skin pigment greatly reduces the ultraviolet-mediated synthesis of vitamin D_3.

A final word of caution is in order: Blum (1961) pointed out that Eskimos are more exposed to ultraviolet radiation than are some inhabitants of tropical forests. Accordingly, Eskimos should have dark skin. They do not.

RETINAL PIGMENTATION AND VISUAL ACUITY

Hoffman (1974) investigated the hypothesis that the degree of retinal pigmentation in the human eye is adaptive because retinal pigmentation is associated with the maintenance of visual acuity in environments that are stressful to the eyes, such as deserts and snow fields.

The degree of retinal pigmentation was estimated by ophthalmoscopic examination and binocular visual acuity was tested over ten levels of brightness in eighty-four subjects. The general level of retinal pigmentation did not influence visual acuity, and pupil size did not differ at various levels of illumination between individuals who were grouped by degree of retinal pigmentation.

SOLAR HEAT LOAD

Blum (1961) reviewed the differential effect of the solar heat load in people with light and with dark skin. This analysis is important because if heat load cannot be dissipated, the body temperature may rise to dangerous levels. Blum found that even though there are no climatic conditions so severe that death results *directly* from raising the body temperature, circulatory failure and collapse do occur under environmental conditions that lead to excessive water loss—such as in a desert climate. Under extreme conditions of high radiant energy, individuals who could most readily reduce heat load would have an advantage and thus survive. Apparently, dark skin absorbs about 30 percent more sunlight than does white skin (Blum, 1961). Accordingly, dark skin brings about greater local heating. This, in turn, leads to more profuse sweating in individuals with dark skin than is experienced by people with light skin. On the other hand, although increased sweating may have a cooling effect, increased water loss is disadvantageous. But this further complicates matters. If dark skin is disadvantageous in the tropics—because of the likelihood of high heat load—dark skin should be of some advantage in a cool climate, where heat preservation is important. It would follow that a light skin should be disadvantageous in a cool climate. But dark skin is found in hot climates and light skin is found in cool climates. Accordingly, differential heat load cannot be significantly important for survival in extremes of climate.

Baker (1958) studied heat tolerance of blacks and whites. Since there are differences between hot humid and hot dry climates, the experiments were performed under hot wet conditions in Virginia and under hot dry conditions in the Yuma Desert of Arizona. Under the hot humid conditions, forty pairs of men who were matched for body fat, weight and stature walked around a course at 3.5 miles per hour (6 kilometers per hour) for one hour. Under the hot dry conditions, eight pairs of men who were also matched for body fat, weight, and stature were studied under eight different conditions—combinations of variations in clothing, sun, shade, walking, and sitting. These experiments produced the the following results. (1) Under hot wet conditions, when both blacks and whites were clothed and walking, blacks had higher physiological tolerance. (2) Under hot dry conditions, when both groups were clothed and walking or sitting, tolerance was equal. (3) Under hot dry conditions, when both groups were nude and exposed to the sun, suntanned whites had higher tolerance. Baker concluded that these experiments fit the spatial distribution of present-day populations.

Austin and Ghesquiere (1976) gave a heat acclimatization test and a test of tolerance to humid heat stress to Bantu Ntomba and Pygmoid Batwa men of the Zaire River Basin. In the acclimatization test, both groups walked together at 5 kilometers per hour in the afternoon sun. Both groups were also exposed to heat in a water bath. The Batwa were more stressed on the basis of body core temperature and heart rate. Differences in heat tolerance were not caused by the level of acclimatization or by work capacity, but variations in body size were significant. The authors concluded that if this thesis is valid, small-bodied populations—such as the Batwa and Ituri Pygmies—are likely to be more stressed by heat and by working in the sun than are large-bodied populations. Accordingly, both skin pigmentation and body size may be factors in heat tolerance.

Dill et al. (1983) examined the volume and composition of hand sweat of white and black men and women in desert walks. The authors commented that many investigators have failed to detect ethnic differences in the number and regional distribution of active sweat glands. In this study, measurements were made of sweat secreted on one hand and on the whole body of whites and blacks walking in the desert heat. There was no sweat in the gloves of many blacks, but this was true of only a few whites. The volume of body sweat increased in both ethnic groups with the rate of walking, but the volume of hand sweat rose more in whites than in blacks. Concentrations of sodium, potassium, and chloride varied greatly in both groups and were unrelated to race.

CAMOUFLAGE

Cowles (1959) proposed that the concealment factor offers a reasonable explanation of some of the variation in human skin pigmentation. As the coloration of birds and of mammals is more frequently associated with camouflage in a variegated environment, there is precedence for this thesis. Cowles offered several objections to the theory that dark skin is advantageous in a tropical climate. As any traveler to the tropics knows, equatorial climates are not necessarily extremely hot. And some of the highest temperatures are found in the desert, not in tropical forests. Heavily pigmented skin also radiates no better in the infrared spectrum than does white skin. (Cowles objected to the association of ultraviolet radiation and skin cancer, but these objections are no longer supportable in view of our present knowledge.) On the other hand, Cowles's observations on the effect of solar energy in the visible spectrum are convincing. He stated that about 50 percent of the incoming solar energy is in the visible spectrum. Since heavily pigmented skin absorbs 85 percent and white skin absorbs 55 percent of this energy and converts it to heat, the heat burden from the visible spectrum is a greater burden to blacks than to whites. Cowles then developed his thesis of concealing coloration.

Cowles reported that the skin of blonds and redheads reflects 45 percent of the incident light, the skin of a brunette reflects 35 percent, and black skin reflects only 16 percent. As the amount of light decreases, first the black, then the brunette, and last the white individual will approach "invisibility" in the environment. The reverse order of visibility follows increasing amounts of illumination. Under diminished light, the diameter of one's pupil expands in response to the amount of incoming light, but it will neither accommodate to permit examination of a small dark object nor allow objects to be seen in dark shadows. Combat troops frequently darken their skins for battle. The advantages of concealment during hunting are also self-evident. Accordingly, black-skinned hunters have an advantage over white-skinned hunters in tropical forests. Cowles believed that human pigmentation—like numerous other variants—has persisted long after its utility has disappeared.

COLD INJURY

Post et al. (1975) reviewed epidemiologic data from World Wars I and II, from the Korean War, and from Alaska to show that darkly pigmented individuals may be more susceptible to cold injury than lightly pigmented ones. The epidemiologic data are supported by in vitro and in vivo laboratory experiments and by the geographic distribution of peoples with the lightest skin color in temperate climates. The experiments showed that melanin pigmentation is a factor in the extent and the amount of cellular and tissue damage after freezing and then thawing of guinea pig skin. After freezing injury, pigmented epidermis—when compared with unpigmented epidermis—showed greater flattening and thinning, more pyknotic nuclei, poorer cellular definition, more edema, more loss of nuclear staining, and a greater amount of cellular infiltrate. These and other experiments demonstrated that cold injury was always more severe in pigmented than in nonpigmented skin. The differential susceptibility of the skin of blacks to cold injury as compared to whites has been shown in studies of Senegalese and Ethiopian troops, in U.S. soldiers during World II, in Korea, and in Alaska (Post et al., 1975).

Post and colleagues (1975) postulated that cold weather may have been one of several factors responsible for the evolution of white skin. Crippling of the extremities—which is common in frostbitten victims in hunting and gathering bands—would reduce reproductive fitness, and manual dexterity could be permanently impaired.

OTHER FUNCTIONS OF PIGMENTATION

It may be that population differences in skin color may have resulted from the action of natural selection on other functions of pigmentation genes. Deol (1975) pointed out that pigmentation loci in the mouse function independently of color. It is thus possible that color may not be among the more important traits governed by these loci. For example, the products of pigmentation genes are involved in other metabolic pathways. Deol surmised that since there is strong evidence for interspecies homology of pigmentation loci in mammals, the situation in humans may not be radically different.

The possible roles of vitamin D and ultraviolet light, retinal pigmentation and visual acuity, the solar heat load, camouflage and dark pigmentation, cold and the evolution of white skin, and other functions of pigmentation have been reviewed here. No one theory explains the differential distribution of skin pigmentation in human populations. In fact, we surmise that multiple mechanisms, including some of the foregoing, are probably involved in the evolution of skin color.

ALBINISM

Albinism is a hereditary defect in the metabolism of melanin. In albinos, melanin is decreased or absent in skin, mucosa, hair, or eyes (Fitzpatrick and Quevedo, 1966; Witkop, 1973). Most forms of albinism are recessive, but ocular albinism is X-linked. Albinism is found in all human populations, in most if not all vertebrates, and in many other animals and plants. According to Sorsby (1958), Noah was an

Table 3.1
Classification of Albinism

Type	Features
Oculocutaneous albinism	Tyrosinase test negative
	Tyrosinase-negative oculocutaneous albinism
	Albinism-hemorrhagic diathesis (Hermansky-Pudlak syndrome)
	Tyrosinase test positive
	Tyrosinase-positive oculocutaneous albinism
	Chediak-Higashi syndrome
	Hypopigmentation-microphthalmus oligophrenia syndrome
	Tyrosinase test variable
	Yellow mutant
	Tyrosinase test not reported
	Brown albinism
Ocular albinism	X-linked
	Autosomal recessive
	Forsius-Erikson type (?)
Cutaneous albinism	Without deafness
	Piebaldism with white forelock
	Questionable isolated white forelock
	Questionable isolated occipital lock
	Menkes syndrome
	Miscellaneous
	With deafness
	Waardenburg syndrome
	Ziprkowski-Margolis syndrome
	Questionable cutaneous albinism with deafness

Source: Modified from Witkop (1973).

albino, and albinism has been known at least since the first century A.D. Every major language in Nigeria has a name for albinos, for example *afin* (Yoruba), *anyali* (Igbo), *mbakara-Obot-Ikot* (Efik), *eyaen* (Bini), *ogobu* (Idoma), *zebia* (Hausa), according to Okoro (1975). The Efik name is derogatory and literally translates as "a white man from the bush." Okoro also stated that albinos are taunted with names such as "D.O."—for district officer—a reference to colonial days when the district officer was European.

There are many classifications of albinism. Each new category is refined further as additional variants and methods for testing are discovered. Here we will use the classification of Witkop (1973), to which we add other variants reported since this classification was introduced. See Table 3.1.

Oculocutaneous Albinism

There are four major forms of oculocutaneous albinism, tyrosinase-negative, tyrosinase-positive, brown, and yellow mutant. The comparative biochemical, morphological, and clinical features are shown in Table 3.2. Unfortunately, most of the population surveys for albinism discussed here were completed before the four variants of oculocutaneous albinism were known. The different forms of albinism are modified by the population in which the disorder is found. Obviously, albinism is more easily detected in Africans or in Amerindians than it is in blond Swedes, for instance. On the other hand, brown albinism (King et al., 1980) may be easily overlooked in Africans.

The highest frequencies of albinism are recorded in Amerindians, followed by West Africans and their descendants, and then Europeans. Although albinism is more frequent in blacks than in whites, tyrosinase-negative albinism is an exception. According to Witkop (1971), tyrosinase-negative albinism occurs in the United States with approximately the same frequency among blacks and whites— 1 : 34,000 and 1 : 36,000, respectively. Tyrosinase-positive albinism, however, is more frequent in blacks (1 : 14,000) than in whites (1 : 40,000).

Before we continue our analysis of population differences in albinism, we should examine an important facet of this disorder in Northern Ireland (Froggatt, 1960). Froggatt estimated that 80 percent of the population was studied and found a frequency of albinos of 10.9 per 100,000. (Tyrosinase-negative albinism was discovered after Froggatt's survey.) Witkop (1971) calculated that the frequency of tyrosinase-negative albinism in the study was 68 percent, because Froggatt found that 68 percent of the obligate heterozygotes in the survey had translucent irides. (Translucent irides are found in carriers for tyrosinase-negative albinism in whites, but not in blacks.) Accordingly, 68 percent of the albinos in Northern Ireland had tyrosinase-negative albinism, and 32 percent had tyrosinase-positive albinism. The frequency of tyrosinase-negative albinism in Northern Ireland is thus 1 : 15,000—a much higher prevalence than reported in U.S. whites or blacks.

Since oculocutaneous albinism is inherited in a recessive manner, an increase in the number of consanguineous marriages among par-

Table 3.2
Clinical Biochemical and Morphologic Patterns of Four Forms of Oculocutaneous Albinism

Feature	Tyrosinase-Negative	Tyrosinase-Positive	Brown	Yellow Mutant
Hair color	White	White to yellow-white in infancy; yellow blond or red with age. Darker in blacks than in whites	Light to medium light brown	White at birth. Yellow by 2 yrs. In childhood: whites, yellow-red to yellow-brown
Skin color	Pink to red	Pink-white to cream	Light brown: hypo-pigmented spots on shoulders and arms	White in infancy. By 2 yrs., cream to normal. Tanning; little photosensitivity
Pigmented nevi and freckles	Absent	May be present and numerous	Numerous and confluent	Present
Susceptibility to skin cancer	Present, severe	Present, severe (particularly in Africa)	Unknown	Unknown—none seen
Eye color	Gray to blue; little to no visible pigment	Blue, yellow, brown. Pigment increases with age. Blacks darker than whites	Blue to hazel brown	Blue in infancy. Changes gray-blue to yellow-brown in adults. Blacks darker than whites

Transillumination of iris	No visible pigment	Cartwheel effect with pigment deposit at iris borders	Moderate radiating pigment	No visible pigment in infancy. Cartwheel effect in adults. Translucency marked
Red reflex	Present at all ages and in all races	Usually present until 5 yrs. May be present in adult whites; usually absent in adult blacks and Amerindians	Not described	Marked retinal hypopigmentation. Red reflex retained to adulthood. Normal macula. Red reflex in black adolescent adults
Nystagmus	Severe	Mild to severe	Present	Mild to moderate
Photophobia	Severe	Mild to severe	None	Mild to moderate
Visual acuity	Frequently legally blind. Defect constant or worse with age	Frequently legally blind in childhood; visual acuity may improve with age and pigment accumulation	Not described	Frequently legally blind in childhood but vision better than 20/100 in many adults
Salivary tyrosine	Unknown	Half normal	Not described	Unknown
Melanosomes	Stage II early premelanosomes	Mostly Stage III late premelanosomes. Some Stage IV	Not described	Mostly Stage III premelanosomes. Some Stage IV
Incubation of hair bulbs in tyrosine	No pigment	Pigment	Not described	Essentially no pigment increase

Source: Tyrosinase-positive, tyrosinase-negative and yellow mutant from Witkop (1973); brown albinism from King et al. (1980).

ents of individuals with albinism should be found. Of the eighty-nine matings that produced one or more affected children, at least ten were consanguineous; four (4.5 percent) were between first cousins; four (4.5 percent) were between second cousins; and there was a distant kinship in two others. Previous estimates of the rate of consanguineous marriage in Northern Ireland averaged 0.9 percent. (It was not mentioned whether this analysis included the total population, or Protestants only. We would expect that consanguineous marriage would be rarer among Catholics.) At any rate, the consanguineous rates for the parents of albinos were well above the consanguineous estimates for the general population, which one would expect in recessive inheritance.

With recessive inheritance, a phenotype frequency of q^2 and a first cousin marriage rate of α in the population, the theoretical frequency, F, of first-cousin matings of heterozygotes that results in homozygous affected offspring is $F = \alpha/(\alpha + 16q)$, if $\alpha = 0.9$, F is 5.7 percent when the phenotype frequency is 8.75 per 100,000, and 5.3 percent when the phenotype frequency is 10.9 per 100,000 (Froggatt, 1960). These figures compare favorably with the observed 4.5 percent matings of first cousins of parents with albino children. Bell (cited in Froggatt, 1960) estimated an α of 0.4 percent in a general hospital population in England and in Wales. If this F value is used, the F values are 2.6 and 2.4 percent for the different phenotype estimates, which are smaller than the observed 4.5 percent.

Froggatt (1960) also calculated the mutation rate for albinism in the Northern Ireland study. He first outlined two main problems in the calculation of the mutation rate. The largest potential error is in the calculation of the relative fertility (f) of the homozygote. Second, if the gene is not completely recessive (and no genes may be), then the heterozygote may have a selective advantage or disadvantage. Froggatt proposed that even a small selective advantage will cancel out a large homozygote disadvantage because of the high ratio of heterozygotes to homozygotes affected. Further, if the heterozygote is at a selective disadvantage, for the gene frequency to be maintained, a higher mutation rate must be postulated than if the gene had no effect in heterozygous form. Calculation of the mutation rate in the albino population used the classical formula: $m = (1 - f)q^2$, where m = mutation rate, f = relative fertility, and q = gene frequency for albinism.

In a study of 989 albinos in Nigeria, Okoro (1975) reported "significant differences between the number of albinos in the six northern states where the population is more nomadic (72 albinos) and the six southern states where the population is more settled (917 albinos)"

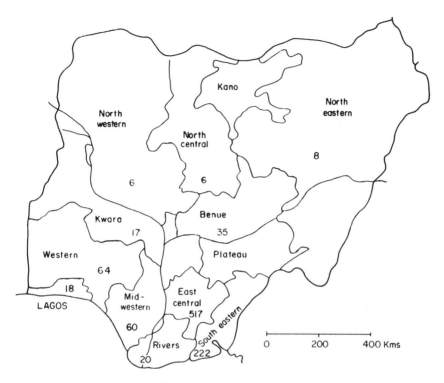

Figure 3.2. The distribution of albinos in Nigeria
Source: Okoro, 1975. Reprinted by permission of Blackwell Scientific Publications, Ltd., Oxford, England

(p. 490). The distribution of subjects with albinism is shown in Figure 3.2. The only section of Nigeria where the frequencies of albinism were estimated was in the East Central Region. Here the prevalence of albinism was 1 : 15,000.

The distribution of albinism in some Amerindian populations is shown in Table 3.3. According to Woolf (1974), the frequency of albinism among the Hopi, Jemez, and Zuni is in the order of the prevalence in the Cuna Indians of San Blas Province, Lower Panama. The frequency of albinism in this group ranges from 1 in 145 to 1 in 213. The reason for the high frequency of albinism in these groups is not known. Selection for the heterozygote, cultural selection in past generations, gene flow, and genetic drift have all been postulated. Albinos do have reduced reproductive fitness, and in many societies there is social selection against them.

Table 3.3
Frequency of Albinism in Southwest Indian Populations

Population and Region	Estimated Population Size	No. of Albinos	Estimated Prevalence
Acoma—New Mexico	2,400	0	
Apache, Mescalara—New Mexico			
Apache, Jicarilla—New Mexico	13,900	0	
Apache—Arizona			
Cochiti—New Mexico	600	0	
Hopi—Arizona	5,000	22	1 in 227
Isleta—New Mexico	2,000	0	
Jemez—New Mexico	1,400	10	1 in 140
Laguna—New Mexico	4,000	2	1 in 2,000
Navaho—Arizona, New Mexico, Colorado, Utah	90,000	24	1 in 3,750
Papago—Arizona	9,000	0	
Picuris—New Mexico	150	0	
Pima—Arizona	7,000	0	
San Felipe—New Mexico	1,300	0	
San Ildefonso—New Mexico	290	0	
San Juan—New Mexico	1,000	2	1 in 500
Sandia—New Mexico	200	0	
Santa Ana—New Mexico	400	0	
Santa Clara—New Mexico	800	0	
Santo Domingo—New Mexico	1,900	0	
Taos—New Mexico	1,300	0	
Zia—New Mexico	425	0	
Zuni—New Mexico	4,200	17	1 in 247

Source: Modified from Woolf (1964).

Ocular Albinism

Ocular albinism is usually X-linked. It apparently occurs in all populations, but relative frequencies are lacking (Witkop, 1979). In whites the eye is the only organ affected. In males there is slight pigmentation of the iris and hypopigmentation of the fundus. Nystagmus, head nodding, and color vision defects may exist alone or in various combinations. Heterozygous females have translucent irides and a mosaic pigmentation of the fundus. Witkop (1973) stated that about 1 percent of all people attending the North Carolina School for the Partially Sighted are ocular albinos, compared to 9 percent who are oculocutaneous albinos.

Ocular albinism was reported for the first time in blacks by O'Donnel et al. (1978a). Notable differences were found between blacks and whites with this disorder. In the black kindreds, almost all of the affected males had pigmentation of the fundus, and their irides were not translucent. O'Donnell and colleagues preferred to call the condition "ocular albinism cum pigmento," because of the paradoxical pigmented appearance. The visual acuity of the black ocular albinos was also better than that of white albinos. O'Donnell and co-workers listed three characteristics of this condition: (1) if mosaicism of retinal pigment is found in the albino's mother, a diagnosis of ocular albinism in the male is confirmed (but some female carriers do not have mosaicism of their retinal pigment); (2) in darkly pigmented whites and in blacks, hypopigmented cutaneous macules and patches indicate X-linked ocular albinism; and (3) biopsy of clinically normal skin reveals giant pigment granules.

Autosomal Recessive Ocular Albinism

O'Donnell et al. (1978b) also described a form of ocular albinism in a white family in which females were as severely affected as males. These individuals have impaired vision, translucent irides, congenital nystagmus, photophobia, albinotic fundus with hypoplasia of the fovea, and strabismus. Obligate heterozygotes lack the ocular abnormalities that are found in females who are heterozygous for X-linked ocular albinism. Skin and hair bulbs are normal. This form of albinism has not been described in blacks.

SKIN CANCERS

It has long been known that squamous cell carcinoma and basal cell carcinoma of the skin are more common in light-skinned individuals than in dark-skinned individuals. The differential effects of ultraviolet light on dark and light skin are the likely explanation.

In the 989 albinos that were studied in Nigeria by Okoro (1975), 95 percent had oculocutaneous albinism. This report did not further characterize the group. The effects of many years of exposure to ultraviolet radiations resulted in an assortment of destructive skin lesions. These lesions consisted of sunburns, blisters, solar elastosis on the neck, arms, and other exposed parts, ephelides (freckles), centrofacial lentigenosis, solar keratoses, chronic superficial ulcers, and, ultimately, squamous cell and basal cell carcinomas (Figures 3.3 and 3.4). The most exposed skin was mainly affected. And, most important: no albino over the age of twenty years was free from malignant or premalignant lesions. Further, the number of albinos rapidly diminished after the third decade, a fact that indicated their relatively short life.

Kromberg et al. (1988) studied 111 patients with tyrosinase-

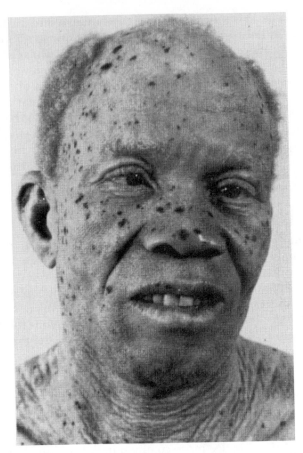

Figure 3.3. A thirty-year-old albino Nigerian man with ephelides and solar elastosis
Source: Okoro, 1975. Reprinted by permission of Blackwell Scientific Publications, Ltd., Oxford, England

positive albinism and the risks for skin cancer in South Africa. The subjects ranged in age from one to seventy-two years of age, with a mean age of eighteen. According to these authors, albinism occurs in about 1 in 3,900 individuals in the South African black population. The study ascertained the risk for skin cancer in terms of age, ethnic group, and the presence of ephelides (freckles), identified the type of cancer, and compared the risk in two different geographical areas. The rate of skin cancers in the albino population was not as high as that in Nigeria but the skin of albinos in South African blacks was significantly lighter than that of whites, and was thus more prone to skin cancer (Kromberg et al., 1987). The prevalence rate for malig-

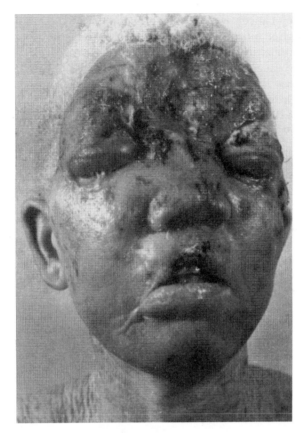

Figure 3.4. A twenty-eight-year old Nigerian albino woman with squamous cell carcinoma
Source: Okoro, 1975. Reprinted by permission of Blackwell Scientific Publications, Ltd., Oxford, England

nant or premalignant skin lesions was 23 percent and positively correlated with age, rising from 6 percent in subjects younger than ten years of age to 100 percent in those older than age fifty. Ephelides were age-dependent and were found in 50 percent of subjects. The absence of ephelides in subjects older than ten years of age was correlated with a significantly higher risk of developing skin cancer. The Sotho-Tawana ethnic group had a higher prevalence of cancer and a lower rate of freckles than the Zulu population, but the differences were not statistically significant. Squamous cell carcinoma was far more frequent than basal cell carcinoma, and both cancers were significantly increased over that of the general population. Melanomas were not found. The head was the most common site of cancer. The incidence of cancer was significantly higher in the population that lived at higher altitude and lower latitude than the population that lived at lower altitude and higher latitude.

With the exception of melanomas on the sole of the foot, melanomas are more frequent in light-skinned peoples than in dark-skinned peoples (Crombie, 1979). Crombie investigated the incidence of melanomas that were recorded in fifty-nine population-based cancer registries. Whites had a wide range of population incidence, and females usually had a higher incidence than did males. If the soles of the feet are excluded, the highest frequencies were found in Northern Europeans and the lowest frequencies in Africans and their descendants. Asians had an intermediate incidence. The highest frequencies of melanomas of the foot, however, were found in Africans. The reason is unknown, although it may be significant that the palms of the hands and the soles of the feet of Africans are relatively unpigmented. Crombie believed that there might be a relationship between pigmented patches—which are commonly seen on the soles of the feet of Africans—and melanomas.

4

Dermatoglyphics

The skin of the palms and soles of humans differs from skin on other parts of the body. Neither hair nor sebaceous glands exist on the palms of the hands and the soles of the feet, but sweat glands are numerous. Pigmentation on the palms and soles is reduced, and the ridges of skin are separated by narrow sulci. Since sole and toe prints are difficult to collect for population studies, our discussion is confined to fingerprint and palm patterns, based on the work of Cummins and Midlo (1961), Penrose (1963), and Holt (1968). For a detailed discussion of the development of epidermal ridges, see Penrose and Ohara (1973). The characteristics and general classifications of dermatoglyphics are discussed first. Next, we will review the special categories that have been developed for anthropological, pathological, and genetic investigations. Finally, population differences in fingerprint and palmar patterns will be analyzed. Unfortunately, several methods of classification of fingerprint and palm patterns exist; some are geared toward clinical investigations, others are more suitable for anthropologic studies or genetic analysis.

DISTAL PHALANGES

The ridges of the distal phalanges form arches, loops, and whorls. A *triradius* is located at the intersection of three opposing ridges, as shown in Figure 4.1. A triradius provides one of the two points for ridge counting, and it is also a focus for tracing. A *core* is the internal feature of a pattern: this is the other landmark for ridge counting. The core may resemble an island, a short straight ridge, a circle, or any of several other forms.

Arches, Loops, and Whorls

The ridges of arches pass from one margin of the finger to the other with a gentle distal curvature. There is no triradius in a plain

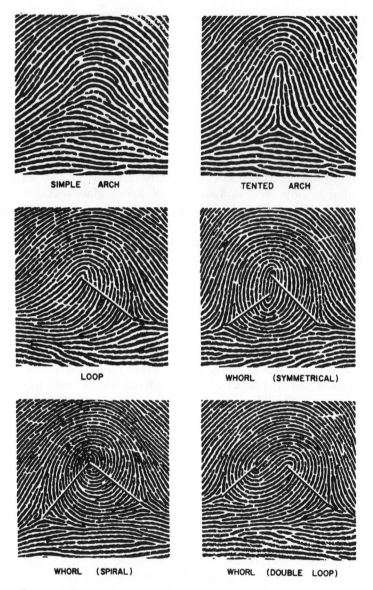

Figure 4.1. Fingerprint patterns. The straight lines crossing the ridges in the lower four patterns are the lines used to determine total ridge counts.
Source; Holt, 1968. Courtesy of Charles C Thomas, Publisher, Springfield, IL

(simple) arch. A tented arch (see Figure 4.1) has a triradius located in or near the midaxis of the digit.

A loop (Figure 4.1) possesses only one triradius. In forming a loop, the ridges curve around one extremity of the pattern and form the head. Ridges flow to the margin of the finger from the opposite extremity of the pattern. This segment of the pattern is described as *open*. If a loop opens toward the ulnar side, it is an ulnar loop; if it opens toward the radial side, it is a radial loop.

A whorl has a concentric design, usually with two triradii (Figure 4.1). A whorl always consists of two loops. The majority of the ridges encircle a central core.

Total Ridge Count

The total ridge count (TRC) is strongly inherited (Holt, 1968). The total ridge count is usually described as TRC = ri, where ri (i = 1, 2, . . . 10) is the ridge count on the i-th finger. Accordingly, the total ridge count is the sum of the counts on all of the fingers.

The number of ridges that intercept or touch a straight line is counted from the triradium to the core of the pattern (Figure 4.1). The two terminal points are excluded. The two ridges that result from a bifurcation close to the line are both included, but ridges that run close to the line without touching it are excluded. Arches have a ridge count of zero. Whorls have two triradii. It is the practice to count the higher one of the two or greater counts. (An absolute ridge count is one in which the sum of the counts in whorls is taken.)

Pattern Intensity Index

The pattern intensity index (PII) is defined as PII = ti, where ti (i = 1, 2, . . . 10) is the number of triradii in the ten finger patterns. Arches—as noted—usually have no triradii, loops have one, and whorls, two. Some arches are tented; these have one triradius. Whorls with three triradii are also encountered. Since PII is dependent on the type and the size of the pattern, PII may be estimated from the pattern type by counting the number of loops plus twice the number of whorls.

PALMAR PATTERNS

The first part of this discussion on Palmar patterns is summarized from Cummins and Midlo (1961), and from Holt (1968). We will then

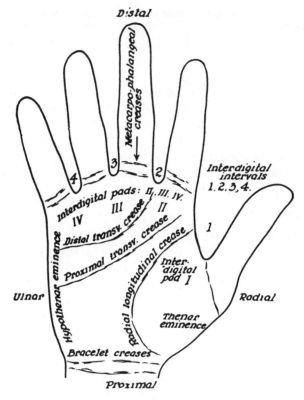

Figure 4.2. Anatomical landmarks in palmar dermatoglyphics
Source: Cummins and Midlo, 1961. Reprinted with the permission
of Dover Publications, Inc., New York, NY

present more recent methods of analysis. All of the methods mentioned here are in current use.

Landmarks

Standard anatomical references or *landmarks* are necessary to characterize palmar dermatoglyphics (Figure 4.2). The principal anatomical regions of the hand are proximal, distal, radial, and ulnar. Beginning with the cleft between the thumb and the index finger, the intervals between the digits are numbered 1, 2, 3, 4. There are six elevations or pads around the center of the palm. The four interdigital pads are made more prominent by flexing the fingers. The pads are in close proximity to an interdigital interval and are numbered in Roman numerals by the number of the neighboring interval: I, II, III, IV. The thenar eminence is prominent in a large region of the proximoradial quadrant of the palm. The hypothenar eminence is an elongated elevation in the ulnar segment of the palm.

The major flexion creases of the palm are also useful landmarks (Figure 4.2). The most distal crease, called the bracelet crease, coincides with the proximal boundary of ridged skin. The metacarpalphalangyl creases separate the digits and palm. The palmist's "line of life"—the radial longitudinal crease—curves to include the thenar eminence and the region of interdigital pad I. The "line of the heart"—the distal transverse crease—and the radial segment of the proximal transverse crease—the "line of the head"—constitute an incomplete proximal boundary of the region that is occupied by interdigital pads II, III, and IV. Zones for formulating palmar main lines are defined, in part, by the levels of termination at the ulnar border of the distal, proximal, and transverse creases.

Palmar Dermatoglyphic Areas

The palms are divided into six dermatoglyphic areas, or configurational fields: the hypothenar, the thenar, and the four interdigitals (Figure 4.3). The configurations of these palmar areas are classified according to the nomenclature of the patterns of fingerprints.

Triradii

Four digital triradii (Figure 4.4) are proximal to the bases of digits II, III, IV, and V. They are named as follows: a, b, c, d, in radioulnar sequence. The digital region is embraced by the two—distal or digital—radiants of each triradius. The proximal radiant—the palmar main line—is most important. It is directed toward the interior of the palm. Four palmar main lines originate from the distal triradii and are named in radioulnar sequence: A, B, C, D. Each letter corresponds to the designation of the respective triradius.

Axial triradii are usually found at or very near the proximal margin of the palm between the thenar and hypothenar eminences, or as far distal as the center of the palm. There may be one, two, or three axial triradii or, rarely, none. A palmar main line is traced from the distal radiant of the axial triradius. It is designated T from the symbol of the axial triradius, t. Occasionally, some triradii are neither axial nor distal and occur in association with certain pattern formations. Accessory triradii are also found. These triradii are related to interdigital regions and represent origins of supplementary main lines.

COLORADO COLLEGE LIBRARY
COLORADO SPRINGS
COLORADO

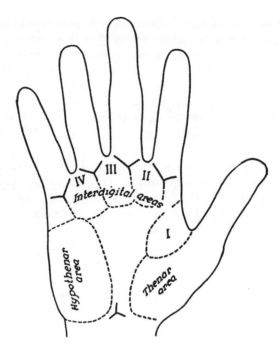

Figure 4.3. The six dermatoglyphic areas of the palm
Source: Cummins and Midlo, 1961. Reprinted with the permission of Dover Publications, Inc., New York, NY

Main Lines

It is usually not difficult to trace a main line from a triradius on the palm, but occasionally the configuration is such that the proximal radiant could be either of two ridges. In this situation, the ridge on the radial side is chosen as the beginning of the main line. The classification of main lines is developed from a numbered sequence around the periphery of the palm (Figure 4.5). Each traced line is described by the position in which it terminates. The sequence of numbers begins with the proximal part of the thenar eminence and continues around the proximal, ulnar, distal, and radial boundaries of the palm. The axial triradius is number 2; the digital triradii are labeled d, c, b, and are numbered, respectively, 6, 8, 10, and 12. Fusion of two main lines occurs when a tracing continues from one digital triradius to another digital triradius (Figures 4.5 and 4.6). A line that is traced to an interdigital interval is formulated by the number of that interval. A line D, for example, extends to the second interdigital interval and is formulated as 11. A line that terminates along the proximal margin of

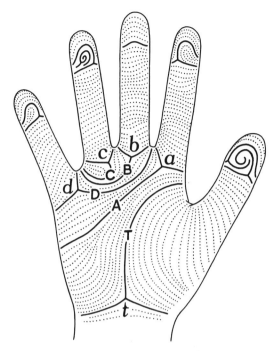

Figure 4.4. The main lines of the palm (A, B, C, D, and T) and the type lines of the fingers. The patterns on digits I and IV are whorls; digit II has a radial loop, digit III has an arch, and digit V has an ulnar loop
Source: Holt, 1968. Courtesy of Charles C Thomas, Publisher, Springfield, IL

the thenar eminence at any point along the radial side of the axial triradius is formulated as 1.

The four remaining intervals along the proximal margin of the hypothenar eminence and the ulnar border of the palm are not limited by definite anatomical landmarks. Position 4 is the approximate midpoint of the ulnar border. The proximal flexion crease frequently reaches this point, and the termination of this crease is a landmark that estimates the extent of the proximal and distal halves of this border. Position 3 is the interval between positions 4 and 2. The distal half of the ulnar border is position 5, which in turn is divided into two halves; the proximal half is numbered 5' and the distal half, 5". The distal flexion crease often ends at a level that corresponds to the point that separates positions 5' and 5".

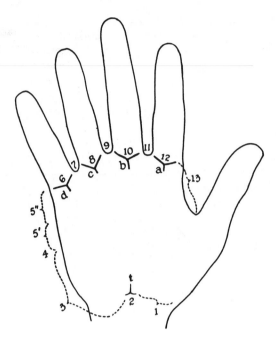

Figure 4.5. The scheme of the numbers for formulating palmar main lines
Source: Cummins and Midlo, 1961. Reprinted with the permission of Dover Publications, Inc., New York, NY

Special Palmar Formulations

Palmar configurations are diverse.

Absence of Distal Triradius and Abortive Main Lines

When the distal triradius is absent, the digital region is not discretely bounded. This variant is mainly confined to triradius c. When the triradius is absent, the corresponding main line is also absent. This situation is formulated by the symbol 0 in the formula in place of the symbol for the main line that is affected. Allied to the absence of a main line is an abbreviated main line. Although the triradius is present, there is practically no main line. This configuration is formulated as x. More frequently, main lines may be aborted to such an extent that the proximal radiant can be traced for only a short distance. This condition is formulated as X.

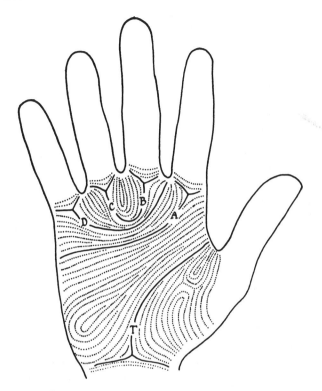

Figure 4.6. The identification of the main lines (A, B, C, D, and T) in a palm having the formula 11.10.8.5'.13-t-L'.L/L.O.L.V. *Source:* Cummins and Midlo, 1961. Reprinted with the permission of Dover Publications, Inc., New York, NY

Formulation of Axial Triradii

A single axial triradius is usually at or close to the palmar margin in the depression between the thenar and hypothenar eminences. Occasionally, a palm may have two or even three triradii at different intervals in the longitudinal axis. There may be no axial triradius. Axial triradii are usually found within a narrow field aligned with the axis of the fourth digit in the palmar formula. If the axial triradius is at or very near the proximal margin, it is named t. The most distally situated position of an axial triradius is near the center of the palm and is named t″; an axial triradius in an intermediate position is t′. If the determination of an axial triradius cannot be made, the symbol ? is used. In palms where there is no t, the ridges of the thenar and hypothenar regions diverge to form a parting, labeled p. If there are two or three axial triradii, they are named in proximal, distal order: t, t′, t″.

Formulation of Configurational Areas

The configuration of each area has a descriptive symbol, and the series of symbols is set as the pattern formula in this order: hypothenar; thenar/interdigital I, interdigital II; interdigital III; interdigital IV. If a configurational area has two configurations, it is formulated with two symbols (Cummins and Midlo, 1961). The only configurations that are formulated in the hypothenar system are those in the proximal portion—the area of the proximal hypothenar pad. The region is bounded on its radial side by a line drawn from the base of the ring finger to position two of the palmar border (see Figure 4.3). If t′ is present, its ulnar radiant divides the hypothenar configuration into distal and proximal elements. Whorls, loops, and tented arches are the three primary types of true patterns in the hypothenar region. Plain arches, open fields, multiplications, and vestiges are not true patterns. Usually there are three triradii in palmar whorls—unlike the usual two whorls in fingerprints. Occasionally, the triradius on the ulnar margin may be outside the area of the ridged skin. The symbol for the whorl is W. Whorls that have spiral or double-looped centers and are enclosed by a concentric periphery are qualified in the formula by adding superscript s, W^s. Whorls that have no more than eight ridges from the core to the nearest triradial point are formulated as small by a lower-case w, instead of W. Hypothenar loops are similar to loops of fingerprints; however, in the hypothenar area there are three instead of two directions of opening: the radial margin, the ulnar margin, and the proximal or carpal margin. The symbol for loops is L, and superscript letters are added as abbreviations of the directions of opening: L^u, T^r, T^c, which indicates the direction of the base as ulnar, radial, or carpal.

An A is the symbol for the plain arch, and superscript letters are used as in tented arches to indicate the direction of the concavity. Open fields (O) are configurations in which the ridges are practically straight and are very rare.

Vestiges are variable. They may lack the sharp recurvatures of ridges, but well-developed vestiges may show some resemblance to a pattern. The symbol for a vestige is V, and when it is close to a definable pattern, the symbol of the pattern type is prefixed (for example, L^cV). There may also be patterns within patterns.

Second, Third, and Fourth Interdigital Areas

The area between digital triradii a and b is designated interdigital II. The area between digital b and c is interdigital III, and the area between triradii c and d is interdigital IV. There may be two patterns in one interdigital area, each of which is entitled to formulation as in other areas.

The C Line

Plato (1970) introduced a new classification of the C line terminations, and pointed out that line C is the only main line of the palm that is truly polymorphic. The paths of the A lines and of the B lines usually follow an ulnar direction; the D line has a radial pathway. On the other hand, the C line may have an ulnar, radial, or proximal direction. The C line may be absent when the c triradius is not present. Plato proposed that the termination of the C line be classified into four modal types that would depend on the direction of its path (Figure 4.7). The following modal types were proposed: first is the ulnar—which includes the termination 4, 5, 6, and 7; second, the radial—represented by terminations 9, 10, 11, 12, or 13; third, the proximal—as shown by X, x, and 8 terminations; and fourth, absent—when there is no c triradius.

Plato (1970) analyzed the following populations from the literature: "Orientals, Oceanians, Negroes, American Indians (North and Central), Caucasians, and Asian Indians." Orientals had the highest frequency of ulnar types (with a maximal frequency of almost 80 percent), followed by Negroes, Oceanians, and American Indians. An abrupt decrease in the frequency of ulnar types was found among the Caucasians and even more so among the Asian Indians. The opposite was found in frequencies of the radial types. The Oceanians and the American Indians show the highest incidence of absence of the c triradius. Since the data were heterogeneous—that is, pooled from the literature and somewhat arbitrarily classified—no statistical analyses were attempted.

The Classification of Penrose and Loesch

Penrose and Loesch (1970) maintained that the traditional methods of palmar dermatoglyphic classification that are used by anthropologists are unsuitable for pathological comparisons and genetic studies. These investigators introduced a method that is more appropriate for such investigations. The principle is to describe all loops. (A whorl always consists of two loops.) All triradii are enumerated, and the most important ones are specified. Arch formations, cusps, multi-

CHRIS C. PLATO

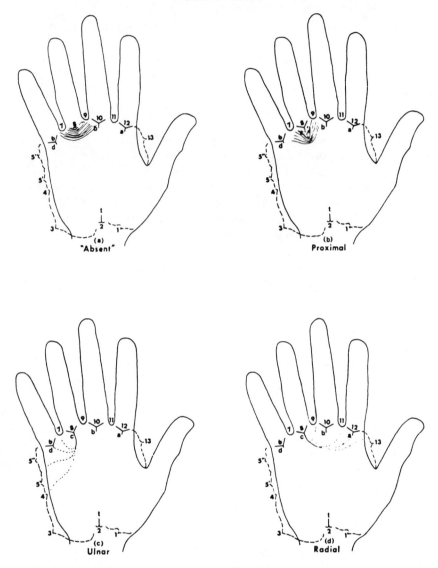

Figure 4.7. The classification of the C line of the palm into four modal types according to its direction and degree of transversality: (a) absent, (b) proximal, (c) ulnar, and (d) radial

Source: Plato, 1970. Reprinted with the permission of Alan R. Liss Inc., publisher and copyrights holder

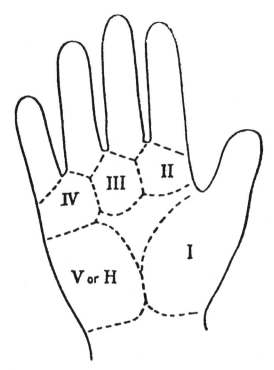

Figure 4.8. Configurational areas on the palm, which correspond to fetal mounds
Source: Penrose and Loesch, 1970. Reprinted with the permission of Blackwell Scientific Publications, Ltd., Oxford, England

plications, and vestiges are neglected because they are not topologically significant. The exits of main lines are also not critical; therefore they are not included. This information may, however, be added to the description.

In Penrose and Loesch's system (1970), a loop is categorized according to the configurational area in which it occurs and the direction of its core. The areas that are used in this classification are shown in Figure 4.8 and roughly correspond to fetal mounds. The thenar and first interdigital areas are combined to form area I. The hypothenar area—which could be described as area V—is called H. Loops are termed peripheral—I, II, III, IV, and H—when the cores point away from the center of the palm (Figure 4.9a) and central—II, III, IV, and H—when they point toward the center of the palm (Figure 4.9b).

Three particular loops are separately specified (Figure 4.9c). The

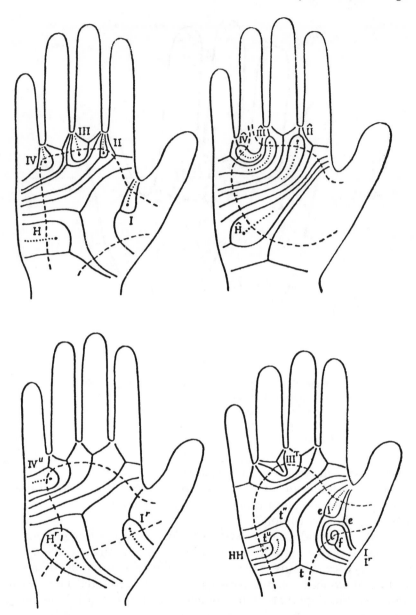

Figure 4.9. *Top left,* Peripheral loops I, II, III, IV, and H. Their relationship to a line drawn at right angles to the ridges is shown. *Top right,* Central loops II, III, IV, and H. Note their relationship to a line drawn at right angles to the ridges. *Bottom left,* Loops that are separately specified, IVu, Hr, and Ir. *Bottom right,* IIIr, IIIᵀ H H e e f t tʺ tu 4 (4)

Source: Penrose and Loesch, 1970. Reprinted with the permission of Blackwell Scientific Publications, Ltd., Oxford, England

first is a peripheral loop on the fourth interdigital area, the core of which exists on the ulnar border of the palm, called IV^u. The second is a loop on the hypothenar region with its direction and exit toward the radial side of the wrist, called H^r, equivalent to the traditional carpal loop. The third is a peripheral loop on area one with exit on the radial-thenar border, called I^r. Tented loops can also be distinguished. It is usually possible to assign a direction to the loop core, but if this is not possible, the tented loop is classified according to the configurational area upon which its center lies. As shown in Figure 4.9, an interdigital peripheral tented loop most commonly is found in the area III and is termed III^T. In the hypothenar area, tented loops are described as T^c, T^r, and T^u, according to the nomenclature of Cummins and Midlo (1961).

In Penrose and Loesch's system (1970), on area I, a triradius is indicated by the letter e when it lies in the distal half of the area; otherwise it is called f. Axial triradii are called t, t', t'', and t'''. A border or extralimital triradius is termed t^b. A triradius that is situated near the center of the hypothenar area is called t^u. Rarely, a triradius deviates sufficiently to the radial side to be called t^r. Occasionally, one of the two interdigital triradii subtends two digits and is the equivalent to the fusion of two triradii. The fused triradii *ab* are termed *z*; fused triradii *bc* are called *z'* and *cd z''*. Interdigital triradii—including those that may be considered accessory but not z, z', or z "—are recorded by giving their number, which is usually 4 (Figure 9d).

The principle of classification of the hypothenar region is to list the loops and triradii.

Penrose and Loesch (1970) studied 250 male and 250 female Europeans. For the sake of simplicity, tented loops were not distinguished separately; triradii e and f were grouped together. Males had an excess of pattern on area I. Females had an excess of z triradii. There was a great excess of area I and IV patterns on the left hand and an excess of II and III patterns on the right. Pattern intensity was not significantly greater in males than in females and was almost the same for the right and the left hands.

Interdigital Patterns

Plato and Wertlecki (1972) proposed another method of subclassifying the interdigital patterns by a method similar to that of Cummins and Midlo (1961). This method considered the termination of the main lines, the presence of accessory triradii, or the combination of the two features. The following patterns were described:

Pattern	Notation
Open/Arch	O
Vestige	V
Ulnar (loop)	U
Radial (loop)	R
Adjacent (loop)	A
Ulnar/Radial (loop)	U/R
Accessory Triradius (loop)	Y
Ulnar/Accessory (loop)	R/Y
Whorl	W

Types O, V, and W need no further explanation; open arches, vestiges and whorls have already been described (Cummins and Midlo, 1961). See Figure 4.10 for an illustration of the various types of loop patterns. The ulnar and radial types were formed by the ulnar or radial paths of the main lines involved. The adjacent pattern is the loop formed by the joining of the terminations of the two side-by-side palm lines. In the U/R type both radial and ulnar loops are found in the same interdigital area. The Y type is formed by the presence of an accessory triradius not directly connected to any of the main lines. The U/Y and R/Y loops are formed by the close association of a main loop and an accessory triradius. These patterns were analyzed in Caucasians, Africans, Micronesians, and peoples of New Guinea. General population differences were found, but the large groups were not delineated. Plato and Wertlecki (1972) believed, however, that more data are needed before definite population frequencies can be established. This appears to be a useful method of classification, particularly since it is easily adaptable for computer entry and analysis. The applicability of this approach is, however, yet to be determined.

DOWN SYNDROME DERMATOGLYPHICS

Holt (1964) studied a sample of 310 individuals with Down syndrome and confirmed that the frequencies of dermal ridge patterns on the fingers differed from those of the general population. Their fingerprints showed less variation, with fewer whorls and arches. (These differences were unusual because the usual tendency is for the frequency of arches to increase as whorls diminish.) The frequency of radial loops was also decreased, and they occurred mainly on digits IV and V, instead of digit II. Interestingly, the deficiency of whorls and

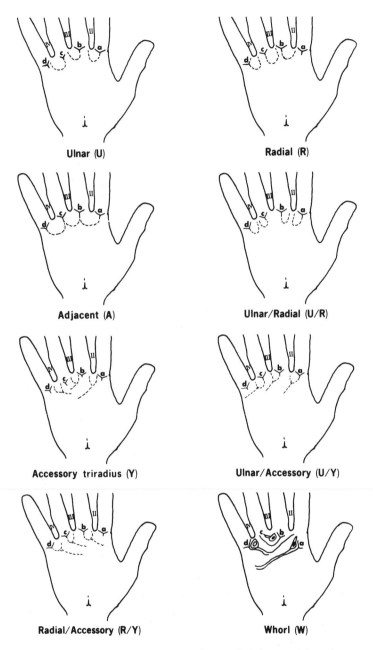

Figure 4.10. Types of "loop" pattern in the interdigital areas of the palm
Source: Plato and Wertelecki, 1972. Reprinted with the permission of Alan R. Liss
Inc., publisher and copyrights holder

arches was accompanied by an increase in ulnar loops, which tended to be high and L-shaped.

Davee, Reed, and Plato (1989) confirmed a study by Fang (1950) that people with Down syndrome had lower total $a–b$ ridge counts in palmar interdigital area II (ID II) than a group of controls in a study of 603 white Down syndrome individuals, 93 black Down syndrome individuals, and 668 white and black controls. Davee and co-workers also found that the white and black Down syndrome individuals had lower total $a–b$ counts than the controls. Additionally, the black controls and those affected by Down syndrome had lower $a–b$ ridge counts than the white counterparts. The total $a–b$ ridge counts in the four populations progressively decreased, with the white controls having the highest ridge counts and the black Down syndrome individuals the lowest. The lower $a–b$ counts were attributed to a pattern in the form of a loop or vestiges in the ID II region.

POPULATION STUDIES

The following sections discuss population studies in Africa, the United States, and Australasia.

Africa

Vecchi (1981) reviewed the literature on geographical distribution of digital dermatoglyphics from surveys of forty-two different male and forty-three different female samples of African populations. Previous studies had reported geographical clines of digital dermatoglyphic patterns in the form of a decrease in the frequency of whorls and an increase in the frequency of loops from north to south in West Africa (Lestrange, 1953; Sunderland and Coope, 1973), and an increasing frequency gradient of loops from west to east in North Africa (Chamla, 1962). Only some of the geographical gradients that had been described in Africa could be confirmed. Vecchi found that in males there is an increase in L frequency and a decrease in W frequency in a north-south direction for West Africa and in an east-west direction for North Africa. The decrease in W frequency from west to east that was found by Sunderland and Coope was not substantiated. Vecchi postulated that the differences between his study and that of Sunderland and Coope occurred because he, Vecchi, used different criteria for the selection of samples. Additionally, Vecchi found that if the entire Sub-Saharan region is considered, the distribution of digital dermatoglyphics is far more complex than a simple north-south gradient of increasing L frequency and decreasing W frequency that Gessain (1957) described.

The Pygmies are of special interest in dermatoglyphics. Glanville (1969) was the first to describe total digital ridge counts in this group. Previous studies on pattern types had shown that there are marked pattern variations among Pygmy populations. The frequency of whorls among the Babinga of the Central African Republic is among the highest in Africa (41–44 percent) (Dankmeijer, 1947; Barrai, 1968); however, Glanville's study of the Efe (Bambuti) Pygmies of the Ituri Forest confirmed the low frequency of whorls in this group. In his sample of 153 males and 54 females, the mean total ridge count was 96.4 in the male and 96.6 in the female—the lowest value recorded in any population.

In a study of Moroccans, Vrydagh-Laourex (1979) found that the observed values agree with the few published data for frequencies of whorls and digital pattern intensity. A trend toward the black African values was observed, with lower frequencies of arches and radial loops on index fingers, higher patterning on thenar areas, and more frequent termination of the D line in the 7 and 7–5–5 types. This led to a lower mean D line termination, a scarcity of distal and triradius t″, and a lower a–b ridge count. The higher patterning in the interdigital areas of the palm, which is characteristic of black populations, was not observed.

Blacks and Whites in the United States

Most of the studies of dermatoglyphics in blacks in the United States are from Steinberg et al. (1975) and Qazi et al. (1977). In a comparison of U.S. blacks with African groups, Steinberg et al. (1975) found similarities, as one would expect. The dermatoglyphics of U.S. blacks were closer to those of other African blacks than to those of Pygmies or San. The modal types of the C line of U.S. blacks were similar to those of black Africans, as were the palmar pattern frequencies. Qazi et al. (1977) found that both males and females had a higher frequency of whorls, a lower frequency of ulnar loops, a lower frequency of arch patterns, and a higher pattern intensity index (the PII was 13.04 for males and 12.39 for females, versus 11.9 for males and 11.2 for females) than those in the group studied by Steinberg and colleagues. In the study by Qazi and co-workers, however, the means of the TRC for fingers were similar to those found by Steinberg and co-workers (the TRC was 119 for males and 106.5 for females). Qazi and co-workers discovered higher frequencies of patterns in thenar/I and III interdigital than in the Steinberg et al. survey (1975).

Qazi et al. (1977) then compared their dermatoglyphic data for African-Americans with those of European-Americans. African-Americans had higher frequencies of whorls and lower frequencies of radial loops on digits of both sexes; lower frequencies of ulnar loops in females and lower frequencies of arch patterns in males; lower total mean ridge count in males; higher frequencies of II interdigital and IV interdigital patterns in both sexes and lower frequency of hypothenar patterns in males; lower frequencies of ulnar direction of modal C line in females and lower frequency of radial direction of C line in males; lower mean a–b count in both sexes; and higher *atd* angle in males.

Quazi and co-workers (1977) confirmed the racial and sexual variations in the dermatoglyphics of African-Americans and showed that statistically significant differences in the type and frequency of patterns occur not only between European-Americans and African-Americans but also among peoples of similar racial backgrounds— such as the group of black Americans studied by Steinberg and colleagues (1975). These differences complicate the evaluation of dermatoglyphics in clinical situations.

Australasia

In a study of Papuans and New Guineans, Plato et al. (1975) reported that the dermatoglyphic distribution of the New Guinea peoples falls within the range of other Australasian peoples, such as Micronesians, other Pacific Islanders, Australian aborigines, and Southeast Asians. Australia appears to be the focus of high whorl frequency and pattern intensity index. Plato and colleagues emphasized that these values decrease steadily as one proceeds to the Amerindians, as well as westward to the Oriental populations, Europeans, and, finally, blacks. In their palmar patterns, Australasians have relatively high frequencies of hypothenar, thenar/I, II, and IV interdigital area patterns, with low frequencies of patterns in the II interdigital area. These peoples also have the highest frequency of accessory triradii of any group, as well as the highest frequencies of both simian and Sydney palmar creases—which exceed 40 percent and 25 percent in some populations of Australia.

5

Polymorphic Variation

The concept of genetic polymorphism grew out of the studies of E. B. Ford (1945, 1964), who derived the concept from his observations that the phenotypes of some characteristics in insects, which appeared to be under the control of variant alleles, were rather more frequent than would have been expected. A *polymorphism* is defined as the occurrence in natural populations of two or more clearly defined phenotypes or variants (forms) of the same trait in which the least frequent form constitutes 0.01 or 1 percent of the population. These traits are under the control of two or more alleles at the same locus. As H. Harris (1980) pointed out, determining this level of frequency is rather arbitrary, even though the 1 percent level has been established to eliminate the possibility that the phenotype or allele might be maintained in the population by mutation alone. He restricted the use of this concept to those genetic loci in which the most common identifiable allele is no more frequent than 0.99. This means that the frequency of heterozygotes in the population is at least 2 percent. Even though the heterozygote frequency in many populations will be less than 2 percent and is, therefore, called nonpolymorphic, this lower degree of variation should not be ignored. Such loci have been called *monomorphic*, but many are not. For the purposes of this discussion the generally accepted definition of polymorphism as clarified by Harris will be used to define the presence of genetic polymorphism in the populations that have been studied for a given trait.

METHODS OF IDENTIFYING PROTEIN POLYMORPHISMS

Population geneticists have used many biochemical and immunologic techniques to identify polymorphisms. Gel electrophoresis using a variety of supporting media is the most common method of demonstrating variation among proteins from a number of different tissues. In a list of seventy-three proteins and enzyme systems that have been shown to be polymorphic, the polymorphism could be demonstrated in whole or to a major degree by electrophoresis in fifty-nine (80.8 percent) (Harris, 1980). A variety of media are used for different biochemical systems, including cellulose acetate, agar, agarose gel, starch gel, and polyacrylamide gel. Commonly, electrophoresis is performed in a continuous buffer system at a specific pH (or within a very narrow range of pH), which has been chosen because of the isoelectric point of the protein, or has been determined by trial and error to reveal variation in a particular group of serum or red cell proteins or enzymes. Other variations in technique can demonstrate even more variability. These maneuvers include the following:

1. Electrophoresis in a vertical direction and/or at high voltage.
2. Electrophoresis in two dimensions in the same or different media (for example, electrophoresis in one direction in cellulose acetate and at right angles to the first direction in starch or acrylamide gel).
3. A discontinuous buffer system in which the buffer in the gel is different from that in the electrode vessels (different buffer systems may also be used in succession as in two-dimensional electrophoresis).
4. Different concentrations of supporting medium in the same gel (for example, a high concentration of polyacrylamide is used in the initial portion of the gel followed by a lower concentration).
5. Adding chemicals to the gel to separate structural components from one another (for example, adding mercaptoethanol to the gel to separate polypeptide chains within each protein molecule).
6. The buffer system may be made up to produce a range of pH within the same gel. In this system, protein molecules move to positions within this pH gradient according to their specific isoelectric points. The gradient can be very narrow or wide by design. This method, called isoelectric focusing, is especially sensitive in identifying molecular heterogeneity due to electrostatic differences.

Two other methods have been used to identify protein polymorphisms. These are quantitative assays, alone or in combination with electrophoresis, and immunological techniques including antibody neutralization and immunodiffusion and precipitation.

ORIGINS AND SIGNIFICANCE OF THE CONCEPT
OF POLYMORPHISMS

Polymorphism has been recognized in human populations since the discovery of the ABO blood groups and the differences in the frequencies in distribution of the ABO blood groups. Polymorphic variation has been found in all natural populations that have been studied extensively. That the occurrence of polymorphism is of interest goes without saying. But what factors are responsible for the origins and maintenance of polymorphisms in natural populations?

The amount and frequency of polymorphism in human populations far exceeded expected levels. Estimates of the amount of polymorphism are reflected in the frequency of heterozygosity that has been identified in individuals in different ethnic populations. More than 104 different genetic loci that code for enzymes have been examined electrophoretically for detectable polymorphisms. H. Harris reported (1980) that 33 (31.7 percent) of these loci had been found to be polymorphic in at least one major ethnic group. This rate of polymorphism almost certainly underestimates the true amount of genetic variability that exists, because the electrophoretic method of screening the proteins is limited to those enzymes that are soluble and only detects those differences that are based on electrostatic charge variations. It may also be true that the distribution of differences among soluble proteins is not representative of the amount of variation in other classes of proteins, such as those involved in the structure of cell membranes. In one attempt to estimate the frequency of heterozygosity, McConkey et al. (1979) used two-dimensional polyacrylamide gel electrophoresis and double-label autoradiography under conditions in which allelic products that differed by a single charged amino acid would be distinguished. From these experiments the rate of heterozygosity was estimated at slightly less than 1 percent for changes involving charged amino acids.

Among different species the range of heterozygosity ranges from 0.04 in a large group of mammals to 0.14 in forty-three groups of *Drosophila*. Population geneticists attempt to explain why some variants are common in certain populations and not in others. There are two competing explanations for this observation. The first is championed by those who believe that the predominant force producing this kind of variation is natural selection. This argument, the "selectionist" theory, is based on the concept that the tremendous amount of heterozygosity represents a reservoir of genetic differences that has been used or is available to help the species adapt to changes in its environment which would adversely effect the survival of the species.

The degree to which the observed variation is a consequence of past environmental challenges, and the adaptation of a particular population to those challenges, indicates the likelihood that selection is a predominant force in promoting the spread of the genetic variants responsible for the polymorphism.

The opposing theory is based on the concept that most of the great variability observed in natural populations is present because it is selectively neutral. In other words, the polymorphism has no particular adaptive value nor does it confer any selective disadvantage on the possessor. The variability may be a consequence of a genetic phenomenon called the "founder effect" or "founder principle." According to this notion, an otherwise rare or uncommon trait occurs by chance in small groups of individuals who are a nonrepresentative sample of a much larger population and who originate or found new populations. This trait is then spread to subsequent generations of this population with a frequency that depends on the subsequent reproductive history of those individuals who carry the trait or variant. Another population genetic phenomenon is called "random genetic drift," in which random changes or chance fluctuations of gene frequency occur in small populations because of the wide variations in combinations of gametes in the individuals in that population. These two phenomena seem, in many instances, to explain variation that has not been shown to be a consequence of selection by a specific environmental agent. For example, sickle cell trait seems to have been selected for by the disease falciparum malaria (see Chapter 8). The process by which a genetic trait achieves polymorphic frequency in a population without the apparent or demonstrable influence of selection has been called "non-Darwinian evolution" (Kimura and Jukes, 1969; Boyer et al. 1969). At the moment, neither the "selectionist" theory nor the "founder effect" theory satisfactorily explains all polymorphism.

The process of genetic drift predicts that allele frequencies in small subpopulations will diverge from each other. Measurements of the degree to which these populations differ genetically show that the genetic divergence between human subpopulations is rather small. A mere 7 percent of the total genetic divergence among what are commonly classified as the Caucasian, African, and Mongoloid races can be ascribed to genetic differences *among* these racial groups. A most important corollary to this observation is that 93 percent of the genetic divergence occurs *within* these groups (Nei, 1975).

The genetic polymorphisms that have been identified in a wide variety of species and in human populations offer a remarkable storehouse of information. Unfortunately, the true significance of most polymorphisms is unknown, and may never be known. Some are undoubtedly relics of the past. For now, polymorphisms are used to assist in mapping the human genome, to identify paternity, to trace the origin of different ethnic groups and other special populations, and to indicate possible susceptibility to disease.

TYPES OF POLYMORPHISM

Polymorphisms have been found at each functional level of the organism. The most obvious human polymorphism is gender. Variations of body structures may achieve polymorphic frequencies in different populations, for example preauricular sinus, supernumerary nipples, metatarsus adductus, polydactyly, branchial cleft anomaly, and café-au-lait spots. Several of these anomalies have achieved polymorphic frequency in blacks, such as preauricular sinus, supernumerary nipples, and polydactyly (see Chapter 11). It is difficult to believe that any of these variants could have achieved polymorphic frequency through some selective mechanism.

Another level of polymorphism is manifested in the quality of the cerumen of the ear canal. Human ear wax is either sticky or dry and flaky. The dry, flaky type behaves as an autosomal recessive trait and is polymorphic only in certain Japanese populations. Still another inherited trait is the ability to taste phenylthiocarbamide (PTC). The inability to taste solutions of PTC is inherited as an autosomal recessive trait. Individuals who are PTC nontasters have an increased risk of developing goiter.

In recent years, improved methods of chromosome analysis have made it possible to identify variations in chromosomal structure which do not appear to have any definitive or identifiable phenotypic effect and are inherited. These usually involve heterochromatic regions that occur in certain chromosomes.

The most extensively studied polymorphisms involve enzymes and serum proteins. These tests usually employ electrophoretic techniques. Protein variation in different body secretions has also been examined, for example, salivary proteins and human milk proteins. Polymorphism in immunoglobulins has been identified by techniques of immunodiffusion and/or immunoelectrophoresis. The most complex molecular polymorphism identified thus far is the HLA histocompatibility gene system, a group of antigens originally defined to provide the basis for organ transplantation. More recently, using a family of DNA cleaving enzymes called restriction endonucleases, it has been possible to define a new type of polymorphism of the genetic material itself, DNA. These DNA polymorphisms are called restriction fragment length polymorphisms (RFLPs). The red cell antigens that constitute the ABO, Rh, MNS, Kell, Duffy, and other red cell antigen polymorphisms were among the first polymorphic systems identified.

In this chapter, each system in which polymorphisms exhibit differences between the frequencies in blacks or populations with significant African ancestry and other major ethnic groups will be described. The possible significance of these differences is considered. It is important to note, however, that Excoffier et al. (1987) cautioned that the selection of study populations introduces some practical problems, particularly in Africa. First, the choice of the population is determined by the amount and quality of the data, and many reports are not representative of the largest African groups. Second, in small samples, only the highest gene frequencies may be recorded, and rare genes, though present, may not be detected. And as has been pointed out (Bowman, 1966), samples are often biased: patients, students, blood donors, armed forces recruits, and military personnel are common sources. Random blood sample ascertainment is rare in human population studies: blood samples are understandably limited to those who agree to give blood.

CHROMOSOMAL POLYMORPHISM

The first chromosomal polymorphism described was in the variation in the length of the Y chromosome. Cohen et al. (1966) noted that there were ethnic differences in the length of the Y chromosome, with the mean length of the Y being greater in Japanese males than in other groups. In later studies, Shapiro et al. (1984) found that pericentric inversion of the Y chromosome is more common in Hispanic and Asian populations than in others.

One of the unexpected results of the introduction of chromosome banding techniques was the discovery of variation in the structure of the chromosomes that was not associated with phenotypic abnormality in the individual (Craig-Holmes and Shaw, 1971; McKenzie and Lubs, 1975; Patil and Lubs, 1977). Lubs and Ruddle (1971) first reported a significantly increased frequency of pericentric inversion of chromosome 9 (Inv 9) in U.S. blacks compared with whites. In other studies (McKenzie and Lubs, 1975) it was noted that the lengths of the secondary constriction heterochromatic regions of chromosomes 1, 9, and 16 are variable. Other polymorphisms that have been identified are qh+ and qh- in these same chromosomes. Hsu et al. (1987) looked for polymorphisms in chromosomes 1, 9, 16, and Y in cytogenetic records involving 6,250 prenatal diagnoses. These investigators found significant differences in the frequency of pericentric inversions for different ethnic groups. Inv (9) was most frequent and polymorphic in blacks (3.57 percent) and in Hispanics (2.42 percent). Whites (0.73 percent) and Asians (0.26 percent) had much lower frequencies of this variant. Pericentric inversion of chromosome 1 was rare. There did not appear to be any association between the presence of the inv (9) and fetal loss, as had been previously suggested. No association was found between the presence of Y chromosome of variable length and fetal loss or an increased risk of later criminal behavior. It was concluded that there is no basis for suspecting possible clinical abnormality based on the finding of inv (9), 9qh+, 1qh+, 16qh+, Yq+, or Yq−.

Chromosomal Q polymorphism has been studied in other groups. In a group of 116 Turkoman individuals of the Kar-Kum Desert of Central Asia (sixteen to twenty years of age), 109 (94 percent) had Q polymorphic variants, while only 7 (6 percent) showed a complete absence of Q bands (Ibraimov and Mirrakhimov, 1983). In another investigation, the chomosomes of 80 "normal" U.S. blacks were evaluated by the Giemsa (CBG) banding technique to estimate size and inversion heteromorphism of chromosomes 1, 9, and 16 (Verma et al., 1981). The sizes of inversion heteromorphisms were classified into five levels. The frequencies of size heteromorphisms of chromosomes 1 and 16 were 10.63 percent and 6.88 percent, respectively, which were not significantly different from those of a "normal" population of whites. Size heteromorphisms for chromosome 9 were higher in whites; 47.5 percent versus 30 percent in blacks. The frequencies of inversion heteromorphisms were chromosome 1, 17.5 percent; 9, 21.9 percent; and 16, zero. Overall, sixty-one chromosomes were found to have an inversion (twenty-eight in chromosome 1 and thirty-three in chromosome 9). A higher incidence of inversion heteromorphisms of chromosomes 1 and 9 was found in U.S. blacks. No inversions were found in chromosome 16 in either group. There appeared to be a significant increase in the size of the h region in association with inversion ($r = 0.99$, $P < 0.01$). In general, enlarged h regions have a higher frequency of inversions. A similar study was carried out in 400 black and white children using Q and C banding to detect heteromorphisms (Lubs et al., 1977). Overall, 6.7 polymorphisms per child were found. The number of polymorphisms was only 3.7 per child when the quality of both banding studies was poor and 12.5 per child when the quality of both banding studies was good. In most instances, the difference between blacks and whites was statistically significant: the polymorphism was more frequent in the children.

Even where the difference was insignificant, the frequency of polymorphisms was greater in black children. On the average, 7 polymorphisms were found in black and 5 in white children. The Q banding polymorphisms were found in chromosomes 3, 4, 13, and 22. The greatest differences in C polymorphisms were seen in chromosomes 4, 12, 18, 19, 20, and 22. In each sample, large centromere regions were more common in black children. The most striking difference was seen in chromosome 19. The highest frequency of inversions was found in a sample in which combined Q and C banding was used. The 1qh partial inversion was eight times more frequent in white children (1 in 200), while the complete 9qh inversion was six times more frequent in black children than in whites.

Racial differences were observed in the size of the short arm of acrocentric chromosomes by Verma and Dosik (1981). These investigators used the technique of chromosomal staining called RFA (R bands by fluorescence using acridine orange), which distinguishes each human chromosome with certainty. With this method, the short arms of human acrocentric chromosomes are well delineated, and minor differences can be readily detected. Verma and Dosik studied the structure of the acrocentrics in 100 blacks and whites—all healthy, unrelated—between the ages of twenty-five and sixty-five. Half of each ethnic group was female. The entire short arm from band p11 to p13 was used to record the size. A system of five levels of size was used, ranging from very small to very large. The short arm of chromosome 18 (18p) was used as the reference for average size. The extremes of size were rare. The size heteromorphisms for chromosomes 13, 14, 15, 21, and 22 were evaluated. Based on the frequency with which a short arm was larger or smaller than the average size, size heteromorphisms occurred 1.5 times as often in blacks for chromosomes 13, 14, and 15, and 2 times as often for chromosomes 21 and 22. No difference between the sexes was determined.

The significance of this variation is a matter of some conjecture. Constitutive heterochromatin is not thought to contain any structural genes. This chromatin material may be associated with the organization of the nucleolus, which is involved with the production of ribosomal RNA. It will be difficult to arrive at a definitive hypothesis without knowing more about the role of this chromosomal material. One cannot rule out the possibility that this heterochromatic material is selectively neutral or that the racial differences are due to the founder effect and/or random genetic drift.

XX TRUE HERMAPHRODITISM

Ramsay and colleagues (1988) reported a high incidence of 46, XX true hermaphroditism among southern African blacks. Pedigree analysis excluded a simple inheritance pattern, and no environmental factor was implicated. With the exception of three whites and two persons of mixed black-white ancestry, all of the individuals were Bantu-speaking blacks, from two main ethnic subdivisions, Nguni and Sotho.

A true hermaphrodite is a person in whom testicular and ovarian tissues are present, but external genitalia are frequently ambiguous. More than half of the reported patients have a 46,XX karyotype and the remainder have a 46,XY karyotype or sex chromosome mosaicism (Ramsey et al., 1988).

Varying numbers of hermaphrodites were studied for specific anomalies. Thirty-eight of thirty-eight had an enlarged clitoris (or small phallus). Twenty-four patients were examined for a separate vaginal opening with a formed vagina, and four had this abnormality. Twenty of twenty-two patients had uterine and fallopian tube differentiation. A gonad was palpated in thirteen of twenty-nine patients so examined.

Although maleness in XX males is frequently the result of the transfer of Y material to the paternally inherited X chromosome, no such material was found in any of these individuals. The investigators were not able to elucidate the etiology of true XX hermaphroditism, and the Y-DNA hybridization studies show that the true XX hermaphrodites are different from XX maleness.

BLOOD GROUP POLYMORPHISMS

The oldest and probably the most extensively studied polymorphism is the ABO blood group antigen system, which was described by Landsteiner in 1900. The antigens for this system can be detected by the natural antibodies in the serum of different individuals, which are present or absent in accordance with the red cell antigenic type of the individual. (Blood group A contains anti-B antibodies and vice versa.) This serum is used to determine another person's ABO type. Using the appropriate antisera, blood types A, B, AB, and O are readily identified. Blood group A can be further subtyped into A_1 and A_2 groups using antisera, although individuals of genotype A_1A_2 cannot be distinguished from genotypes A_1A_1 or A_1O without family studies.

Other subgroups of A are found. Phenotype $A_{1,3}$, for example, is believed to be more common in blacks.

Among Europeans, approximately 47 percent are group O, 42 percent are group A ($A_1 + A_2$), 8 percent are group B, and 3 percent are group AB. These relative frequencies vary from population to population.

In 1927, the MN and P blood group antigen systems were identified by using antisera produced by immunizing rabbits against human red cells (Race and Sanger, 1975).

Another major system is the rhesus or Rh system. Levine and Stetson (1939) reported a pregnancy complicated by stillbirth and the presence of antibodies in the mother's serum that agglutinated the red cells of 85 percent of U.S. whites. Landsteiner and Weiner (1940) found an antibody that detected this polymorphism by exposing human red cells to antisera made by immunizing rabbits with the red cells from rhesus monkeys, hence, the name of the system, rhesus. It was subsequently found that this antigenic system is responsible for hemolytic disease of the newborn. This antibody was found to have a similar pattern of reaction in the population of whites as the antibody described by Levine and Stetson. Reactors to the antibody are Rh positive, and nonreactors are Rh negative. Although a number of other specificities have been shown to be related to the Rh system, the positive phenotype, which is autosomal dominant, accounts for about 90 percent of the patients with hemolytic disease of the newborn. The genetic makeup of this system is very complex. Two systems of description of the antigens of this system exist: those of Fisher (CDE/cde) and Wiener (R;r).

In addition to the ABO and Rh blood group systems, several others have been identified that are polymorphic in one or more populations; for example the MNSs, P, Dombrock, secretor, Lutheran, Kell, Lewis, Duffy, Kidd, Cartwright, Auberger, Diego, and Xg red cell antigens systems.

The antigenic differences in the ABO system are due to differences in the structure of carbohydrates attached to high-molecular-weight glycoproteins, which constitute the blood group substances found in normal body secretions, such as saliva and gastric juice. The sugar moieties responsible for the antigenic specificity are added sequentially by enzymes under specific genetic control (Hakomori, 1981). An alpha fucose molecule is added, under the control of the H gene. This is followed by the addition of an N-acetylgalactosamine molecule, which confers type A specificity. If a galactose molecule is added the specificity will be type B. If no carbohydrate moiety is added, the H substance persists and the blood type will be O. The enzymes directing these reactions are glycosyl transferases, either an A transferase or a B transferase, depending on which carbohydrate molecule the enzyme attaches to the basic core structural macromolecule. It appears that the carbohydrate determinants that determine group A specificity are qualitatively the same in groups A_1 and A_2, but there is a quantitative difference in the sugar determinants attached. This appears to be a consequence of a difference in the effectiveness of the N-acetylgalactosaminyl transferases, which is determined by the respective group A genes (Schachter et al., 1973). The enzymes can be biochemically distinguished by isoelectric focusing and by their pH activity curves (Topping and Watkins, 1975; Watkins, 1980). It is, therefore, possible to make the phenotypic distinctions previously discussed in the A group genotypes which would otherwise require family studies. None of the many other well-defined blood group systems have been characterized.

The biochemical basis for the antigenic differences in these other systems remains largely unknown, except for the P and MN systems. In the P system the antigenicity is apparently located in the carbohydrate moieties of glycoproteins and glycolipids, with at least two loci involved (Watkins, 1980). In the MN system, antigenicity occurs in the sialoglycoprotein portion of the red cell membrane (Marchesi et al., 1972), and the M and N reactive substances are differentiated by at least two amino acid substitutions in the NH_2-terminal portion of the polypeptide chain (positions 1 and 5) (Dahr et al., 1977; Blumenfeld and Adamany, 1978).

BLOOD GROUPS IN POPULATIONS

The ABO System

The composite *gene* frequencies of the ABO system worldwide are O, 0.623; A, 0.215; and B, 0.162. By way of comparison, the highest frequency of blood group O is found in Central and South American Indians, where almost all are of this type. The frequency of group O in Northwest Europe is about 0.68 and in U.S. blacks is 0.713.

The gene for blood group A is common in Europe, especially in central and eastern regions. Armenians in Iran and other regions also have high frequencies of the A allele (Bowman and Walker, 1963). The aborigines of southwest Australia and the Blackfeet Indians in western North America also have high frequencies. The A_2 gene is found in high frequency in Europe (0.10), the Near East, Africa, and western Asia. Unusual forms of the *A* gene are seen in special populations, for example the *Aint* and *Abantu* genes in blacks.

In contrast to the A blood group gene, the *B* gene is almost totally absent from the aboriginal tribes of Australia and the Americas. There is a very high frequency of this gene in central Asia and north-' ern India, while West Africa and Egypt have the lowest frequencies. Variations in the *B* gene have been recognized in molecular studies of the B group enzymes. Three levels of activity have been recognized, one of which (the highest) belongs to the B group black populations. The BI and BII levels of activity are found in European populations. Quantitative estimates of red cell A, B, and H receptors reveal differences between white and black populations.

Other Polymorphic Blood Systems

The other blood systems that we will discuss are the secretor, P, Rh, Duffy, Kidd, MNSs, Dombrock, Cartwright, Diego, and the Xg groups.

The secretor blood group system (Sese) is a balanced dimorphism in most populations. The frequency of the *Se* gene is about 0.5 throughout Europe. The San of South Africa, the Australian aborigines, Eskimos, and Amerindians have a frequency of about 0.85.

The P system is another polymorphic variation. The frequency of the *P1* gene is high in black Africa (0.90), intermediate in Europe (about 0.50), and lowest in the Far East.

The Kell blood group system is complex, especially important because it is often responsible for transfusion reactions and the development of antibodies after pregnancy. It is made up of four (4) groups of closely linked alleles: *K, k; Kp^a, Kp^b; Js^a, Js^b, K^{11}, K^{17}*. The first of each pair is uncommon while the alternative is frequent. When red cells are missing the common antigen they will always have the uncommon antigen. Cells without k will have K. The two main antigens are K (Kell) and k (Cellano). The k antigen is much more frequent than the K. Very rare individuals have no Kell antigens. They are designated K$_o$. They express a K$_x$ reactivity which is probably the substrate on which Kell antigens are manufactured. When blood is routinely tested for transfusion usually only the Kell antigen is tested for.

The K antigen is more frequent in the white population (0.09) than in blacks (0.035) where it is relatively infrequent. On the other hand *Js^a* is much more frequent in black populations (0.04 to 0.20) than in white (0.001).

The Rh blood group system is extremely complex, as noted earlier.

Table 5.1
V Antigen Frequency in Selected Populations

Population	No. tested	Positive	%
Whites in London	407	2	0.5
Whites in New York	444	2	0.4
Whites in Seattle	514	1	0.2
Blacks in New York	168	45	26.8
Blacks in Seattle	327	94	28.8
W. Africans in Lagos & Accra	150	60	40.0
Bantus in South Africa	511	170	33.3
Orientals	272	1	0.4
Amerindians	174	2	1.2

Source: Modified from Race and Sanger (1975).

A detailed description of Rh genotypes and genetics is beyond the scope of this book (see Race and Sanger, 1975; Mourant et al., 1976d). The distribution of these genotypes among European populations is similar, except for the Basques of Spain and France and the Bedouins of the Sinai Desert, in whom the frequency of r (*cde*) is approximately 0.5. The principal characteristic of black African populations is the very high frequency of *cDe* (R^0) ranging from 0.50 to 0.90. This genotype has a frequency of only 0.03 in European populations. The V antigen is also a distinctive marker in black African groups, especially in the Sudan. The V antigen accompanies *cDe* twice as frequently as it is associated with *cde*. This antigen is found only in individuals possessing the gene complexes *cDe*, *cDue*, or *cde*. The V antigen also occurs in Arabs. Table 5.1 shows the frequency of the V antigen in several populations.

The genes in Duffy blood group system are designated Fy^a, Fy^b, and Fy. The frequency of the Fy^a gene is about 0.40 in England and throughout Europe. The frequency is much higher in the Far East (among Japanese and Koreans) and India, and is virtually 100 percent in most Melanesian populations. The frequency of this gene is very low in American and African blacks (0.053). The Fy gene has a very high frequency and is an especially good marker of black populations (U.S. blacks—0.825; West Africans—over 0.90) and in estimating racial admixture between blacks and whites. The Fy gene is also frequent in the Near and Middle East (Saudi Arabia, Yemen, and Kurdistan).

The Duffy blood group system figured prominently in admixture estimates in the State of Puebla, Mexico (Lisker et al., 1988). Lisker and co-workers analyzed ABO, MN, Rh-Hr, Duffy, and Diego blood groups in addition to serum haptoglobins, albumin factor Bf, hemoglobin, and glucose-6 phosphate dehydrogenase, and found that the proportions of black, Amerindian, and white genes were 10.7, 56.3, and 33.0 percent, respectively. Although Lisker and co-workers noted historical evidence that the black population was high in the sixteenth century, they cautioned that further studies are needed, because if Fy^a is excluded from the calculations, the black ancestry is reduced to 1.6 percent.

The Kidd blood group system has two alleles identified as Jk^a and Jk^b. Black African populations have the highest frequency of Jk^a (0.75), the frequency is intermediate in European populations (0.50) and low in Chinese groups (0.30).

This MNSs system polymorphism is distributed worldwide. The M gene has a frequency averaging between 0.50 and 0.60. It is of anthropological importance because of the identification of a pair of allelic genes U/u that are detected by an anti-U antibody. Essentially all Europeans react to this antibody, but only a minority of American blacks do (0.012). Among blacks who are phenotypically S-s-, 84 percent are U- and 16 percent U+. The symbol Su indicates the S-s- phenotype, which is associated with the U- status. The U- type is highest in Congo Pygmies (35 percent) (Fraser et al., 1966). The frequency of the Su gene is about 0.60 percent. By contrast, no U- individuals were found in tests of 1,000 Bantu (Marsh et al., 1974). Yet another antigen, M1+, in this system is able to distinguish between whites (4 percent), U.S. blacks (24 percent), and certain Africans (47 percent). The alignment of the $M1$ gene with S,s or Su does not differ from the rest of the M genes.

Two additional antigens, Hunter (Hu) and Henshaw (He), were found to be associated with the MN system. The Hu antigen was found in 7 percent of U.S. blacks and 22 percent of West Africans. The He antigen was found in 2.7 percent of a West African population, but not in Europeans. Most of the positives had the N and S antigens also. About 2–3 percent of U.S. blacks are positive for the He antigen. It is associated with Ns in Papua, with MS in Congo blacks and in the Khoikhoi, and with Ms in Cape Coloureds. The evidence is strong that in most populations in which it segregates, the He gene is linked with the $MNSs$ genes.

The alleles in the Dombrock blood group system are Do^a and Do^b. The Do^a gene has a frequency of about 0.42 in Northern Europe. Black Africans may have the lowest frequency. In Thais the gene frequency is as low as 0.07.

In the Cartwright blood group system the alleles are Yt^a and Yt^b. The Yt^a frequency is very high in Northern Europeans at 0.96. The

Yt^b gene has a low frequency in all populations tested, with the lowest frequency probably found in black populations.

The main gene in the Diego blood group system, Di^a, is characteristic of Mongolian populations. The frequencies of this gene vary from one population to another. The highest frequencies are found in Amerindian groups in Venezuela (about 0.40). Diego frequencies are lower in the Far East, with intermediate prevalences in Central or South America. The frequency of Di^a is about 0.05 in Korea and Tibet. The Diego antigen is not found in black populations and is uncommon in Europeans.

The Xg blood group system is unique because it is X-linked. The gene frequency of this system ranges from 0.50 in Finland and Norway to 0.76 in Sardinia. In the Amerindians of North America and in the Australian aborigines, the frequency of Xg^a is about 0.80. In blacks the frequency of this gene is about 0.55.

Blood Groups as Population Markers

Some of the blood group genes are distinctive enough to be considered specific markers of a particular population. In blacks these are V (Rh), Js^a (Kell), and Fy (Duffy). These might be called true African markers. In general, Africans have high blood group O frequencies, while A and B groups range from 0.10 to 0.20. The frequencies of M and S genes are lower in blacks than in Europeans, and the Su gene is common in blacks.

In the Rh system, the distribution of certain of the genotypes is unique to these groups. The cDe (R^0) haplotype has a relatively high frequency in Sub-Saharan Africa, ranging from 0.60 to 0.90 or higher. The V antigen also exists in these populations and is found in individuals who are cDe (R^0) or cde (r) haplotypes. This latter haplotype is lower in Africans than in Europeans. The cDe (R^0) haplotype reaches its maximal frequency in tropical Africa. The K gene is rare, but the Kell system has a variant that is peculiar to Africans called Js^a.

The Duffy system in blacks has a very high frequency of a silent gene Fy, which is responsible for the Fy(a-b-) phenotype. When the distribution of blood groups in Africa is examined, the high values of the typically African markers seem to be found in West Africa, with an admixture with European genes in East Africa.

Table 5.2 summarizes some of the gene frequencies for the polymorphic blood groups in European and black populations.

Genetic Variation and Disorders in Peoples of African Origin

Table 5.2
Frequencies of Polymorphic Blood Group Genes in European
and Black Populations

System	Gene		Whites (%)	Blacks (%)
			Frequency	
ABO	O		0.660	0.713
	A^1		0.209	0.135
	A^2		0.070	0.038
	B		0.061	0.114
P	P^1		0.540	0.836
	P^2		0.460	0.164
Lutheran	Lu^a		0.039	0.020
	Lu^b		0.961	0.980
Duffy	Fy^a		0.421	0.053
	Fy^b		0.549	0.122
	Fy		0.030	0.825
Kidd	Jk^a		0.486	0.725
	Jk^b		0.514	0.275
Xg^a	Xg^a		0.659	0.550
MNSs	M		0.532	0.495
	N		0.468	0.505
	S		0.327	0.171
	s		0.673	0.716
	*		0.001	0.113
Rh	D		0.590	0.713
	d		0.410	0.287
	C		0.433	0.179
	c		0.567	0.821
	E		0.155	0.107
	e		0.845	0.893
	V		0.001	0.157
Kell	K	(K^1)	0.045	0.005
	k	(K^2)	0.954	
	Kp^a	(K^3)	0.012	
	Kp^b	(K^4)	0.988	
	K^u	(K^5)	0.999	
	Js^a	(K^6)	0.000	0.144
	Js^b	(K^7)	1.000	0.856

Source: Modified from Giblett, 1969.

*Gene with no S or s expression; U usually not expressed.

ENZYME POLYMORPHISMS

To determine the molecular structure of human proteins, biochemists have isolated purified and concentrated protein with a single peak of activity or a single antigenic specificity of greatly increased concentration as the evidence of purity. As more sensitive and sophisticated methods of analysis and protein fractionation were introduced, such as column chromatography and immunoelectrophoresis, it became apparent that for some proteins no amount of purification or protection during purification procedures could produce "pure" proteins by the usual criteria. Despite extensive purification, some proteins still showed more than a single zone of activity or immunoreactivity. Eventually it was found that these multiple peaks of activity were different structural forms of the same enzyme with the same or very similar substrate specificity. These forms of the enzyme also had different antigenic determinants. Although it could not be proven, apparently these different forms existed in vivo, and were not artifacts of the experimental technique.

Isozymes

The multiple forms exhibited by enzyme proteins are called isozymes or isoenzymes (Market and Moller, 1959). Isozymes or isoenzymes are generally defined as "multiple molecular forms of a given enzyme occurring in a single individual or in different members of the same species" (H. Harris, 1980, p. 98).

H. Harris (1980) summarized the major concepts of the molecular origins of isozymes and provided much of the impetus to progress made in this field of study. It appears that different kinds of molecular relationships are likely to be involved in different types of isozymes produced in different ways. The application of zone electrophoresis in different media under varying conditions has been responsible for the swift discovery of a variety of examples of isoenzyme systems in recent years. This powerful investigative tool has been extended to a wide range of enzyme systems through the development of sensitive and specific staining methods, and through innovative electrophoretic methods that permit one to identify a pattern of specific enzyme activities in a mixture of enzymes taken from a simple, crude homogenate of fresh tissue. Only relatively small amounts of material are required for isozyme analysis, which makes it feasible to examine material from large numbers of individuals.

A group of proteins that make up a family of isozymes are very much alike, but not identical, in their enzymatic properties. They may catalyze the same reactions but they have different kinetics. The members of the isozyme family can exist in similar tissues, but there may be differences in the isozyme patterns among cells of different tissues. The pattern range of isozymes may be attributed to one or more of the following three mechanisms:

1. There may be two or more gene loci coding for structurally different polypeptide chains of the enzyme.
2. Multiple alleles may occur at the same genetic locus, which codes for the distinctly different structures of particular polypeptide chain.   .
3. Secondary isozymes may form because of posttranslational modification of the enzyme structure. Because multiple allelic differences are most likely the result of one or a maximum of a few amino acid differences, polypeptide chains that make up isozymes that are generated from multiple gene loci are more likely to show much greater differences.

Aconitase

The enzyme aconitase (ACONS) catalyzes the conversion of cis-aconitate to isocitrate. There are two forms of this enzyme, one soluble (ACONS) and one mitochondrial (ACONM). Electrophoretic analysis has demonstrated two alleles of *ACONM* and seven alleles of *ACONS*. Variant forms are found in both mitochondrial and soluble ACON enzymes. The electrophoretic patterns in heterozygous individuals are consistent with monomers of both the soluble and mitochondrial isozymes. Polymorphic variation was observed in ACONS in a sample of a Nigerian population at the three most common alleles (Slaughter, Hopkinson, and Harris, 1975). This same isozyme system was examined in three samples from northeastern Brazil. Two were collected in the State of Bahia and one in the State of Sergipe. The samples differed in degrees of black admixture: the greater the degree of black admixture in the population, the higher the frequencies of the alleles *ACONS 4*, *ACONS 2*, and *ACONS 6*. The findings fit with the known *ACONS* gene frequencies in present-day Nigerians and with the past history of Yoruba slaves in Bahia (Azevêdo et al., 1979).

Red Cell Acid Phosphatase

Six different phenotypes of red cell acid phosphatase are found in European populations. The designation of the types and their frequencies are 13 percent for type A, 43 percent for type BA, 36 percent for type B, 3 percent for type CA, 5 percent for type CB, and 1 in 600 for type C. These phenotypes are determined by three autosomal alleles, ACP_I^A, ACP_I^B, and ACP_I^C. Types A, B, and C are homozygous genotypes. The ACP_I^C allele is much less frequent in blacks than in whites. Several other red cell acid phosphatase isozyme phenotypes have been recognized in addition to the six already mentioned (Giblett and Scott, 1965; Karp and Sutton, 1967). These are called RA, RB, and RC. Another, allele ACP_I^R, in combination with the previously mentioned alleles produces these heterozygous combinations. This allele is vary rare in white populations, but has a gene frequency of 0.01 in U.S. blacks, and is especially common in South African blacks, where it may be as frequent as 0.20 (Jenkins and Corfield, 1972). In the Babinga Pygmies the frequency of the ACPR allele (17 percent) is comparable to that seen in the Khoikhoi and San people (Santachiara-Benerecetti et al., 1977). The frequency of this allele is higher in the Khoisan people than in any other population in the world. In the Obamba and Bateke populations, the frequency of Pr is 0.013 (1978).

Not only does each of the ACP alleles determine a structural form of the red cell acid phosphatase isozyme system, but also the structural differences seem to determine quantitative differences in acid phosphatase activity. There appear to be significant differences in the average level of activity between ACP types. On the average, cells of type B individuals show about 50 percent more activity than do those of type A individuals. Red cell activity of type CB individuals is greater than in persons of type CA or B. Studies of acid phosphatase activity were performed on three populations in the Sudan and one in Nigrilis, India (Saha and Patgunarajah, 1981). In these populations, the relative activity of the acid phosphatase genes was ACP^A—1.0, ACP^B—1.2, and ACP^C—1.3. Red cell acid phosphatase activity also seemed to be higher when HbA_2 was increased (15 percent) and the subject had sickle cell anemia (21 percent). People with haptoglobin type Hp 2 had an 18 percent higher level of acid phosphatase than did those with Hp 1. People with glucose-6-phosphate dehydrogenase (G-6-PD) deficiency had a lower level of acid phosphatase activity (20 percent) than did those with normal G-6-PD activity (see Chapter 6).

Adenylate Kinase

Variants of adenylate kinase (AK) enzyme were characterized using horizontal starch-gel electrophoresis by Fildes and Harris (1966). Three autosomal inherited isozyme forms were named AK 1, AK 2-1, and AK 2-2. The heterozygous form, AK 2-1, occurred in about 10 percent of a British population. Bowman et al. (1967) described adenylate kinase polymorphism in European-Americans, African-Americans, West Africans, and Lacandon Mayans of Chiapas, Mexico.

Nine percent of 1,315 European-Americans were polymorphic for AK 2-1, but only 1 percent of 1,063 African-Americans had AK 2-1. Eight hundred blacks from Ghana and Nigeria were examined for adenylate kinase polymorphism, and none had AK 2-1. Further, 150 subjects from inbred Lacandon Mayan families were all AK 1. Bowman et al. (1967) performed intermixture calculations on the African-American sample and found that with adenylate kinase, European intermixture was estimated at 13 percent, rather than the usual 25–30 percent predictions.

Even though adenylate kinase is monomorphic or infrequently polymorphic in Africans and their descendants, some other populations are similar. The Lacondones have already been mentioned. Bowman et al. (1971) studied this enzyme in Vietnamese populations and found AK 2-1 to be present in only 1 percent of the Vietnamese, 2 percent of the Khmer, 3 percent of the Malo-Polynesian Rhade, and 0 percent among the Mon Khmer Sedang and Stieng. Among Iranians in Shiraz, 11 percent had AK 2-1 (Bowman and Ronaghy, 1967).

Peptidase

The peptidases (PEP) are a group of enzymes in red cells which hydrolyze dipeptides and tripeptides into their constituent amino acids. Five enzymes (A, B, C, D, and E) have been described. Each is produced by separate, unlinked, genetic loci. They are readily detected by gel electrophoresis.

Three types of Pep A, under the control of two autosomal alleles (*PEP A¹* and *PEP A²*), are called Pep A 1 (the most common type in all populations studied thus far), Pep A 2-1, and Pep A 2. With the exception of black populations, the latter two phenotypes are quite rare. Pep A 2-1 has been found in 15 percent and Pep A 2 in 1 percent of black populations. The gene frequency for *PEP A²* is 0.07 in blacks. The isozyme pattern is consistent with that of a dimer.

Peptidase C (Pep C) is polymorphic among the Babinga Pygmies (Santachiara-Benerecetti, 1980). The electrophoretic patterns were consistent with three alleles, one of which is silent. The silent allele has a gene frequency of 0.208 in this Pygmy population. A deficiency of this isozyme does not appear to have any adverse effects. It is also polymorphic in European and American whites.

Peptidase D (Pep D) specifically splits iminodipeptides with C-terminal prolineor hydroxyproline, for example, prolylglycine. It has been named prolidase, imidopeptidase, and proline dipeptidase. Prolidase deficiency is associated with a massive excretion of di- and tripeptides, all containing proline with another amino acid that is amino-terminal (Powell et al., 1973). A number of electrophoretic variants have been identified. These are under the control of three alleles, *PEP D¹*, *PEP D²*, and *PEP D³*. This isozyme group is polymorphic in Africans and U.S. blacks and in East Indians and Pakistanis. The *PEP D³* allele has its highest frequency in blacks.

Phosphoglucomutase

Phosphoglucomutase (PGM), a phosphotransferase enzyme, catalyzes the interconversion of glucose-1-phosphate and glucose-6-phosphate; PGM is found in most tissues. The complex isozyme patterns observed with electrophoresis suggest that these patterns are produced by at least three gene loci. The individual isozymes are monomers. Three common phenotypes have been identified: $PGM_1 1$, $PGM_1 2$–1, and $PGM_1 2$. In English populations the frequencies of these phenotypes are $PGM_1 1$—58 percent, $PGM_1 2$—1 to 36 percent, and $PGM_1 2$—6 percent (H. Harris, 1980). Phosphoglucomutase is determined by two autosomal alleles ($PGM_1{}^1$ and $PGM_1{}^2$). The technique of isoelectric focusing in thin acrylamide gels used in PGM electrophoresis showed that the three PGM_1 phenotypes could be resolved into ten different phenotypes. The phenotypes can be explained by four common alleles: $PGM_1{}^1+$—0.62 frequency; $PGM_1{}^1$—0.17; $PGM_1{}^2+$—0.14; $PGM_1{}^2$—0.07. (the + and − symbols refer to differences in isoelectric points of the isozymes). A second PGM genetic locus (PGM_2) was identified. Most individuals are homozygous for the $PGM_2{}^1$ allele. Other, much less frequent phenotypes of this locus are 2-1, 3-1, 4-1, and 5-1. Yet a third genetic locus, PGM_3, has been identified with three common phenotypes, 2, 2–1, and 1. In a majority of tissues, 80 to 95 percent of the PGM enzyme activity is derived from the PGM_1 locus and the remainder derives from the PGM_2 locus. Figure 5.1 shows the allelic patterns of the three loci of PGM (H. Harris, 1980).

The $PGM_2{}^2$ allele has been found only in blacks (Vogel and Motulsky, 1979). Studies of PGM subtyping reveal other variants of PGM_1 in certain African groups (Santachiara-Benerecetti et al., 1982). A unique PGM_1 variant called 1 Twa was detected in the Twa Pygmies from North Rwanda. A $PGM_1{}^{2S}$ allele was found in 9.6 percent of the North Twa Pygmies. A low frequency of PGM_1 1 Twa (1.2 to 3.6 percent) was found in all other groups except the Hutu people (6.4 percent). A high frequency of PGM_1 (1F) was found in the South Twa Pygmies (20 percent) and the Tutsi people (18 percent).

In another study by Tipler et al. (1982), eleven South African populations were typed for the PGM_1 locus using isoelectric focusing (pH range 5–8) in acrylamide gels. The gene frequencies of four alleles were compared to those of other European and black populations.

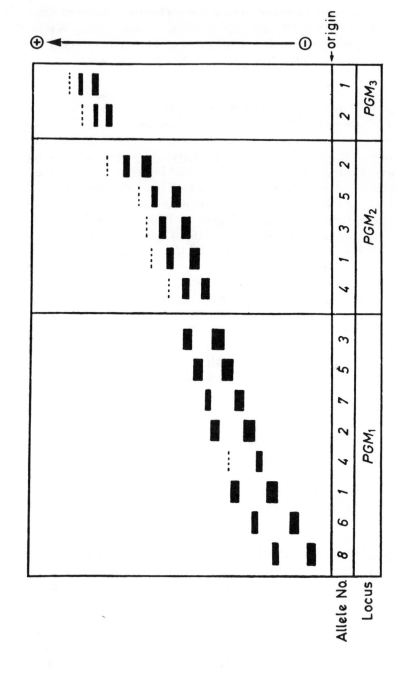

Figure 5.1. Allelic patterns of three variants of phosphoglucomutase (PGM)

Source: Harris, 1980. Reprinted with the permission of H. Harris and Elsevier Science Publishing, Inc., New York, NY

Marked differences were found. Blacks have a lower $PGM_1{}^{2-}$ frequency. Asiatic Indians, on the other hand, have a relatively high frequency of $PGM_1{}^2$ due to a higher frequency of the $PGM_1{}^{2+}$ allele. White South Africans of Dutch ancestry and Ashkenazim groups have PGM_1 alleles similar to those of Europeans. More rare $PGM_1{}^6$ and $PGM_1{}^7$ alleles were also found. The PGM_2 2-1 "Atkinson" phenotype was also found.

Alcohol Dehydrogenase

Alcohol dehydrogenase (ADH) with the coenzyme nicotinamide adenine dinucleotide (NAD), catalyzes the reversible conversion of organic alcohols to ketones in a variety of tissues. The major locus of this activity is the liver. Its physiological function is thought to be the degradation of ethanol produced by microbes in the gastrointestinal tract. Von Wartburg et al. (1965) identified an "atypical" form of this enzyme which had a pH optimum (8.8) that was markedly different from that of the usual enzyme (pH 10.8). Approximately 6 percent of Europeans were found to have this form of the enzyme (Smith, Hopkinson, and Harris, 1971).

Most Japanese (approximately 90 percent) appear to have ADH activity consistent with the "atypical" pH optimum (Yin et al., 1984). The ADH isozymes with "atypical" activity are thought to consist structurally of beta 2 subunits, while typical subunits have beta 1 subunits, both of the ADH_2 gene. Using isoelectric focusing to distinguish ADH_2 (2-2 phenotype) from ADH_2 (2-1 phenotype), the "atypical" livers could be separated into two groups, A_1 and A_2. Type A_1 livers were found to be homozygous ADH_2 (2-2 phenotype) (found in 53 percent of Japanese liver samples), and type A_2 were found to be heterozygous ADH_2 (2-1 phenotype) (found in 31 percent of Japanese livers). Livers from black individuals show the usual frequency of this enzyme.

Smith, Hopkinson, and Harris (1972; 1973) studied the ADH isozyme patterns using starch gel electrophoresis. They found evidence for the operation of three autosomal loci, ADH_1, ADH_2, and ADH_3. These loci code for the corresponding polypeptide subunits, alpha, beta, and gamma. All three loci are active in the liver, but only beta and gamma are active in the kidney. Only the beta subunit is active in the lung, while only the gamma subunit is synthesized in the stomach. The isozyme pattern is consistent with a dimeric pattern. Hybrid patterns are produced in the liver, but not in other tissues. A developmental sequence of this enzyme system has been found in the sequence $ADH_1 \rightarrow ADH_2 \rightarrow ADH_3$, which is not activated until the postnatal period. At each of the *ADH* loci it appears that there are at least two common alleles.

New molecular forms of ADH were identified in 29 percent of the liver homogenates from U.S. blacks (Bosron et al., 1983). These have been collectively designated ADH Indianapolis (ADHInd). Three different ADHInd phenotypes were identified on starch gel electrophoresis, and four forms were isolated using affinity and ion-exchange chromatography (Bosron et al., 1983). The most anodal form has a single pH optimum at 7.0 for ethanol. It appears to be a homodimer and is a new subunit. The other three have two pH optima at pH 7.0 and 10.0. They appear to be heterodimers of the new subunit, with the usual alpha, beta, and gamma subunits. It appears that this new subunit is a variant of the ADH_2 locus and so would be called ADH_2Ind. It codes for the beta Ind subunit. The frequency of this newly discovered gene was 0.16 in U.S. blacks. The frequency of ADH_3 *1* and ADH_3 *2* alleles is also different in black Americans and Japanese.

Debrisoquin Hydroxylation

Debrisoquin is a drug whose metabolism seems to differ from individual to individual. Metabolic oxidation seems to play a central role in its detoxication. Subjects to whom the drug is administered excrete unchanged drug together with five oxidation products: 4-, 5-, 6-, 7-, and 8-hydroxy-debrisoquine. The 4-hydroxylation reaction is polymorphic in relation to whether subjects are able to carry out this reaction (Woolhouse et al., 1979). The alicyclic 4-hydroxylation of debrisoquin is polymorphic in American and British whites and in Ghanaians (Woolhouse et al., 1979; Mbanefo et al., 1980). Individuals are poor metabolizers or extensive metabolizers in relation to their ability to carry out the 4-hydroxylation reaction. Extensive metabolizers may be homozygous or heterozygous. British whites are primarily monomorphic extensive metabolizers (92 percent), and homozygous poor metabolizers make up 8 percent of the population. Three phenotypes were observed in the Ghanaian population: homozygous extensive metabolizers (58 percent), heterozygous extensive metabolizers (36 percent), and homozygous poor metabolizers (6 percent). In a study of 123 Nigerian volunteers, 10 (8.1 percent) were unable to carry out the 4-hydroxylation reaction effectively. The incidence of the allele governing impaired 4-hydroxylation among Nigerians was calculated at 0.28. It appears that there was an association between the ability to carry out 4-hydroxylation and 6- and 7-hydroxylation of debrisoquine. This suggests that the alleles controlling alicyclic oxidation also influence aromatic hydroxylation. In another study of this drug-metabolizing polymorphism in cancer patients (Ritchie and Idle, 1982), it appeared that poor metabolizers were less vulnerable than extensive metabolizers to relative cancer risk. Investigations in animals suggest that there is a relationship between vitamin B_1 metabolism and DNA binding and debrisoquin oxidation.

Red Cell Glutathione Peroxidase

An inexpensive method has been devised to detect glutathione peroxidase (GPX1) after electrophoresis on Cellogel (Board, 1983; Meera Khan et al., 1984). Glutathione peroxidase is a selenoenzyme that catalyzes the oxidation of reduced glutathione (GSH) by peroxides (Destro-Bisol and Spedini, 1989). Surveys of several ethnic groups revealed variant phenotypes at polymorphic frequencies in African and Lebanese populations. In 398 Dutch persons no GPX1 variants were found; however, 1 of 72 Afro-Jamaicans and 3 of 116 Punjabis had a GPX isozyme pattern consistent with a GPX1 2-1 produced by alleles *GPX1*1* and *GPX1*2*. The 2-1 variant was found to be identical to the "Thomas" variant described by Beutler and West (1974). Thus far African-Americans, Ashkenazi Jews, the Punjabis of the Indian subcontinent, and African-Jamaicans have the *GPX1*2* allele in polymorphic frequencies. These data have been interpreted by some authors to indicate that the *GPX1*2* allele was originally an African variant and that present-day Punjabis, like Ashkenazi Jews, are predominantly of Mediterranean origin with a significant proportion of "African ancestry" (Mourant et al., 1976d).

Red Cell Glutathione Reductase

Erythrocyte glutathione reductase (GSR) is a red cell enzyme that catalyzes the conversion of oxidized glutathione to the reduced form of this molecule. Analysis was made of 414 blood samples from Sudanese males in Khartoum for GSR in relation to G-6-PD and hemoglobin types (Saha, 1981). The frequencies of the *GSR* alleles were *GSR1*—0.9493 and *GSR2*—0.507. The frequency of glutathione reductase deficiency (GSR0) was only 0.0241. The frequency of *GSR2* was higher in the G-6-PD-deficient men than in the normal men. There was no relation to hemoglobin types, but there was an excess of GSR0 in those who had hemoglobin AS.

Frischer et al. (1973) analyzed the prevalence of glutathione reductase deficiency in populations of the United States, South Vietnam, and Ethiopia. Decreased glutathion reductase activity was found in 0.3 percent of European-Americans, 1.9 percent of African-Americans, 7.3 percent of Ethiopians, 14.6 percent of Iranians, and 22.0 percent of Vietnamese. The authors suggested that decreased glutathione reductase activity in apparently healthy persons is far more common than hitherto believed. Decreased activity may reflect differences in the prevalence of malnutrition with riboflavin deprivation, the type and distribution of glutathione reductase alleles, and other genetic or environmental factors.

Alpha-N-acetyl-D-glucosaminidase

Alpha-N-acetyl-D-glucosaminidase (NAG) is the enzyme that is deficient in cases of the Sanfilippo syndrome type B. Serum and plasma samples from 360 whites and 126 blacks were examined for their levels of NAG (Vance et al., 1980). Samples from the blacks had higher mean levels of the enzyme when compared to the whites. Log transformation and admixture analysis demonstrated three distributions of NAG activity in blacks and two in whites. Segregation analysis of the NAG activity of twenty-nine white half-sibling twin families suggested a genetic model for the inheritance of NAG activity. There appear to be alleles for high and low NAG activity. The results suggest a genetic polymorphism for NAG activity in black and white populations.

Acetyltransferase

Acetyltransferase is active in the liver. The enzyme functions in the transfer of an acetyl group from acetyl-coenzyme A to the drug isoniazid. Measurement of acetyltransferase activity in liver samples shows marked difference in enzyme levels between "rapid" and "slow" inactivators of the activity of isoniazid characterized by the rate at which they are able to acetylate this drug. Not only do "slow" inactivators have higher and more prolonged levels of isoniazide in the bloodstream but also they are prone to develop peripheral neuropathy in long-term therapy. Fortunately, this problem can be avoided by the administration of vitamin B_6 (pyridoxine). About 50 percent of whites and blacks are slow inactivators, while only about 10 percent of Oriental populations fall into this category (H. Harris, 1980). This same enzyme appears to be involved in the acetylation of other drugs, such as sulfamethazine, hydralazine, sulfapyridine, and dapsone.

Acetylator phenotyping can be carried out by means of a sulphadimidine test (Karim et al., 1981). This test revealed that 65 percent of Libyan Arabs were slow acetylators. The frequency of the allele controlling slow acetylation is estimated at about 0.81, which is similar to European (0.72–0.75), black (0.73), and nearby Middle Eastern frequencies.

Paroxonase

Paroxon is an organophosphorus anticholinesterase compound used topically in the treatment of glaucoma. Microsomal oxidation of the insecticide parathion produces this compound in mammals. Parathion is inert until it is transformed into paroxon. Paroxonase is an arylesterase able to hydrolyze paroxon to p-nitrophenol. A polymorphism of the activity of this enzyme has been found (Geldmacher-von Mallinckrodt et al., 1983) consisting of two phenotypes controlled by two alleles at a single autosomal locus. Homozygotes of the low-activity allele (*PON¹*) have a gene frequency of 0.716–0.777. Homozygotes of the high-activity allele (*PON²*) have a gene frequency of 0.223–0.284. Heterozygotes also have high activity. They can be distinguished by determining the ratio of paroxonase to arylesterase activity (Eckerson et al., 1983). In African and Oriental populations, less than 10 percent of the population is in the low-activity group.

HEREDITARY ELLIPTOCYTOSIS

Hereditary elliptocytosis is a hereditary disorder of the erythrocytes characterized by an elliptical or oval shape of the erythrocytes. Preliminary studies in France showed a high frequency of hereditary elliptocytosis in West Africans and individuals from the Antilles (Dhermy et al. 1989). These authors confirmed this finding by searching for hereditary elliptocytosis in subjects from the Peoples Republic of Benin, Burkina-Faso, the Ivory Coast, and Togo. They found 22 people with hereditary elliptocytosis out of 3,450 subjects. Hereditary elliptocytosis was thus found to be about ten times more frequent in West Africa than in whites from Europe or the United States. Hereditary elliptocytosis is related to mutations of spectrin; spectrin was characterized in the 22 positive subjects and 17 showed the $Sp\alpha^{I/65}kD$ variant. A second variant, $Sp\alpha^{I/46}$ occurred in the other samples.

SERUM PROTEIN POLYMORPHISMS

A variety of proteins are polymorphic.

Haptoglobins

Haptoglobin is an α_2-globulin that is synthesized in the liver. A major property of this protein is that it binds free hemoglobin tightly and forms a complex that promotes the degradation of hemoglobin while allowing the retention of iron. This hemoglobin-haptoglobin complex is also easily detected after gel electrophoresis because it has peroxidase activity.

Three distinct phenotypes of haptoglobin (Hp) can be recognized after starch gel electrophoresis at alkaline pH (Smithies, 1955). These are called Hp 1-1, Hp 2-1, and Hp 2-2. They differ in electrophoretic mobility and the number of bands of peroxidase activity which represent the number of proteins. Hp 1-1 has only a single band, whereas Hp 2-2 has a series of less rapidly moving bands. Type Hp 2-1 is a combination of both of these patterns, similar in appearance but qualitatively different. These patterns appeared to be under the control of two alleles Hp^1 and Hp^2 (Smithies and Walker, 1955). They are highly polymorphic in most populations. The multiple bands in Hp types 2-1 and 2-2 are a series of polymers of larger size. All the bands contain two nonidentical polypeptide chains called alpha and beta. The differences between the three types are the result of differences in the α-chains: the β-chains remain the same.

Additional complexity can be identified when electrophoresis is carried out in 8.0 molar urea (denaturation) and reducing conditions with mercaptoethanol to separate the polypeptide chains. Two kinds of α-chains can be identified from type 1-1 phenotype by this method, called hp1F alpha and hp1S alpha (Smithies et al., 1962, 1966). Persons who are type 1-1 may be homozygous for the hp1F or hp1S α-chains or heterozygous for these chains. Family studies are consistent with a two allele system Hp^{1F} and Hp^{1S}. The α-chain form, type 2-2 (hp2 alpha), is a single zone under conditions of reducing denaturing electrophoresis. This form is about twice the size of the hp1F and S α-chains and is thought to represent an end-to-end fusion of an hp1F alpha and an hp1S alpha. The F and S chains differ by a single amino acid. This means that the Hp^2 is almost a duplication of the Hp^1 allele. Smithies (1966) asserted that the α-chain form, type Hp 2-2, developed by mispairing and unequal crossing over. A consequence of this theory is that four different kinds of hp2 alpha chains would be expected. A triplicated allele might also occur by the same mechanism. An unusual haptoglobin type, called the "Johnson" phenotype, appears to be consistent with this kind of structural variation. The "null" allele at the Hp locus is called Hp^0. In this phenotype it is presumed that no α-chains are synthesized and therefore no haptoglobin is produced. The molecular basis for the null allele is unknown. One phenotype that is polymorphic in many black populations is Hp 2-1M, which occurs in association with the Hp^0 allele (Giblett and Steinberg, 1960). An Hp^{2M} allele is responsible for this phenotype. It is likely that there are a number of different null alleles, each having arisen from a different mutation.

In a survey of individuals in Gambia, no haptoglobin could be demonstrated in 203 out of 857 subjects. In the remaining individuals the Hp frequencies were Hp^1—0.651; Hp^2—0.349 (Welch et al., 1979). In another study of 825 persons in the Keneba population in the Gambia (Jenkins and Dunn, 1981), it was found that 22.5 percent were ahaptoglobinemic (Hp0). The frequency of Hp0 was higher in the 209 persons with malaria parasites in their blood (38.8 percent) than in nonparasitic individuals (17.5 percent). This finding is consistent with a low level of hemolytic activity in the blood of individuals with parasitemia. Children in this population had a low incidence of Hp0 except in the first two months. The incidence of Hp0 rose after two years, then leveled off in the older age groups. A positive correlation was found between Hp0 and malaria indicators (Jenkins and Dunn, 1981).

In a series of haptoglobin surveys in Iran, Bowman (1964a) found that serum samples that were not reasonably fresh, often appeared to be Hp0, but on retesting with fresh serum, the samples were Hp2-2. This problem was not encountered with Hp1-1 and Hp2-1 sera. Haptoglobin also disappeared during a fava bean-induced hemolytic crisis. Accordingly, the Hp0 phenotype in some populations should be viewed with suspicion.

The geographic distribution of the Hp^1 gene worldwide has been summarized by Giblett (1969). The Hp^1 gene frequency is similar throughout the European continent (0.35–0.43). The Swedish Laplanders and the southernmost Italians have the lowest frequencies: 0.32. In Africa the Hp^1 frequencies are much higher than in Europe (0.47–0.87). The exceptions to this finding are the Ethiopians and the Pygmies of the Ituri Forest (0.40) and the San of South Africa (0.29–0.31). These low Hp^1 frequencies are suspect because of the frequent occurrence of Hp0 in these populations (30 percent). The Hp^1 gene frequency in U.S. blacks is about 0.55. The proportions of the Hp phenotypes in U.S. blacks are 0.28 for Hp1-1; 0.38 for Hp 2-1; 0.10 for Hp 2-1M; 0.19 for Hp 2-2; and 0.05 for Hp0. The major difference between U.S. blacks and whites is in the frequency of Hp 1-1 (blacks have twice the frequency of whites) and Hp 2-2 (whites have twice the frequency of blacks).

In another study, eleven African groups living in an area from Algerian Sahara to Central Africa were surveyed (Constans et al., 1981a). The Hp^1 gene frequency was higher in Saharan samples than any other groups. The frequency of this gene declined from north to south. In samples from a group of Pygmies, the Hp^2 gene frequency was higher than the Hp^1 frequency. By using a sensitive radioimmunoassay, Hp phenotypes can be detected when there appears to be a deficiency of haptoglobin as with Hp0. A relative absence of Hp 1-1 and a significant excess of Hp 2-2 individuals was found in the Pygmy and San groups. In contrast, although a similar frequency of Hp0 persons was found in some of the Nigerian tribes, their Hp^1 frequencies were generally higher than those in most other African regions. More Hp2-1M phenotypes were found in the Hp0 population than in the non-Hp0 population. The large polymers of Hp related to the presence of the α_2-chain (Hp^2 gene product) appears to be primarily involved with the mechanism of Hp 0 and Hp 2–1M phenotypes in African groups. Polyacrylamide gel electrophoresis of serum samples from the Sara, Bantu, and Peulth peoples revealed a high frequency of the Hp^2 gene. The Hp^{1F} and Hp^{1S} gene frequencies are similar in the Hp^1 group of individuals in African populations or Hp^{1F} is somewhat greater than Hp^{1S}. The Hp^{1FF} gene is almost never found in Mongoloid populations. In Europeans the ratio of Hp^{1S} : Hp^{1F} is about 2 : 1. In all of the populations studied thus far the Hp^2 gene is Hp^{2FS} even when the Hp^{1F} gene is lacking in the population.

Transferrin

Transferrin (also called siderophilin) is a protein found in the β-globulin fraction of serum. Transferrin (Tf) has a molecular weight of 77,000 and carries iron from the plasma to cells of the bone marrow and tissues where it is stored. Transferrin is a glycoprotein consisting of a single polypeptide chain containing about 5.5 percent carbohydrate with eight galactose, eight glucosamine, four mannose, and four sialic acid residues, and one residue of fucose. Sialic acid appears to be a vital factor in determining Tf electrophoretic mobility. In addition to its iron-binding function, Tf may also be involved in the removal of certain organic materials from the circulation (Sass-Kuhn et al., 1984).

Smithies (1957, 1958), using two-dimensional electrophoresis, demonstrated that transferrins exhibited inherited variation. These variants were subsequently shown to be codominant alleles. The most frequent form of Tf is called C. Eighteen electrophoretic variants of Tf have been described. Eight variants migrate more rapidly and nine move more slowly than Tf C. Most of them occur in the heterozygous state. Only a few of these variants achieve polymorphic frequency in any population, for example, B0-1 in the Navajo (8 percent), B2 in some European populations (1 percent), DCHI in Oriental populations (6 percent), and D1 in black and Australian aboriginal populations (12 percent). Wang et al. (1965, 1966, 1967) showed that the electrophoretic difference between TfC, Tf D1, TfDCHI, and TfB2 was due to a specific single amino acid substitution in each case. The variants do not appear to differ in iron-binding capacity or other functions.

In other studies the technique of isoelectric focusing in the pH gradient of 4.0–6.5 has shown that TfC has two subtypes, TfC_1 and TfC_2 (Kueppers and Harpel, 1980). Three Tf bands are seen in homozygotes. Double bands are seen in heterozygotes and probably represent different degrees of iron saturation. Family studies are consistent with a codominant mode of inheritance. The allele frequencies in U.S. whites are 0.8 for TfC_1 and 0.19 for TfC_2; in U.S. blacks they are 0.84 for TfC_1 and 0.11 for TfC_2. In the Bi Aka Pygmies, the two subtypes TfC_1 and TfC_2 have been identified. The TfD_1 frequency in this population is one of the highest observed in African groups (Constans et al., 1981).

An association has been found between $TfD1$ and the $GcAb$ genes. The frequency of individuals with the TfD_1 allele in Africa ranges from 0.051 in South African blacks to 0.073 in North Africa. In the Australian aborigines the highest frequency of TfD_1 has been found (Kirk et al., 1964): 37 percent of this population had the Tf CD1 phenotype. Almost 4 percent of the population was homozygous.

Group-Specific Component

The group-specific component (Gc) was discovered by Hirschfeld (1959) using immunoelectrophoresis. It is an α_2-globulin that has been shown to be a vitamin-D binding globulin by Daiger et al. (1975). This protein is highly polymorphic and, like that of the haptoglobins, Gc occurs world-wide. It is a single polypeptide chain with a molecular weight of 52,000. It is a glycoprotein with about 4 percent carbohydrate. The phenotypes Gc 1–1, Gc 2–1, and Gc 2–2 can be distinguished by agar or starch gel electrophoresis (Bearn et al., 1964) and are controlled by codominant alleles, Gc^1 and Gc^2. It has been hypothesized that the frequency of the Gc^2 allele is inversely related to the amount of sunshine. Low levels of sunshine are associated with an increased frequency of the Gc^2 allele (Mourant et al., 1976). There are, however, exceptions to this thesis.

A number of techniques have been devised to demonstrate a large number of variants of Gc. Subtypes of Gc, Gc^{1F} and Gc^{1S} have been found to be due to carbohydrate differences that follow protein synthesis. About eight-four different variant forms of Gc have been identified. Some of the polymorphic variants are found in the Chippewa

Amerindians (Gc^{Chip}, 8 percent) and the Australian aborigines (Gc^{Ab}, 10 percent).

Subtyping has most recently been carried out by isoelectric focusing. New variants have been found in the Fula people of Senegal. Among Saharan, Middle Eastern, and African populations, five variants, including one previously unknown, were identified. The lowest frequency of Gc^2 occurs where the sunlight is strongest (Constans et al., 1981b; Constans and Viau, 1977). Subtypes are found in all populations that were studied. In a village in Gambia, West Africa, the fast variant Gc^{Ab} (aborigine) was found with a frequency of 0.013. The allele frequencies of these genes differ in U.S. whites and blacks. No difference in the amount of Gc protein has been demonstrated as measured by radial immunodiffusion (Kueppers and Harpel, 1979). The frequencies are as follows:

	Gc^{1F}	Gc^{1S}	Gc^{Ab}
Whites	0.149	0.572	0.000
Blacks	0.732	0.147	0.015

There are marked differences in the frequencies of Gc^{1F} and Gc^{1S} between U.S. blacks and whites. This finding is consistent with the north-south gradient in Gc^{1F} gene frequency found in Africa that appears to parallel the gradient seen in skin pigmentation (Constans et al., 1980). The presence of the Gc^{Ab} allele is consistent with its presence in Central African populations. This is a very rich polymorphic protein system that should have wide use for genetic and anthropologic studies.

Ceruloplasmin

Ceruloplasmin (Cp) is a plasma metalloprotein in the α_2-globulin region which binds and transports 90–95 percent of plasma copper. It is a blue glycoprotein that carries six to seven copper ions per molecule of protein. In humans Cp consists of a single polypeptide chain (Takahashi et al., 1984). According to Walaas et al. (1967), Cp has oxidase activity and will oxidize dopa, epinephrine, serotonin, and p-phenyl-enediamine in vitro. Whether this function operates in vivo is not known. Because Wilson disease is associated with a deficiency of Cp, it has been suggested that it plays a role in copper storage (Bearn, 1966).

Three variants of Cp were identified by Schreffler et al. (1967) using starch gel electrophoresis at alkaline pH. Five different phenotypes of Cp were found in U.S. blacks (Cp B, Cp AB, Cp A, Cp AB, and Cp BC). Eventually two additional phenotypes were added (Shokeir et al., 1970). The frequency of the polymorphic distribution of Cp for U.S. blacks in comparison with U.S. whites is as follows:

	Cp^A	Cp^B	Cp^C
Whites	0.006	0.994	0.000
Blacks	0.053–0.060	0.994	0.003

Another variant Cp^{NH} (for New Haven) was described with a mobility similar to that of Cp^C. The Cp AB phenotype is found in approximately 19 percent of U.S. blacks (Giblett, 1969).

Two new variants of Cp were found in studies of a population from Ann Arbor, Michigan: Cp Michigan was found in both blacks and whites. A variant of CpC called CpC Ann Arbor was found in one family. Another variant, CpA Wap, was found in one person from each of the Macushi and Wapishana Amerindian tribes.

Alpha-1-Antitrypsin, Protease Inhibitor

Alpha-1-antitrypsin is an α_1-globulin found in the prealbumin fraction of serum. It is synthesized in the liver, and binds to trypsin and other proteolytic enzymes and inhibits their proteolytic activity; α_1-globulin has 90 percent of the total trypsin-inhibiting capacity. It has a molecular weight of approximately 54,000, and contains 12.2 percent carbohydrate. Laurell and Eriksson (1963) reported that this protein was absent in patients with degenerative pulmonary disease of a rather early onset, leading to death in their forties. It was later found that children with this deficiency developed a form of infantile cirrhosis (Gans et al., 1969). Only 15–30 percent of individuals with the deficiency actually develop degenerative lung disease. Others do not seem to be affected by the deficiency. The amount of alpha-1-antitrypsin in the liver seems plentiful but the level in the serum is reduced to 15 percent of normal levels. The most common gene is Pi^M.

Fagerhol and Braend (1965) described polymorphism of the prealbumin protein; subsequently Fagerhol and Laurell (1967) found that the prealbumin protein was the same as alpha-1-antitrypsin. The name Pi was suggested for this system to designate protease inhibitor. At least thirty allelic variants of Pi have been described (Hug et al., 1981). The Pi genotype of individuals with degenerative pulmonary disease may be $Pi^Z Pi^Z$ (most frequent) or $Pi^Z Pi^o$ or $Pi^o Pi^o$ (most rare).

Two percent of the white population are heterozygotes for the Pi^Z allele. The frequency of the Pi^Z allele in white U.S. whites is 0.01. In Europe about 1 in 1,500–3,000 newborn infants have Pi deficiency; in black and white Americans the overall frequency is 1 in 10,000.

The M heterozygote can be subtyped to M_1 and M_2, with the latter being the most frequent in most populations. Isoelectric focusing with thin layer polyacrylamide gels at pH 4.5 was used to study the serum of individuals from a village in Gambia, West Africa. A new variant allele, Pi^{Gam}, was found at polymorphic frequency (0.0642) (Welch, McGregor, and Williams, 1980). Unlike other populations, only the common allele Pi^M (0.9358) is found (Welch et al., 1980). Inheritance of the variant was confirmed by family studies. It is not associated with a deficiency of Pi activity or concentration.

A similar technique was used to study samples from U.S. black and white populations. A common variant $PiM3$ was found with an isoelectric point between $PiM1$ and $PiM2$. The frequency of these alleles is 0.11 in U.S. whites for Pi^{M3} and 0.54 in blacks (Kueppers and Christopherson, 1978). When Pi^{M3} and Pi^{M1} are included in the system, heterozygosity is five times greater in whites and ten times greater in blacks.

Thyroxin-Binding Globulin

Thyroxine-binding globulin (TBG) functions to carry thyroid hormone from the thryroid gland to the tissues. It has a single polypeptide chain with four oligosaccharide units, with an average of ten terminal sialic acid residues. The amount of this protein is reduced in some individuals, but it does not affect their thyroid status. The inheritance of this deficiency is X-linked recessive (Nicoloff et al., 1964). The frequency of TBG deficiency in newborns is 1 in 3,600 (Sorcini et al., 1980). Genetic variants of TBG have been identified (Rivas et al., 1971). An X-linked polymorphism of TBG has been demonstrated in populations of Africans and Melanesians (Daiger et al., 1981). A slowly migrating TBG variant on electrophoresis was shown to be frequent in blacks (0.11). An electrophoretically slow variant (TBG S) is also seen in the Melanesian and Polynesian populations, ranging from 1 to 10 percent (Kamboh and Kirkwood, 1984).

Microheterogeneity can be demonstrated in isoelectric focusing of purified TBG-1 (Grimaldi, 1983). In most whites and blacks it has 4 bands. In black donors two other phenotypes are seen: TBG-2, which has 4 bands at slightly different positions from TBG-1; and TBG-1,2, with all bands present in TBG-1 and TBG-2. Deglycosylation reduces the number of bands in TBG-1 and 2 to two bands, and to three bands in TBG-1,2. The polymorphic variants show no difference in thyroxine-binding ability.

Plasminogen

Plasminogen (PLGN) is a proenzyme that is activated by a protease plasminogen activator to plasmin. This is, in turn, responsible for plasma fibrinolytic activity, the ability to break down or dissolve fibrin. A common genetic polymorphism in plasminogen has been revealed by studying neuraminidase-treated serum or plasma samples with isoelectric focusing in polyacrylamide gel followed by immunofixation or overlaying the gels with casein after activating the urokinase (Raum et al., 1980). From the pattern on isoelectric focusing, it appeared that two alleles *PLGN*A* and *PAGN*B* were most likely. The gene frequencies for these alleles were different in white, black, and Oriental populations. The frequencies are as follows:

	PLGN*A	*PLGN*B*
Whites	0.69	0.30
Blacks	0.80	0.18
Orientals	0.96	0.03

The frequency of the genes in blacks is intermediate between that of whites and Orientals. The distribution of phenotypes fits Hardy-Weinberg expectations. The pattern of variants is consistent with codominant inheritance. There does not appear to be a difference in plasminogen activity related to the different variants.

Salivary Protein

The salivary proteins constitute a very complex polymorphic system. The salivary proteins that have been found to be polymorphic are as follows: salivary amylase, salivary vitamin B_{12}-binding proteins, parotid basic protein, postparotid basic protein, parotid proline-rich protein (PPP), and parotid double-band protein. The PPP complex consists of a minimum of seven genes. These are *Pa*, *Pr*, *Ps*, *PM*, *Db*, *G1*, and *PIF* (Azen and Denniston, 1980). This gene complex controls the important acidic and basic proline-rich proteins that constitute approximately two-thirds of the parotid salivary proteins. Most of these loci have a null allele (a gene that produces no gene product) in addition to others that determine structural proteins (McKusick, 1989).

The PS proteins or parotid size variant proteins are determined by inheritance of two expressed alleles, Ps^1, Ps^2, and one unexpressed allele, Ps^0. The electrophoretic variation appears to be due to differences in molecular weights between the proteins (Goodman and Karn, 1983). The Ps proteins are glycosylated with an isoelectric point (pI) of 8.1. Gene frequencies in whites (N = 150) and blacks (N = 101) are as follows:

	Ps^1	Ps^2	Ps^0
Whites	0.598	0.101	0.301
Blacks	0.185	0.126	0.689

The null allele frequency in blacks is more than twice that in whites.

Another salivary protein, PmS, has a strong positive correlation with the presence of a smaller salivary protein Pm (PmF) (Azen and Denniston, 1980). The Pm (parotid middle band protein) salivary protein was found by Ikemoto et al. (1977) on acid-urea starch gel electrophoresis migrating between Pa and Pb in the Japanese population. This finding suggested that PmS is probably part of the Pm polymorphic salivary protein system.

	$PmF+$	$PmS+$
Whites	0.15 (N = 140)	0.12 (N = 150)
Blacks	not determined	0.24 (N = 101)

This complex polymorphic system was also investigated in 200 Kenyan schoolboys (Pronk et al., 1984). The frequencies for the polymorphic salivary proteins are as follows: Pr^1—0.66, Pa^+—0.18, Db^+—0.55, Pb^2—0.12, $AMY1^{A2}$—0.008, and $AMY1^E$—0.03. When we compare these frequencies to those of West Africa and U.S. blacks, there appears to be a deficiency of the Pb^2 alleles and the $AMY1^3$ allele is missing. A new phenotype, AMY1E, was found, which suggested that there was an allele present with an estimated frequency of 0.02.

Two new genetic protein polymorphisms of salivary proteins have been identified in parotid saliva, called CON 1 and CON 2 (Azen and Yu, 1984). They are detected by immunologic and concanavalin A reactions on nitrocellulose. The proteins are determined by the autosomal dominant inheritance of one expressed and one unexpressed (recessive) allele. Gene frequencies are as follows:

	CON^{1+}	CON^{1-}	CON^{2+}	CON^{2-}
Blacks	0.581	0.419	0.007	0.993
Whites	0.396	0.604	0.034	0.966
Chinese	0.580	0.420	0.000	1.000

The CON 1 protein shows a strong association with the Ps protein system; the CON 2 protein system has a strong association with the PmF system proteins; and CON 2 has a strong association with G1. These alleles appear to be linked to the salivary protein gene complex (SPC). There appear to be significant differences in the allele distributions of some of these parotid systems, which makes them useful for population studies.

Milk Protein

Polymorphism occurs in the milk proteins and seminal fluid of cattle and humans. Bell et al. demonstrated variation in the β-lacto-globulin of bovine milk (1970). Studies of the milk casein in milk from women of the Kikuyu population using urea-starch gel electrophoresis revealed an α- and β-casein polymorphic pattern (Ponzone et al., 1975) similar to that in whites; however, the gene frequencies for the β-casein locus were the reverse of that in the white population. A new β-casein variant called β-E was identified.

Properdin Factor B

Properdin factor B (Bf) is a genetically controlled polymorphism demonstrated by agarose gel electrophoresis. The *Bf* gene frequencies were determined for 194 blacks from southeastern United States (Budowle et al., 1981). The frequencies are as follows: *BfF*—0.626, *BfS*—0.034, *BfF1*—0.327, and *BfS1*—0.013. The frequencies in blacks from the southeastern United States and South African blacks are quite similar but are different from Boston blacks. These differences might be due to racial admixture. In another study, properdin factor B phenotypes were identified in 1,258 members of several South African groups including seven Bantu-speaking black populations, one Asiatic Indian, and one Coloured population (Mauff et al., 1976). The allele frequencies were distributed as follows:

	Bf^F	Bf^S	B^{FRARE}
Black	0.655	0.282	0.063
Indian	0.322	0.645	0.033
Coloured	0.513	0.435	0.052

In the course of these studies, two new F alleles and one new S allele were found. In yet another group of 918 Saudi Arabs, an unusually high frequency of the "rare" $Bf^{S0.7}$ allele (0.1514) was found. The frequencies of the common alleles were Bf^S, 0.5174 and Bf^F, 0.3213 (Klauda et al., 1984).

The fourth component of complement, C4, is composed of two tightly linked genes ($C4A$ and $C4B$) in the major histocompatibility complex of chromosome 6, as demonstrated by agarose gel electrophoresis. Seven alleles at the $C4A$ and five alleles at the $C4B$ loci were examined in 169 black individuals from the southeastern United States (Budowle et al., 1983). Phenotypic frequencies of C4A6, C4A5, C4A4, C4B4, C4B3, and C4BQ0 were significantly different between black and white Americans.

A combination of isoelectric focusing and a hemolytic assay to detect bands delineated extensive structural polymorphism in human C8. Inheritance patterns in families appear to be codominant and the distribution of phenotypes fits Hardy-Weinberg expectations (Raum et al., 1979). The $C8$ allotypes were determined for two previously studied families, each of which has a homozygous C8 deficient propositus. The analysis suggests that C8 deficiency is a "silent" or "null" allele of the $C8$ structural locus. It appears that no gene product at all or none resembling C8 can be demonstrated. It would seem, therefore, that half-normal levels of C8 cannot be used as a single criterion for establishing heterozygous C8 deficiency. The $C8$ structural locus does not appear to be linked to the human histocompatibility (HLA) complex. The distribution of $C8$ alleles is given in the following table. It appears that blacks have a significantly higher frequency of the $C8^{A1}$ allele than whites or Orientals. The frequencies are as shown:

	$C8^A$	$C8^B$	$C8^{A1}$
Orientals	0.655	0.345	0.000
Blacks	0.692	0.259	0.049
Whites	0.649	0.349	0.003

Apolipoproteins

Approximately twelve proteins have been described as components of lipoprotein particles and have been called "apoliproteins." Apoliproteins participate in lipoprotein synthesis, secretion, processing, and catabolism. Sepehrnia et al. (1988) screened a large number of serum samples from Nigerians to investigate structural variation at six apoliprotein (APO) loci: A-I, A-II, A-IV, C-II, E, and H. The APO A-I

and *A-II* loci were monomorphic, with the exception of a putative APO A-I variant. On the other hand, marked variation was noted at the *APO A-IV, APO A C-II, APO E*, and *APO H* loci. The alleles were markers unique to blacks. Specific black alleles were observed at the *APO A-IV* and *APO H* loci, and the *APO E* locus was characterized by having the highest *APO E*4* allele frequency (0.296) in world populations. Further, a comparison of mean heterozygosities (H) at the *APO A-IV, C-II, E*, and *H* loci between Europeans (H frequency, 0.18) and Nigerians (*H* frequency, 0.24) reflected the high proportion of genetic variation among blacks at the apolipoprotein loci.

Sepehrnia and co-workers (1988) emphasized that the contrast between the occurrence and frequency of apolipoprotein alleles in Nigerians and Europeans uncovered an opportunity to compare apoliprotein gene effects in diverse populations. *APO E*2* and *Apo E*4* alleles, for example, are associated with high and low cholesterol levels, respectively. Accordingly, it might be predicted that populations with high *APO E*4* allele frequencies would have high mean cholesterol values (as do the Finns); however, although Nigerians have the highest frequency of the *APO E*4* allele, their adjusted mean cholesterol level is 161 milligrams per deciliter, which is low compared to most European populations. Evidently the effects of the *APO E*4* allele are modified by diet. Finally, Sepehrnia et al. suggested that the common *APO H* and the unique *APO C-II* and *APO A-IV* alleles observed in blacks may be useful in studies of ethnic differences in lipid metabolism and cardiovascular risk.

HLA POLYMORPHISMS

The HLA (human leukocyte-system A) antigens are part of the major histocompatibility complex (MHC). The MHC is a group of genetically linked loci located on the short arm of chromosome 6 (Francke and Pellegrino, 1977), which determines the structure of a rich variety of cell surface antigenic determinants. The primary function of these antigens is to enable the host to discriminate between "self" and "nonself" (which involves recognition) and to mount and direct effective immunologic surveillance (which involves response). These are the functions involved in the viability of a foreign tissue graft: the survival of a "nonself" graft depends on the closeness of the "match" between the MHC antigens of the recipient and the donated organ (Snell et al., 1976). The MHC appears to have the greatest diversity and variability of any system of proteins and antigens yet described. This tremendous degree of polymorphism has made it the single most useful system in determining paternity. There appears to be an association between certain of the HLA antigens and particular diseases (Svejgaard et al., 1983; Mittal, 1984), which aids in diagnosis and prognosis. The loci of the MHC have been divided into three classes (I, II, and III), based on differences in structure and function.

Class I loci include three loci of the HLA region: *HLA-A*, *HLA-B*, and *HLA-C*. The antigens produced by these loci are controlled by codominant alleles and are expressed on the surface of nucleated cells. They are glycoproteins and are part of cell membranes. These antigens are released from cell surfaces and can be found in normal serum, seminal fluid, and breast milk (Mittal, 1976, 1979). These molecules are made up of a heavy polypeptide chain with a molecular weight of 44,000 and a light chain with a molecular weight of 12,000. This latter chain is the same as the β_2-microglobulin and is controlled by a gene on chromosome 15 and does not contribute to the HLA polymorphism. The heavy chain has five regions or domains with three outside the membrane, one within the membrane, and one inside the cell. It also has an oligosaccharide carbohydrate moiety of about 3,000 molecular weight (Bodmer, 1983). The antigens in this group seem to be required when a cytotoxic cell kills an infected target cell.

Class II loci include *HLA* loci *D*, *DP* (*SB*), *DQ* (*MB*), and *DR*. The gene products of these loci are glycosylated polypeptide chains called alpha (molecular weight of 33,000) and beta (molecular weight of 29,000). Each chain has four domains: two ouside the membrane, one within it, and one extending into the cytoplasm. The regions of each of these loci have a different number of α- and β-chain genes, which are as follows: *HLA-D—6 α, 7 β*; *HLA-DR—1 α, 2 or 3 β*, *HLA-DP* and *HLA-DQ—2 α* and *2 β*. The DQ and DR specificities are present on B cells, monocytes, and macrophages but not on T lymphocytes. These antigens seem to be regulators of the lymphocyte subgroups of cells that cooperate in manifesting the immune response. Cells that cooperate must have the same class II antigens.

Class III loci determine serum proteins, which are called complement components C2, C4, and properdin factor B. They seem to be important in antibody-mediated cytolysis. The current system used to name the components of the HLA region is to designate the gene loci by HLA plus one or more letters. The alleles of each locus are designated by numbers after the locus symbols. Provisionally identified specificities carry another letter, a lower case w, inserted between the locus letter and the allele number. Loci of the *HLA* region are called A, B, C, D, DP, and DQ (Nomenclature, 1984). The number of antigens currently recognized in the various loci are: A—23; B—49; C—8; D—6; DP—3; and DR—16 (Nomenclature, 1984). Each HLA antigen is a multivalent cluster of several antigenic determinants. Each cluster is different from the others in the number and type of determinants of which it is composed. The loci that code for the *HLA* genes are so closely linked that the set of alleles present on each of the chromosomes of parents is usually transmitted intact to their children. This set of closely linked genes is called a haplotype. One of each pair of the haplotypes from each parent is passed on to each child.

There are several other significant genetic features of the *HLA* region: (1) frequencies of different HLA antigens and genes differ within an ethnic group as well as between ethnic groups; (2) certain *HLA* genes of different loci show high "gametic association" or "linkage-disequilibrium"; (3) different allelic HLA antigens show varying degrees of "cross reactivity" due to sharing of common antigenic determinants; (4) quantitative expression of HLA antigens varies from tissue to tissue and from individual to individual (Mittal, 1984).

Studies in different populations reveal that at almost every locus there is a difference in *HLA* gene locus frequency in one or more of the populations that have been studied. Table 5.3 is a partial listing of *HLA* locus gene frequencies for *HLA-A, -B, -C,* and *-DR* in white, black and Japanese populations.

Striking differences were noted when 1,465 healthy white and 128 healthy black individuals were compared for the genetic distribution of twenty-five different HLA antigens. Whites had significantly higher frequencies of A1, A3, B8, and Bw16, and blacks had higher frequencies of A28 and Aw30 (Mittal, 1976). Among whites the *A1-B8* haplotype had the highest incidence, whereas the *A2-B12* was highest among blacks. Black patients with renal failure had a twofold increase in the frequency of *Bw17* compared to healthy black individuals. This difference was not found among whites.

When 141 blacks, of which 135 were unrelated, from Kenya and Tanzania were studied for B-related HLA antigens, it appeared that several B15-related HLA antigens as a group occurred with a relatively high frequency (30 percent) in unrelated individuals. The antigens SV and possibly Bu, frequent in the black population, are found in fewer than 1 percent of whites (Hall et al., 1980). These variants have a strong association with C locus antigens in blacks. In a 1987 study, other unique characteristics of the HLA-D region in the black population were reported by Dunston et al. These investigators found that U.S. blacks and whites were significantly different in serologically detected and lymphocyte-defined HLA-D region antigen frequencies and in linkage relationships seen in DR and DQ specificities. The most frequent DR specificity in U.S. blacks, DRw13, occurred in 30

Table 5.3
HLA Locus Gene Frequencies

	HLA-A Frequency (%)				HLA-B Frequency (%)		
Gene Loci	European Causasoids (N = 228)	African Blacks (N = 102)	Japanese (N = 195)	Gene Loci	European Caucasoids (N = 228)	African Blacks (N = 102)	Japanese (N = 195)
A1	0.16	0.04	0.01	B5	0.06	0.03	0.21
A2	0.27	0.09	0.25	B7	0.10	0.07	0.07
A3	0.13	0.06	0.007	B8	0.09	0.07	0.002
A23	0.02	0.11	—	B12	0.17	0.13	0.07
A24	0.09	0.02	0.37	B13	0.03	0.02	0.008
A25	0.02	0.04	—	B14	0.02	0.04	0.005
A26	0.04	0.05	0.13	B18	0.06	0.02	—
A11	0.05	—	0.07	B27	0.05	—	0.003
A28	0.04	0.09	—	B15	0.05	0.03	0.09
Aw29	0.06	0.06	0.002	Bw38	0.02	—	0.02
Aw30	0.04	0.22	0.005	Bw39	0.04	0.02	0.05
Aw31	0.02	0.04	0.87	B17	0.06	0.16	0.006
Aw32	0.03	0.02	0.005	Bw21	0.02	0.02	0.02
Aw33	0.007	0.01	0.02	Bw22	0.04	—	0.07
Aw43	—	0.04	—	Bw35	0.10	0.07	0.09
Blank	0.02	0.11	0.04	B37	0.01	—	0.008
				B40	0.08	0.02	0.22
				Bw41	—	0.02	—
				Bw42	—	0.12	—
				Blank	0.04	0.18	0.08

	HLA-C Frequency (%)				HLA-DR Frequency (%)		
Gene Loci	N = 321	N = 101	N = 203	Gene Loci	N = 334	N = 77	N = 164
Cw1	0.05	—	0.11	DRw1	0.06	—	0.05
Cw2	0.05	0.11	0.01	DRw2	0.11	0.09	0.17
Cw3	0.09	0.06	0.16	DRw3	0.09	0.12	—
Cw4	0.13	0.14	0.04	DRw4	0.08	0.04	0.14
Cw5	0.08	0.01	0.01	DRw5	0.15	0.07	0.05
Cw6	0.13	0.18	0.02	DRw6	0.09	0.10	0.07
Blank	0.47	0.50	0.53	DRw7	0.16	0.07	—
				W1A8	0.06	0.07	0.07
				Blank	0.21	0.45	0.45

Source: Modified from Bodner et al., 1978.

percent of a sample of 139 black individuals. A unique variant of this specificity appeared to be due to a β-chain difference.

Other DR/DQ differences were also found: for example DR5, which is associated with DQw3 in whites, is frequently associated with DQw1 in blacks (38 percent). And DRw8, which is usually not associated with a defined DQ specificity in whites, occurs with DQw3 in 83 percent of blacks. Except for DW2, DR2, DW19, and DRw13, the frequencies for HLA-D specificities are significantly lower in blacks. The HLA-DR3 antigen frequencies in U.S. black and white populations are 25 percent and 20.5 percent, respectively (Rosen-Bronson, 1987). Whereas 93.3 percent of DR3 cells from whites type as Dw3, only 7.7 percent of DR3 cells from blacks type as Dw3. The remaining 92.3 percent do not type as any known HLA-D specificity. Another unique feature in the black population is that *DR2* is frequently associated (47 percent) with *DQwa*, a newly defined allele (Kunikana et al., 1987). Three haplotypes are most frequently found among black DR3 individuals, which are characterized by *B* locus specificities, and which are either unique to blacks or found at much higher frequency in black populations (Rosen-Bronson, 1987). These are (1) *A30, Cw-, Bw42, Dw-, DR3, DRw52, DQwa;* (2) *Bw71, Dw3, DR3, DRw52, DQw2;* and (3) *Bw71, Dw-, DR3, DR52, DQw2.*

These and other ethnic differences that have been identified in recent HLA studies underscore the importance of having organs for transplantation (such as kidneys) donated by members of the same family, and if possible, at least by the same ethnic group, to maximize the chance of having an organ match with reasonable chance for long-term survival. (A similar argument can be used for the preference of intraethnic blood transfusions, to reduce the likelihood of sensitization to antigens, such as Duffy, C, E, and Kell—particularly in black patients who need multiple transfusions over many years.)

DNA POLYMORPHISMS

It is possible to analyze DNA structure, gene arrangement and the location of genes to specific chromosomes using a large specialized family of enzymes known as restriction endonucleases. These are bacterial enzymes that "cut" DNA, not into random segments, as is usual for DNAases, but at specific places. These enzymes, which are responsible for the advent of recombinant DNA technology, cleave DNA only where a specific small sequence of bases occurs in the molecule (Nathans and Smith, 1975). Usually they recognize a sequence of six bases, for example, *Bam*HI recognizes the sequence GGATCC, while *Hind*III recognizes the sequence AAGCTT. The cleavage may occur at the same place in both DNA strands leaving flush ends, but more often it is acentric, leaving "sticky ends" available for base pairing with other fragments of DNA from the same or other species. A tremendous array of different-sized fragments of DNA is produced, which can be demonstrated using agarose gel electrophoresis of the enzyme-treated DNA followed by transfer to filter paper (a technique called "Southern blotting") and autoradiography using appropriately radio-labeled probes and restriction enzymes. When DNA from different species is joined together using these enzymes, it is called "recombinant DNA."

The remarkable specificity with which the restriction endonucleases act is the basis for another important application of these enzymes to the study of genetic variation. Numerous base changes can be found widely scattered throughout the human genome as well as the genomes of higher organisms. These base changes can have the effect of eliminating restriction sites for particular enzymes and introducing new sites for others. The variability produced in this way is inherited in simple Mendelian fashion. The action of restriction enzymes produces pieces of DNA or fragments that are recognized by the use of a radiolabeled "probe," a piece of DNA containing a particular base sequence, usually a particular gene, which recognizes all the fragments of DNA containing that particular sequence. One of the harmless base changes may be associated with the presence of a mutant gene that produces a different pattern of DNA fragments from that usually found with a particular restriction enzyme. Variability of this sort found in 1 percent or more of the population when they are studied with a particular restriction enzyme is called restriction fragment length polymorphism (RFLP). These have served as a large number of linkage markers for following mutant genes in families and in populations. These RFLPs are called dimorphic because there are usually only two alleles; the restriction site for a given enzyme is either present (+) or absent (−).

Some regions of DNA consist of variable numbers of repeated sequences having groups of core base pairs of ten to fifteen base pairs in length, called *minisatellites*. Different chromosomal segments have differing numbers of these repeated sequences. In these regions the length of DNA-containing minisatellites between particular restriction sites can vary tremendously, producing what are called *hypervariable* regions. The restriction fragments produced vary in size according to the number of repeats in a given region. This variability appears to be inherited in the same way as the restriction polymorphic sites. The amount of variability in these hypervariable regions of DNA is so great that a probe (Jeffreys et al., 1985) from one of these minisatellites produced unique restriction fragment patterns. This constitutes a form of DNA "fingerprinting." The technique can be used to localize or map a specific gene in relation to another gene or genes by using a series of restriction enzymes to produce fragments between and outside the genes being studied using as probe the gene one wishes to localize. This is followed by orienting the fragments produced in the order that best fits the pattern of fragments in relation to the genes being studied. Using this technique, it is possible for investigators to map a gene using only 5 milliliters of whole blood.

One of the first RFLPs found was identified in association with the *HbS* gene (Kan and Dozy, 1978). The RFLPs associated with *HbS* are described in detail in Chapter 7.

Studies of RFLPs associated with human growth hormone-human chorionic somatomammotropin (*hHGH-hCS*) gene cluster in Mediterranean people, Northern Europeans, and blacks (Chakravarti, et al., 1984b) revealed that in these populations on the average 1 out of 500 bases in the *hHGH-hCS* cluster is variant. In another study, a cDNA clone of argininosuccinate synthetase (AS) was used as a probe to screen restriction fragments (Daiger et al., 1983). The structural locus for this enzyme is on chromosome 9. The investigators found AS-like sequences on at least ten human chromosomes, including the X and Y. All the enzymes of three enzymes tested produced high-frequency polymorphisms. These were *Hind*III, *Hind*II, and *Bam*HI. Most of the polymorphic loci were found in blacks, whites, and Orientals with allele frequencies as shown here:

Restriction Enzyme	Allele Frequency	
	p	q
*Hind*II	0.13	0.87
*Hind*III	0.30	0.70
*Bam*HI	0.56	0.44

Similar studies in U.S. blacks demonstrated a large number of restriction fragments when insulin was used as a probe (Lebo et al., 1983). To produce the large number of fragments found, recombination was estimated to have occurred thirty-three times more frequently than expected. The recombination rate between the β-globin gene locus and the structural insulin locus was approximately 14 percent (map distance 14.2 centimorgans). Another study provided data consistent with a mutation rate of 1 out of 240 in hypervariable regions.

A different kind of DNA polymorphism has been found in studies of mitochondrial DNA (mtDNA) using restriction enzymes. Since the mitochondria are inherited from the mother, no meiosis or recombination occurs. Mitochondria do not appear to be influenced by selection pressure, so a study of mtDNA provides an unbiased history of the mutational history of the population being studied. The mtDNAs from 235 individuals representing five ethnic groups were treated with the *Hpa*I restriction enzyme (Denaro et al., 1981). Six different cleavage patterns, called morphs, were identified, each of which could be related to one another by single nucleotide substitutions. Of the differences found in the frequencies of these morphs in different groups, the most striking was found in the frequency of the morph most common in whites and Orientals compared with the frequency found in Africans. These differences appeared to originate by a change in the sequence GTCAAC to GTTAAC. One morph with two fragments was found in 12.5 percent of Orientals and 4 percent of Bantu people. In another study of mtDNA, samples from twenty-one individuals from twenty-one diverse racial and geographic origins were digested with eighteen different restriction endonucleases (Brown, 1980). Eleven produced digests that showed one or more differences between samples. Each of the twenty-one samples could be characterized individually from these digests. The differences could be explained by single base substitutions. Fourteen of the restriction site changes were shared by two or more individuals and seven of these were shared between different races. These data are consistent with group-specific patterns of DNA cleavage.

Mitochondrial DNA was studied in the Wolof and Peuls in Senegal and compared with populations in Africa and elsewhere (Scozzari et al., 1988). The Wolof belong to the group of Senegambians in Senegal along the Atlantic and in the adjacent hinterland. The Wolof comprise about 40 percent of the Sengalese population. The Peuls constitute about 15 percent of the Sengal population and belong to a nomadic group distributed throughout western Africa from Senegal to Cameroon and Chad.

The mtDNA of the Senegalese were analyzed by means of restriction enzymes: *Hpa*I, *Bam*HI, *Hae*II, *Msp*I, *Ava*II, and *Hinc*II, and to a lesser extent, *Hae*II. Important differences were found among the Senegalese, the Bantu of South Africa, and the San. Senegalese mtDNAs showed typical African characteristics; namely, presence and frequency of *Hpa*I morph 3 and high incidence of *Ava*II morph 3. These patterns markedly differentiated the Senegalese from the others. Scozzari and co-workers (1988) asserted that the phylogeny of mtDNA types in Africa illustrates how the three African groups are distinct genetic entities. The San are at one end of the range of variability; Senegalese are at the other end, but still closely related to the Bantu. These investigators also confirmed the work of others in demonstrating that not only blacks are divergent from others but also that mitichondrial differentiation among blacks is greater in Africa than in Europeans and Orientals.

The mtDNA polymorphisms were investigated in a sample of 90 Sicilians (Semino et al., 1989). Interestingly, the *Hipa* I-3/*Ava* II-3 complex that is characteristic of African ancestry was found in the Sicilian population at a frequency of 4.4 percent. Accordingly, for the first time an estimate of the amount of gene flow from Africans to the Sicilian gene pool was obtained.

Mitochondrial DNA has also been used to resolve the question of the geographical origin of humans. Although there is general agreement that humans originated in Africa, all experts do not concur. The following summary of the problem is based on a review by Cann (1987) and an editorial by Lewin (1987).

Lewin indicated that Jon Marks (at Yale) and A. Wilson (of the University of California at Berkeley) asserted that the publicity surrounding mtDNA studies has led to the inference that there was a single African female from which we are all descended who lived about 200,000 years ago. This female was given the catchy name "Mitochondrial Eve." The investigation that led to this speculation was the use of mtDNA to construct family trees. Since mitochondria pass from generation to generation only through the female line, phylogenies derived from mtDNA data trace maternal inheritance. It appears that mtDNA is something of a "passenger" in the events that lead to the development of a new species, and thus it neither contributes to the formation of a species nor reveals what actually happened. And according to Wilson, a Mitochondrial Eve will usually have existed a considerable time before a newly derived species becomes established. Accordingly, the 200,000-year-old African female from which humans derive their mitochondrial DNA was a member of an archaic sapiens species, and not yet anatomically human. Even so, Cann asserted that since mitochondrial DNA undergoes significant change every few hundred thousand years, mitochondrial DNA is a gauge of short-term human evolution, and the best clock available for measuring the time when modern humans emerged from a common ancestor.

Cann (1987) reported that in 1979 she began collecting samples of mitochondrial DNA from the placentas of newborns, and by 1986 she, A. Wilson, and Mark Stoneking had collected 147 samples from children whose ancestors lived in Africa, Asia, Europe, Australia, and New Guinea. Using restriction enzymes, they divided each sample into more than 300 fragments, which were arranged in distinctive patterns by gel electrophoresis. Then, by computer, the number of mutations was estimated that had taken place in each sample since it and the others evolved from a common ancestor. Fourteen of the samples had base sequences virtually identical to other samples in the survey, which left 133 distinct types of mitochondrial DNA. Some samples in this group were closely related in base sequences, indicating divergence from a single female within the past few centuries; other were connected by a common grandmother who lived tens of thousands of years ago. An evolutionary tree was then constructed by computer which placed the 133 mitochondrial DNA types at the tips of the branches, and each cluster, in turn, linked to others at the points at which they emerged from common ancestors. All but 7 mitochondrial DNA types were on limbs that converged on a single branch, and descendants of people from the five global regions were mixed throughout this branch. A second major branch, thinner than the first, contained the 7 remaining mitochondrial DNA types, all derived from peoples of African descent. These 7 were as different one from the other in the composition of mitochondrial DNA types as were any of those on the more widely radiating, multiracial branch. Thus Cann concluded that their common ancestor is just as old as the common ancestor of the 126 on the larger branch. Further, the base of the tree, the region where the two branches split, is the position of the common ancestor of humans: her children diverged and originated the two lines of descent.

In dealing with the important conclusion that modern humans emerged from Africa, Cann related that the emergence of modern humans from Africa is substantiated from two of the tree's characteristics. First, African ancestors can be traced to the base of the tree without encountering any non-African ancestors, but descendants of the other areas have at least one African ancestor. Second, the Africa-only branch has more diverse types of mitochondrial DNA than any other geographic group, which indicates greater evolutionary change among Africans than any other population. As for those who insist that Asia is the birthplace of modern humans, they must explain why Asians of today have little diversity in their mitochondrial gene pool. Cann's studies confirmed the fossil evidence that "Mother Eve" was an African.

But there is dissent. The fossil evidence has long been debated, and now the molecular evidence for the origins of humans may be just as controversial. Excoffier and Langaney (1989) collected all available data on mtDNA RFLP and concluded that of the mtDNA types in ten populations, all types could have evolved from a single common ancestral type. Distribution of shared types among continental groups indicated that Caucasoid populations could be the closest to an ancestral population from which all other continental groups could have diverged and also could occupy an intermediate place between African and Oriental populations. Further, African populations have quite differentiated mtDNA types, most of which appeared only recently. Thus Africans were not likely heirs of a population from which all others could have diverged. Excoffier and Langaney pointed out that the African sample of Cann (1987) consisted of eighteen African-Americans, one Nigerian, and one Bushman (San), and that the sample's heterogeneity must have introduced a strong bias which raised the estimate of its molecular diversity.

Studies of other DNA polymorphisms have been carried out in U.S. blacks, whites, and Hispanics from the New York metropolitan area, using undefined probes called D14S1 (analyzed in *Eco*RI-digested DNA) and HRAS-1 (*Taq*I-digested DNA) (Baird et al., 1986). More than forty alleles were detected with the D14S1 probe, with DNA fragments ranging from 14.3 to 32.5 kilobase pairs (kbp), and eighteen alleles, in the size range 1.85–4.5 kbp, were detected with the HRAS-1 probe. In HRAS-1 the frequency distribution of alleles among three ethnic groups, whites, blacks, and Hispanics, for both of the polymorphisms studied is statistically significant ($P < .01$). The investigators showed how these polymorphic systems can be used in paternity testing. The two polymorphic loci described in these populations yield a power of paternity exclusion comparable to a battery of sixty-three HLA antigens.

Another DNA polymorphic system in which differences have been found between different ethnic groups involves the RFLPs of coagulant factor IX, *Taq*I and *Xmn*I (Lubahn et al., 1987). Two alleles were found. The rarer type *2* allele is significantly higher in Europeans than in U.S. blacks and East Indians. Type *2* alleles are rare in Chinese and Malays.

There is ample evidence that DNA polymorphisms detected as RFLPs provide markers for studies of between-group and within-group variability. Furthermore, studying mtDNA allows us to trace the mutagenic history of a given population. The significance of most of the ethnic differences in RFLP allele frequencies that have been identified to date is unknown. It is likely that the very high degree of polymorphism that we have observed occurred and has been maintained because the changes were selectively neutral. Polymorphic analysis is a very active area of research in genetics and a great deal remains to be learned about the types of variability that exist in the human genome. For the moment, these systems contribute to paternity studies, tracing population movements, determining the origin of mutant genes in populations.

COLOR VISION

The pigments of the three kinds of color vision genes are found in three different classes of retinal cone cells, each containing a different photopigment (Kalmus, 1965; Motulsky, 1988) from which most of this section's discussion is derived. A blue pigment gene is located on chromosome 7 and red and green pigment genes are located at the tip of the long arm of the X chromosome (Xq28). The pigments have maximal absorption spectral peaks of 430 (short wave, blue), 540 (middle wave, green), and 570 (long wave, red). Individuals with normal color vision have the three pigment genes and are trichromats. Dichromats who lack the green pigment gene are termed deuteranopes, and those that lack the red pigment gene are protanopes. About 8 percent of Europeans have defective color vision, approximately 2 percent are dichromats, and the remainder are anomalous trichromats. Deuteranomalous trichromats have defective green visualization, and protanomalous trichromats have anomalous red perception.

There are many techniques for ascertaining color vision; various observers interpret these tests differently. One of the best techniques is anomaloscopy, which is difficult. Thus most surveys are performed with color vision charts, which may differ from one other. Even the same chart may vary from edition to edition, and the edition rarely is recorded in reports on color vision studies.

A review of the biochemical aspects of color pigment genes (Motulsky, 1988), showed that the red and green photopigments resemble each other in 98 percent of their amino acid sequence as deduced from the DNA sequence. Presumably, the red pigment gene is upstream (5′) of the green pigment gene(s). Deuteranopic dichromats lacked green pigment genes; although protanopic dichromats had no normal red pigment gene, there was in its place a hybrid or fusion pigment gene consisting of 5′ segments of the red pigment gene and 3′ segments of the green pigment gene (5′ R–3′ G), and no or variable numbers of green pigment genes. Deuteranomalous trichromats showed fusion genes of type 5′ G–3′R with or without normal green pigment genes. On the other hand, protanomalous trichromat males had a 5′R–3′ fusion gene, no normal red pigment gene, and variable numbers of normal green pigment genes. Motulsky emphasized that these genotypic findings do not necessarily delineate protanopia from protanomaly. He suggested that the molecular basis of defects in color vision may be explained by a variety of unequal crossovers that resulted in deletions of pigment genes or by different fusion genes that consisted of variable segments of the red and green pigment genes. Further, the striking homology between the red and green pigment genes and the variable green pigment genes explained the higher frequency of deuteranomaly (4–5 percent) as compared with protanomaly (0.7–1 percent).

Kalmus (1965) reviewed the occurrence of abnormal color vision in various populations (Table 5.4) and analyzed sources of bias. In general, Africans and their descendants have very low frequencies of abnormalities in color vision. As we mentioned, different techniques of testing undoubtedly account for some of the differences. Ideally, statistical comparisons should be made by testing the scoring efficiency of the investigators before comparisons are attempted. Population frequencies often were derived from other tests done for specific purposes, such as applications for jobs in transport, volunteers for the air force or navy, armed services recruits in peace or wartime, and schoolboys. The bias in such groups is self-evident. Sample size, as in other population surveys, is also crucial.

Kalmus (1965) also introduced and reviewed various selective explanations for population differences. Kalmus postulated that abnormalities in color vision could be detrimental to hunting groups, because hunters may not be able to delineate prey in jungles or forests, and thus may be poor food gatherers or may risk being killed. On the other hand, Kalmus admitted that many persons with defects in color vision may be at an advantage in a camouflaged environment or where game is camouflaged by color. Human-created camouflage is, after all, usually developed by individuals who have normal color vision, and those who have defects in color vision may be able to discern camouflaged targets, and in fact they have been used in wartime for that purpose.

ADMIXTURE IN POPULATIONS

Anthropologists and population geneticists have long been fascinated with admixture estimates in human populations. The requirements for such calculations are parent populations, usually two, and an evolving "hybrid" population over varying periods of time. There are many such studies, but here we will only concern ourselves with admixture estimates of African-American population in the United States, particularly those using biochemical genetic markers and haplotype frequencies. We will not review the extensive literature on the use of morphological characters, such as skin color, because they have been supplanted by more recent methods. Calculations of admixture estimates are complicated by the classical population genetic interactive forces of mutation, migration (initial and ongoing), contin-

Table 5.4
Defective Color Vision in Male Populations

Population	Frequency (%)
Europe	
English	0.068–0.095
Scots	0.075–0.077
French	0.066–0.090
Belgians	0.075–0.086
Germans	0.066–0.078
Swiss	0.080–0.090
Norwegians	0.080–0.101
Czechoslovakians	0.105
Russians	0.067–0.960
Jews (Russians)	0.076
Finns (Leningrad)	0.057
Turks (Istanbul)	0.051
Asia	
Tartars	0.050–0.072
Chinese	0.050–0.069
Japanese	0.035–0.074
Indians (Hindi)	0.000–0.100
Indians (tribal)	0.000–0.090
Israelis	0.021–0.062
Druses (Israel)	0.100
Filipinos	0.043
Fiji Islanders	0.000–0.080
Polynesians (Tongo)	0.075
Africa	
Bechuanas	0.034
Bugandans	0.019
Bahutus	0.027
Batutsi	0.025
Congolese	0.017
North America	
U.S. whites	0.072–0.084
U.S. blacks	0.028–0.039
Amerindians	0.011–0.052
Eskimos	0.025–0.068
Canadian whites	0.112
Mexicans (urban)	0.047–0.077
Mexicans (tribal)	0.000–0.023
South America	
Indians	0.000–0.070
"White" Brazilians	0.069–0.075
"Dark" Brazilians	0.088
Japanese (in Brazil)	0.129

Table 5.4
(*Continued*)

Population	Frequency (%)
Australia	
Whites	0.073
Aborigines	0.020
Mixed	0.032

Source: Modified from Kalmus (1965).

uing admixture, genetic drift, natural selection, linkage disequilibrium, recombination, and sampling errors, to name a few (Reed, 1969; Cavalli-Sforza and Bodmer, 1971; Chakraborty, 1986; Chakraborty and Smouse, 1988).

The role of falciparum malaria, isolation, and admixture in urban and rural blacks of the Charleston, South Carolina, region was studied by Pollitzer (1958). Pollitzer believed that the most likely explanation for the higher frequencies of HbS in blacks of Charleston and the Sea Islands as compared to blacks in other parts of the United States was because admixture in this region amounted to 5 percent as contrasted to higher rates elsewhere.

The choice and the number of the genetic markers or haplotypes is most important. As Chakraborty (1986) pointed out, markers that are markedly influenced by selection, such as sickle hemoglobin, and glucose-6-phosphate dehydrogenase should probably be avoided. Further, a marker that is absent in one of the parent populations and present in high frequency in the other population, such as adenylate kinase electrophoretic phenotypes (Bowman et al., 1967), is obviously preferable over markers in parent populations in which the frequencies are close. Interestingly, admixture estimates are still useful after thousands of years of exchange of genes among dissimilar groups, because considerable morphological and genetic differences exist among populations (Chakraborty, 1986), otherwise this book could not have been written.

Let us return to ancestral populations. Chakraborty (1986) emphasized that the appropriate choice of ancestral populations to be studied has been given little attention. What European frequencies should be used? Which African stock(s) or weighted averages of various African populations should be analyzed? And what "hybrid" black population in the United States, Southern, Eastern, Midwestern, Southwestern, Western, is the best study group? Even the location may be misleading. Chicago blacks were studied by Bowman et al. (1967). The studies suggested that the 13 percent admixture estimates for adenylate kinase polymorphisms were unusually low. On the other hand, admixture estimates for Southern U.S. blacks are generally lower than in some other parts of the country, and there have been massive migrations of blacks from the South to Chicago since the Great Depression and during and after World War II.

As we have mentioned, it is important to separate those genes that are obviously affected by selection and those not so obviously affected. For example, in an exhaustive analysis, Chakraborty (1986) separated such estimates. Admixture estimates of genes infrequently affected by selection ranged from a low of 4 and 7.3 percent, respectively, in Charleston, South Carolina, and in Evans and Bullock Counties, Georgia, to a high of 30.6 percent in Baltimore. Figures using markers affected by selection ranged from a low of 17.5 percent using $G\text{-}6\text{-}PD^{A-}$ in Georgia to highs of 42 to 70 percent using $G\text{-}6\text{-}PD^{A-}$ and Hb^S in other another section of Georgia.

Galactokinase

Galactokinase catalyzes the conversion of galactose to galactose-1-phosphate. Population studies reveal a significant difference between blacks and whites in the distribution of red cell galactokinase (GALK). Based on the shapes of the distributions, it was inferred that whites are essentially all homozygous for one allele, called *GALKA*, whereas blacks are polymorphic with a second allele called *GALKP*, for lower GALK activity. This allele is present in high frequency in blacks and is rare or absent in whites. Three normal distributions can be fit to the distribution of the black data, so the frequency of *GALKA* is approximately 0.217 (Spielman et al., 1978). The current estimate of the admixture of whites with blacks is about 25 percent, but, as we mentioned, an admixture of 13 percent was found using adenylate kinase (Bowman et al., 1967). This implies that virtually all the GALKA genes in blacks were introduced by white admixture and that the original black population was monomorphic for GALKP.

DNA Haplotypes

Admixture estimates using DNA haplotypes were promising addenda to genetic markers (Chakraborty, 1986); however, Chakraborty and Smouse (1988) cautioned scientists in the use of these newer tools in population studies. These authors asserted that a population formed by genetic admixture of two or more source populations may exhibit considerable linkage disequilibrium between genetic loci, and that in the presence of recombination, linkage disequilibrium declines with time. Further, recombination alters haplotype frequencies over time and the haplotype-derived measures of admixture proportions from haplotype frequencies in generations following the admixture event become more biased. These authors also emphasized that even haplotype frequencies that are defined by multiple restriction fragment length polymorphisms should be viewed with caution. Each RFLP site should be considered separately for admixture analysis.

Cystic Fibrosis in African-Americans

Cystic fibrosis (CF) is common in Europeans and their descendants, with a prevalence of 1 in 2500 births in the United States. The disorder is rare in Africans: only eight cases have been reported since the mid-1950s. African-Americans have an estimated incidence of 1 in 17,000 births (Cutting et al., 1989). Presumably, the high prevalence of CF is due to heterozygote advantage (the carrier frequency in European-Americans is about 1 in 25). Cutting et al. queried whether the low frequency of CF in blacks is because blacks have not been subjected to the selection pressure of Europeans, or whether racial admixture is the explanation. On the other hand, these authors pointed out that on the basis of a *CF* gene frequency of 1 in 50 in Europeans and 20 percent racial admixture in blacks, the European *CF* gene frequency in African-Americans should be 1 in 250. But, if the incidence of CF at birth in African-Americans is 1 in 17,000, the *CF* gene frequency is about 1 in 130. If so, European genes account for 50 percent or more of the *CF* gene in blacks, which is unlikely.

Cutting et al. (1989) and Beaudet et al. (1989) found that strong linkage disequilibrium occurs between the *CF* locus and polymorphisms found with DNA probes XV-2C and KM19. In a North American population, 86 percent of CF chromosome occurred with a haplotype that was found on only 14 percent of normal chromosomes (Beaudet et al., 1989). Cutting et al. not only looked at European-Americans but also compared the DNA polymorphism haplotypes linked to the *CF* locus in North American black and European-American families. While strong linkage disequilibrium was detected between the DNA marker XV2c and the *CF* locus ($\delta = 0.46$) and between DNA marker KM19 (8; standardized link low disequilibrium value [LD]) and *CF* ($\delta = 0.67$). Data for these same markers in sixteen African-American patients and their families showed a different haplotype distribution and linkage disequilibrium pattern with the CF locus. The authors concluded that racial admixture alone does not explain the prevalence of *CF* in African-Americans, and that multiple alleles of the *CF* gene may exist in African-Americans.

Beaudet et al. (1989) indicated that it is indeed appropriate to use linkage disequilibrium data for risk calculations of CF, for to do otherwise would neglect genetic information that may considerably alter probabilities for carrier or affected status. And most important, as we have seen, disequilibrium data for one ethnic group may differ from data for others, because a crossover occurred early in the ancestry of a particular ethnic group or because of a separate origin for a *CF* mutation. Beaudet et al. gave examples of genetic isolates of a recent origin, such as the Amish or Mennonite groups in North America. Different linkage disequilibrium data were found in an Italian population; however, the similarity of North American and European data was quite evident. Accordingly, Beaudet and colleagues used the European-American data without dividing them by ethnic origin. It is impractical to do otherwise, because many European-Americans are inextricably mixed and many do not know their ancestral line.

   Let us return to European-American, African-American CF differences. Are the linkage disequilibrium data important clinically? Cutting et al. (1989) cited previous studies that suggested that black CF patients have milder lung disease and more gastrointestinal complications than do European-Americans. Nevertheless, when Cutting and co-workers rated the clinical features of the black and European-American patients in a double-blind fashion, the presence of a particular haplotype or genotype was not associated with severe lung disease or with more gastrointestinal complications in either group of patients. The authors postulated that the lack of clinical heterogeneity may be due to the small sample size or to imprecise ascertainment of clinical status, or perhaps the distance between the DNA polymorphisms and the *CF* locus precluded a significant association with a particular phenotype.

6

Glucose-6-Phosphate Dehydrogenase Deficiency

Beginning in 1945, the University of Chicago–Army Malaria Research Unit conducted intensive investigations of newly synthesized 8-aminoquinolines and of their causal prophylactic and curative properties in severe, mosquito-induced Chesson strain of *Plasmodium vivax* malarial infections (Alving et al., (1960). This unit, which was led by Alving until his death and thereafter by Paul E. Carson, developed most of the present-day standards for prophylaxis and treatment for malaria—including the widespread drug-resistant strains of *P. falciparum*. The early antimalarial drug pamaquine and the later drug primaquine had long been known to be toxic to Asiatic Indian soldiers but not to English soldiers in India (Beutler, 1959); later, in the United States, toxicity was known in blacks but not in whites. The toxicity of primaquine led to a series of investigations by Alving's team and others to understand the mechanism of drug-induced hemolytic anemia, or "primaquine sensitivity," as it was then known.

Dern et al. (1954) demonstrated that primaquine-induced anemia was caused by an abnormality of the red cells and that it was also self-limited in blacks. Carson et al. (1956) discovered that a deficiency of the red cell enzyme glucose-6-phosphate dehydrogenase (G-6-PD) was the common denominator in almost all of the individuals who developed hemolytic anemia when they took primaquine and some other antimalarial drugs. The G-6-PD deficiency was shown to be inherited in an X-linked manner (Childs et al., 1958; Marks and Gross, 1959). Biochemical analysis indicated that the sensitivity of the red cells of primaquine-sensitive people was connected to serum glutathione content and stability (Childs et al., 1958). Glutathione is a tripeptide found in the blood; most of it occurs in the red cells. It is maintained in its functional reduced state (GSH) by an enzyme, glutathione reductase (GSSG). Apparently, GSH stabilizes a number of cell proteins, including enzymes, coenzymes, and hemoglobin, and protects these proteins by being preferentially oxidized, thereby protecting the red cell. The cell is able to regenerate GSH from the oxidized form GSSG to maintain this protective function (Beutler et al., 1957). As Beutler (1983) pointed out, GSSG and protein S-SG disulfides are important regulators of the hexose monophosphate pathway, and red cell GSH is constantly oxidized in the glutathione peroxidase reaction (GSH-Px), in which GSH rids the erythrocyte of damaging hydrogen and organic peroxides.

As shown by Beutler (1983), G-6-PD deficiency may occur in three ways: decreased production of enzyme molecules, formation of enzyme molecules with decreased catalytic activity, or production of enzyme molecules with reduced stability.

The G-6-PD deficiency was also found in individuals with favism (acute hemolytic anemia caused by ingestion of fava beans) and in some individuals with congenital nonspherocytic hemolytic anemia (Szeinberg et al., 1958; Shahidi and Diamond, 1959). Interestingly, although all individuals with favism have G-6-PD deficiency, all individuals with G-6-PD deficiency who eat fava beans do not develop hemolysis (Walker and Bowman, 1960). This seeming paradox has been investigated for many years; the reason is still unknown. Investigators have demonstrated a large number of variants of G-6-PD: one type has increased enzyme activity, other types have very little or no enzyme activity, and some types have severe enzyme deficiency, with less than 12 percent activity. The more common forms of G-6-PD deficiency, particularly that found in blacks, are generally without clinical effect unless certain drugs are ingested. On the other hand, about 30 percent of persons who have congenital nonspherocytic hemolytic anemia have some form of G-6-PD deficiency (Carson and Frischer, 1966).

PATHOPHYSIOLOGIC ASPECTS OF G-6-PD DEFICIENCY

The most frequent type of G-6-PD deficiency found in Mediterranean, Middle Eastern, and Oriental peoples is more severe than that found in blacks, probably because the level of enzyme activity is lower (Carson and Frischer, 1966). Africans and their descendants are generally asymptomatic unless they are subjected to the "stress" caused by administration of certain oxidative drugs or by severe infections, particularly viral hepatitis.

The following compounds are known to cause clinically significant hemolysis of G-6-PD-deficienct red blood cells (Beutler, 1978): the analgesics acetanilid, acetylsalicylic acid, and acetophenetidin; the sul-

fonamides and sulfones sulfanilamide, sulfapyridine, diaphenylsul-
fone, N-acetylsulfanilamide, sulfacetamide, thiazolsulfone, salicylazo-
sulfapyridine, and sulfamethoxypyridazine; the antimalarials primaquine, pamaquine, pentaquine, quinocide, and quinacrine (Atabrine);
the nonsulfonamide antibacterial agents furazolidone, furmethonol,
nitrofurantoin, nitrofurazone, and chloramphenicol; and the compounds of naphthalene, trinitrotoluene, methylene blue, nalidixic
acid, phenylhydrazine, quinine, quinidine, ascorbic acid (in massive
doses), and niridazole.

When individuals of African origin with G-6-PD deficiency are
given 30 milligrams of primaquine daily, hemoglobinuria may appear
after two or three days, and may be associated with fatigue, weakness,
or abdominal and back pain, and jaundice (Dern et al., 1954). The
hemoglobin, red cell count, and hematocrit levels fall rapidly, with a
concomitant increase in reticulocytes. In addition, Heinz bodies can
be found in the red cells during the initial hemolytic reaction. After
approximately six or seven days, the acute hemolytic phase subsides
and the recovery phase begins. This occurs even if the individual
continues to take the drug responsible for the episode. During this
period, the acute phase events are reversed, and the individual feels
almost normal. This process is summarized in Figure 6.1.

The reason for the lack of response to the drug administration
during the "refractory period" appears to be the result of an altered
reactivity of the red cell population in patients deficient in G-6-PD. It
was found that the primaquine-induced hemolysis was related to red
cell age (Beutler et al., 1954). With increasing cell age, the erythrocyte
G-6-PD activity declines in normal as well as deficient cells. However,
in cells that are deficient in G-6-PD, the decline is even more dramatic, so one might say that they undergo accelerated aging or premature
senescence. The degree of primaquine sensitivity is progressively
greater as the G-6-PD-deficient erythrocytes cells age. Further, these
deficient red cells have a shortened life span even when they are not
exposed to an oxidative drug; their life span is approximately 75
percent of normal, that is, about 90 days instead of the normal 120
days (Kellermeyer et al., 1962). The deficient red blood cells constitute
a range of susceptibility to drug-induced hemolysis. This is the reason
for the dose-dependence of primaquine sensitivity: higher dosages of
an oxidative drug will have an effect on more cells than lower dosages.
Earlier studies by Beutler et al. (1954) revealed that erythrocytes less
than 21 days old were not affected by the drug, whereas those more
than 63 days old were hemolyzed by a daily dose of 30 milligrams of
primaquine. Cells ranging in age between 22 and 62 days are variously sensitive, depending on the dosage of the drug administered.

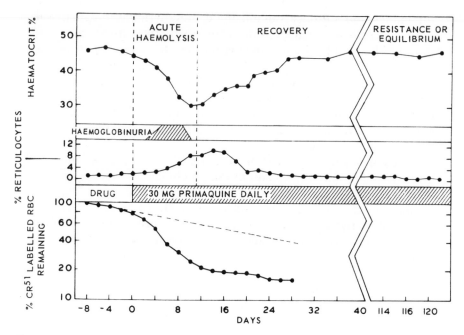

Figure 6.1. Mitigation of the hemolytic effect of primaquine. Hemolysis is self-limited in blacks even though a challenge dose of 30 milligrams of primaquine daily is continued
Source: Alving et al., 1960. Reprinted with the permission of The Office of Publications, World Health Organization, Geneva

Thus the change in reactivity of the erythrocytes of an individual who has G-6-PD deficiency is a consequence of the persistence of a population of relatively young erythrocytes and the continued removal of those relatively few red blood cells as they reach the age of drug sensitivity. The hemolytic anemia induced by a number of other drugs, including acetanilid, sulfanilamide, and naphthalene, is similar to that produced by primaquine. There are some variations in the intensity of the hemolytic reactions produced by some drugs (for example, nitrofurantoin, naphthalene, and thiozolsulfone), which appear to result from differences in the way the drugs are absorbed and/or metabolized (Dern et al., 1955).

BIOCHEMICAL AND MOLECULAR CHARACTERISTICS OF G-6-PD

The enzyme G-6-PD is the first enzyme in the hexose monophosphate shunt. It catalyzes the oxidation of glucose-6-phosphate to 6-phosphogluconate (6-PG). This is accompanied by conversion of the coenzyme nicotinamide-adenine dinucloetide phosphate (NADP) to reduced nicotinamide-adenine dinucleotide phosphate (NADPH). This key metabolic pathway is shown in Figure 6.2. This pathway primarily maintains the intracellular concentration of the reduced form of the coenzyme $NADPH_2$. This compound, in turn, serves to help maintain the levels of reduced glutathione (GSH), which under normal circumstances helps control the levels of hydrogen peroxide and other organic peroxides (Jacob and Jandl, 1966). When the GSH levels are lowered because of reduced enzyme activity, the erythrocytes become vulnerable to the oxidative effects of those drugs that produce hemolysis in individuals with G-6-PD deficiency.

Almost all peoples of African origin who have G-6-PD deficiency have a variant characterized as A− (as compared to the common and most frequent variant with normal enzyme activity called B+). An A+ variant common in people of African origin has no ill effects. These variants were first described by Kirkman (1962) and by Boyer, Porter, and Weilbacher (1962). Another relatively common variant seen mostly in those of Mediterranean, Middle Eastern, and Asian ancestry (Sardinians, Greeks, Oriental and Sephardic Jews, Iranians, Asian Indians) is designated B−. It has up to 7 percent of normal enzyme activity (the average level is 3–4 percent) and is classified as a severe enzyme deficiency. The reaction of individuals with the B− variant to drug-induced hemolysis is more severe than the reactions of people with the A− variant. For example, acetylsalicylic acid and acetophenetidin are only slightly hemolytic in G-6-PD A− deficiency in very large doses. Chloramphenicol is hemolytic in G-6-PD B− deficiency but not in G-6-PD A− deficiency, and quinine and quinidine are hemolytic in G-6-PD B− deficiency but not in G-6-PD A− deficiency. Individuals with the B− variant are also sensitive to hemolysis induced by the fava bean. Population studies in West Africa have revealed that a very small number of Africans have the B− variant.

The molecular weight of G-6-PD has not been established. Estimates have ranged from 105,000 to 240,000 daltons, but in vivo G-6-PD is probably a dimer with a molecular weight in the range of 105,000 to 120,000 (Beutler, 1986). The subunits of the molecule are polypeptide chains about 450 amino acids long. Structural analysis has revealed a single amino acid difference between the common B+ and the A+ variants. An aspartic acid residue in the B+ variant has been replaced by an asparagine in the A+ variant. Based on the genetic code, it is apparent that this amino acid difference can be brought about by the substitution of a single base. Aspartic acid can be coded for by GAU or GAC, but asparagine can be coded for by AAU or

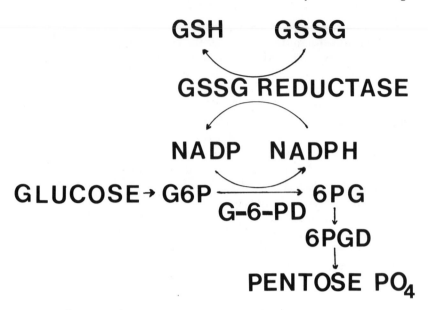

Figure 6.2. The pentose-phosphate pathway of red cell metabolism

AAC. This means that the amino acid change can be brought about by the substitution of guanine in the B+ variant or by an adenine in the A+ variant (GAU → AAU or GAC → AAC) (Yoshida, 1967).

Takizawa and Yoshida (1986) purified glucose-6-phosphate dehydrogenase to homogeneity from human erythrocytes. The amino acid sequence of the subunit consisted of 531 amino acids. A 41-mer oligonucleotide with unique sequences was obtained based on the amino acid findings. Two cDNA libraries developed in phage λgt11—human liver cDNA library and a human hepatoma Li7 cDNA library—were screened with the synthetic nucleotide probe. Two positive clones (G-6-PD-19 contained an insert of 2.0 kilobase pairs (kbp) and encoded 204 amino acid residues that were compatible with the COOH-terminal segment of the enzyme. The insertion of the clone had a 3′ noncoding region of 1.35 kilobase pairs. The other clone (G-6-PD-25) had an insertion of 1.8 kilobase pairs and encoded 362 amino acid residues of G-6-PD. Southern blot analysis of DNA obtained from cells with and without the human X chromosome confirmed that the cDNA hybridized with the sequence in the X chromosome, and thus the clones that were obtained were for X-linked G-6-PD and not for autosomal hexose-6-phosphate or other isozymes.

Examination of the structure of *G-6-PD A(+)* gene (Yoshida, Takizawa, and Prchal, 1988) demonstrated that ^G_C → ^A_T transition was found in the variant gene, resulting in the amino acid substitution Asn

→ Asp at position 142 from the NH_2-terminal of the enzyme. The nucleotide change created an additional *Fok*I cleavage site in the *A(+)* gene. Accordingly, the *Fok*I fragment of the variant *A(+)* DNA differed from that of the common *B(+)* DNA. The *Pvu*II fragment type was also polymorphic in blacks but not in whites. The majority of African-Americans and nonAfrican-Americans have a major hybridization fragment of approximately 4.0 kilobases (*Pvu*II type 2), and about 20 percent of blacks have a major fragment of about 1.5 kilobases (*Pvu*II). The G-6-PD gene with *Pvu*II has another *Pvu*II cleavage site about 0.7 kilobases downstream from the mutation site of *G-6-PD A+*. About 40 percent of *G-6-PD A+* genes have *Pvu*II type 2, but only about 10 percent of *G-6-PD B+* genes are associated with *Pvu*II type 2. A significant linkage disequilibrium was found between the G-6-PD types and the *Pvu*II types at the G-6-PD locus.

A cDNA library was constructed from Epstein-Barr virus transformed lymphoblastoid cells from a subject who had the G-6-PD A− deficiency variant (Hirono and Beutler, 1988). One of four isolated cDNA clones had a sequence that was not found in the other clones nor in the published cDNA sequence. The clone consisted of 138 bases which coded forty-six amino acids. Comparison of the remaining sequences of the clones with the published sequence showed three nucleotide substitutions, C^{33} → G, G^{202} → A, and A^{376} → G. Each change produced a new restriction site. Genomic DNA from five G-6-PD A− subjects was amplified by the polymerase chain reaction. The base substitution at position 376, which was identical to the substitution in G-6-PD A+ was found in all G-6-PD A− samples and no normal controls. Guanine at position 33 was detected in all G-6-PD A− and seven G-6-PD B+ controls. The authors asserted that the finding of the same mutation in G-6-PD A− that is found in G-6-PD A+ suggested that the *G-6-PD A−* mutation arose in a person with *G-6-PD A+*, which added another mutation that causes in vivo instability of G-6-PD. (It is more likely that the *G-6-PD A+* mutant occurred on many occasions.)

Ramot et al. (1959) found G-6-PD activity not only in erythrocytes but in all other tissues in which G-6-PD was studied. All the tissues show the same isoenzymes, indicating that G-6-PD is under the same genetic control in all the cells, but there are minor differences in isoelectric points of erythrocyte, leukocyte, platelet, and fibroblast G-6-PD (Beutler, 1986), possibly because of posttranscriptional changes in the structure of the molecule. Leukocyte G-6-PD activity is normal or only slightly reduced in G-6-PD A − subjects, while it is significantly reduced in persons with Gd Mediterranean (about 30 percent of normal), Far Eastern variants, such as G-6-PD Mahidol (Thai populations), and in

persons with unusual variants that have markedly reduced levels of
G-6-PD in which there may be hemolysis even without exposure to
oxidizing drugs, as in G-6-PD Gd Freiburg.

PROBLEMS IN IDENTIFYING G-6-PD DEFICIENCY

Identifying G-6-PD A− individuals can be difficult after a hemolytic
episode, because the older erythrocytes that manifest the enzyme deficiency may no longer be present. Between hemolytic episodes the
enzyme deficiency can be readily detected with a screening procedure
or a quantitative assay. It may be possible to diagnose the deficiency
even during the refractory period by performing a quantitative assay
on the older, more dense erythrocytes (Ringelhann, 1972). It can also
be diagnosed by combining the assay for G-6-PD with that of another
age-dependent enzyme, such as hexokinase or glutamic-oxaloacetic
transaminase, and as Carson and Frischer (1966) pointed out, an
elevation of glutathione reductase is intrinsic to G-6-PD deficiency.
The elevated glutathione reductase level is particularly useful as an
adjunct in identifying females with heterozygous G-6-PD deficiency.
Although G-6-PD screening tests detect males who are hemizygous
deficient, there is a wide range of variation in heterozygous females
(Tarlov et al., 1962; Beutler, Yeh, and Fairbanks, 1962; Bowman and
Maynard Smith, 1963). Repeated testing of heterozygous females often shows that on some occasions the tests may reveal intermediate
deficiency and on other occasions G-6-PD assays will be in the normal
range in the same individual. These phenomena have also been observed in males with intermediate G-6-PD deficiency, where there is
no possibility of X-chromosome inactivation (Bowman, Carson, and
Frischer, 1969).

Bowman and Maynard Smith (1963) postulated that an autosomal
component (modifying gene pair) may account for the variable manifestation of G-6-PD in females. Interestingly, Kanno et al. (1989)
found two types of subunits in human red cell glucose-6-phosphate
dehydrogenase. The two subunits have the same COOH region that
consists of 479 amino acid residues, but the NH_2-terminal regions are
different in size and sequence. The cDNA and the gene for the NH_2
terminal region of the major subunit were cloned. Southern blot hybridization revealed that the gene for the NH_2-terminal region is on
chromosome 6 and not on the X chromosome. Northern blot hybridization showed two RNA components, one of the COOH-terminal
region and the other for the NH_2-terminal region. Kanno and co-
workers concluded that two separate structural genes, the X-linked

and chromosome 6-linked genes must be responsible for encoding the single-chain subunit.

When family studies of males with G-6-PD deficiency were conducted, it was found that the pedigree pattern of affected persons was consistent with X-linked recessive inheritance (Gross et al., 1958; Childs et al., 1958). Close linkage was also established between the gene for G-6-PD and the genes for color blindness (Adam, 1961; Porter et al., 1962). Additional studies indicated that this gene was probably located between the loci for proteranopia and deuteranopia (Kalmus, 1962). Beutler (1983) reviewed in more detail the X-linkage of G-6-PD. A map of four X-linked genes produced by the fusion of hamster and human genes suggested that the order on the X-chromosome in the region of the G-6-PD locus is phosphoglycerate kinase—alpha-galactosidase—hypoxyxanthine-guanine-phosphoribosyl transferase—G-6-PD.

Affected males have mothers who are heterozygous or homozygous deficient. In about 80 percent of heterozygous affected females, however, it is difficult to detect carriers by quantitative measurement of erythrocytic G-6-PD. A majority of females who are heterozygous for the G-6-PD gene have levels of enzyme activity that fall between normal and deficient. A smaller number of heterozygotes have levels that are very nearly normal or that fall into the range of those that are deficient. The findings from several studies of G-6-PD activity in heterozygous females (Beutler et al., 1962; Davidson et al., 1963; Beutler and Baluda, 1964; Satori et al., 1966; Sansone et al., 1964), along with observations by Lyon (1961), suggested an explanation for this phenomenon. Lyon is given credit for formulating the hypothesis that one X chromosome in each cell is inactivated in early embryonic life and remains inactive thereafter. This is a random process that produces a distribution of enzyme activities ranging from low to intermediate to normal levels. When the X chromosome carrying the gene for G-6-PD deficiency is inactivated, that cell will express normal enzyme levels. When the X chromosome carrying the normal G-6-PD gene is inactivated, that cell will express the deficient phenotype. If, in a given heterozygous individual, significantly more X chromosomes carrying the mutant allele are inactivated, that individual will have normal levels of enzyme. If significantly more X chromosomes carrying the normal allele are inactivated, the individual will have deficient levels of enzyme. In the majority of examinations the heterozygote will have intermediate levels of enzyme. It should also be remembered that some heterozygous affected females (and intermediate affected males—discussed later) often have normal levels of enzyme activity, not because of inactivation but because of intermittent hemolysis.

Although X chromosome inactivation is apparently well established, Frischer, Carson, and Bowman (1965) showed that in intact erythrocyte studies of females heterozygous for G-6-PD deficiency, mixtures of deficient and nondeficient red cells reveal that the mosaic nature of populations of erythrocytes heterozygous for G-6-PD deficiency is a function not only of the genetic variant of G-6-PD deficiency but also of the elapsed time of incubation, when the nitrite-methylene blue methemoglobin reduction test of Brewer (Brewer, Tarlov, and Alving, 1960) is combined with the acid elution technique of Kleihauer and Betke (1963). In fact, in normal males and females, 50 percent red cells and 50 percent "ghosts" will be found if a time between 50 and 100 minutes is selected. Accordingly, a conclusion of mosaicism is unwarranted. On the other hand, if the time of examination is three hours, about 50 percent of the erythrocytes of heterozygous females will be red, and this could lead to similar unsafe inferences of mosaicism.

Family studies are most helpful in establishing the G-6-PD status of females. If a woman has a son with G-6-PD deficiency, then she is at least heterozygously affected, and if a woman's father has G-6-PD deficiency, she is at least heterozygously affected. In fact, one of the advantages of detecting hemizygous G-6-PD in males as part of a screening program or in clinical situations is that the status of three generations may be predicted from one examination. If the male is affected, his mother is at least heterozygously affected, and all of his daughters will be at least heterozygously affected. Further, on the average, one-half of a G-6-PD deficient male's sisters will be heterozygously affected.

CHARACTERIZATION OF G-6-PD VARIANTS

In 1967 a group of biochemists and geneticists met under the auspices of the World Health Organization (WHO) and proposed a system of characterizing G-6-PD variants (Betke et al., 1967). It was suggested that each variant should be subjected to the following studies:

1. Determination of enzyme activity by enzyme assay.
2. Electrophoresis of the enzyme (if needed, after partial purification) in various buffer systems.
3. Determination of substrate specificity, Michaelis-Menten constant (K_m), for glucose-6-phosphate and NADP (and NAD).
4. Utilization of substrate analogues. These examinations were usually made with 2-desoxyglucose-6-phosphate, galactose-6-phosphate, and deamino-NADP. Utilization of such substrate

analogues is frequently used for eliciting qualitative differences in enzyme properties.

5. Determination of thermostability.

6. Determination of pH dependency of enzyme activity.

Following this scheme, it has been possible to identify more than 260 variants of G-6-PD. (See Beutler, 1983, for a listing of the known G-6-PD variants.) If G-6-PD variants are characterized by enzyme activity, at least six classes can be recognized. These are summarized in Table 6.1.

Commonly, males are hemizygous and have a single isoenzyme band. The mobility of the most frequent phenotype in black and other populations in which there is also normal activity has been designated Gd B. The next most frequent phenotype in which there is normal or near normal activity has been designated Gd A+. In black subjects with enzyme deficiency there is a band of reduced activity which has the same mobility as Gd A: this is designated Gd A−. Studies of the enzyme activities in Gd B and Gd A− subjects showed a difference in the in vivo enzyme stability in the erythrocytes. The half-life of the enzyme in Gd B cells is approximately sixty-two days; and in Gd A− the half-life is only thirteen days (Yoshida, 1967; Piomelli et al., 1968). The mean level of enzyme in Gd A individuals is about 10 percent less than that in Gd B people. The enzyme in the Mediterranean type of G-6-PD deficiency, designated Gd Mediterranean, has the electrophoretic mobility of Gd B. These phenotypes are all determined by alleles at the same genetic locus on the X chromosome. The corresponding genes for these phenotypes have been designated as *GdA, GdA−, GdB*, and *Gd Mediterranean*. Where the enzyme activities and the electrophoretic mobilities of two G-6-PD samples are similar, it is necessary to subject them to additional analysis, including determinations of substrate specificity, thermostability, pH dependency, and utilization of substrate analogues.

The availability of electrophoretic variants has been used to provide additional evidence for the X inactivation (Lyon) hypothesis. When erythrocytes from black females heterozygous for G-6-PD deficiency, that is, *GdA−/GdB*, were studied, two distinct populations were identified, Gd A− and Gd B (Beutler and Baluda, 1964). Studies of the electrophoretic patterns of the enzyme from skin biopsies and from uterine leiomyomas in heterozygotes with G-6-PD electrophoretic phenotype Gd A/B showed that the skin biopsies had both phenotypes but the tumors were only either Gd A or Gd B (Linder and Gartler, 1965). This finding is consistent with the postulate that only one of each of the alleles at the G-6-PD locus in the heterozygous

Table 6.1
Classes of G-6-PD Variants

Class	Variant
Enzyme activity increased	G-6-PD Hektoen
Enzyme activity almost normal	G-6-PD A+ (common in tropical Africa)
Enzyme activity moderately decreased	G-6-PD A− (10–60% in hemizygous males, common in tropical Africa, susceptibility to oxidizing drugs increased; no favism)
Intermediate enzyme activity	G-6-PD B
Severe enzyme defect and compensation	G-6-PD B− (Mediterranean hemolysis; susceptibility to oxidizing drugs increased; favism may occur)
Severe enzyme defect and chronic nonspherocytic hemolytic anemia	G-6-PD Albuquerque (hemolysis even without exposure to oxidizing drugs)

Source: Vogel and Motulsky (1979); Bowman, Carson, and Frischer (1969).

female is active in each cell. Further study provides evidence that both alleles are active in oocytes (Gartler et al., 1972).

Thermostability Variants of G-6-PD

The technique of heat denaturation was used in addition to electrophoresis to detect thermostability variants of hemoglobin and G-6-PD in an attempt to measure the amount of genetic variability among populations in the Republic of Cameroon (Bernstein, Bowman, and Kaptue-Noche, 1980). A minimum of three to a maximum of thirteen thermostability variants were estimated for HbA and HbS and a minimum of two to a maximum of ten thermostability variants were estimated for Gd A, Gd B, and Gd A−. Pedigree analysis was used to determine the mode of inheritance of G-6-PD thermostability. Families in which the mother was heterozygous for G-6-PD deficiency were selected, and the male children were compared with their mothers. In each situation, male offspring had thermostability values comparable to those of the same electrophoretic variant of the heterozygous mother, but there was considerable variation between families.

POPULATION GENETICS

A number of populations have been screened for G-6-PD deficiency. The disorder appears to be quite widespread. Some population studies are shown in Table 6.2. The only major population groups that appear to be relatively free (less than 1 percent of the population) of the G-6-PD deficiency are Northern Europeans, Armenians, Zoroastrians, Australian aborigines, Alaskan Eskimos, Amerindians, Ethiopian Highlanders, and Japanese.

Extensive surveys have shown that G-6-PD deficiency is widely distributed. More than 300 G-6-PD variants have been found and at least 100 have polymorphic frequencies in various populations (Vulliamy et al., 1988). One can find the Gd A+ and Gd A− types not only in Africa but wherever African peoples have migrated. The Gd A− phenotype has also been found among Sicilians (Luzzatto, 1973). (In Italy and in Greece the association of a *Hpa*I 13.0 kilobase (kb) DNA fragment with the HbS gene that is common in West Africa is also found; see Chapter 11.) The Gd Mediterranean type is seen commonly in populations in the south of Europe, the Middle East, the Far East, and the Malaysian peninsula. With the establishment of criteria for characterizing variants of G-6-PD variants (discussed earlier), a large number of additional variants have been identified, besides the more common variants. The variants in black or African populations are listed in Table 6.3. The populations are grouped according to the level of enzyme activity. A more comprehensive version of this table contains information on the red cell enzyme activity as a fraction of normal activity, K_m of G-6-P, K_m of NADP, 2-deoxy-G-6-P utilization, deamino NADP utilization, heat stability, and pH optimum (McKusick, 1986).

The relative frequencies of the common phenotypes vary among black populations (H. Harris, 1980). In a sample of Nigerian Yoruba males, the frequency of Gd B was 56 percent; of Gd A, 22 percent; and Gd A−, 22 percent (Porter et al., 1964). Among a comparable sample of African-Americans, the corresponding frequencies were 60–70 percent (Gd B); 15–20 percent (Gd A); and 10–15 percent (Gd A−). The gene frequencies of *Gd A+* and *Gd A−* in seven "anthropological" groups in Africa are shown in Table 6.4.

The geographic distributions of the Gd A− variant and other common deficiency variants appear to conform to the distribution of falciparum malaria (see Table 6.5), which suggests that this gene might have achieved polymorphic frequencies because it confers resistance to this lethal parasitic disease (Luzzatto et al., 1969).

In a 1983 study in Kenya, the frequency of G-6-PD deficiency in men from four tribes was determined by Aruwa and colleagues. The frequency was greatest in the Luo tribe (32 percent) living in the Lake Victoria region. The Mijikenda tribe, who had the second highest frequency of deficiency (16 percent), live in the coastal region, which

Table 6.2
G-6-PD Deficiency among Some Ethnic Groups

Population	Frequency (%)
Jews	
Sephardic (Mediterranean and Asiatic)	2–36
Ashkenazi (European) Jews	0–2
Non-Ashkenazi Jews	0–23.6
Kurdistani	60
Sardinians	14–48
Italians (excluding Sardinians)	0.35
Italians	1.3–2.0
Greeks	3
Iranians	
Moslems (Shiraz)	8
Zoroastrians	0
Armenian	0.6
Ghashghai	11.3
Basseri	13.3
Mamassani	20
Arabs	3–13
Blacks: African-American	12–15
Congolese Bantus	3–21
Other African populations	0–28
Other Groups	
Filipinos	13
Japanese	0
Chinese	2.0–5.5
Thais	11
Vietnam	
Vietnamese	1.4
Khmer (Cambodians)	15.3
Malo-Polynesian	
Cham	9.1
Rhade	2.3
Mon Khmer	
Sedang	0
Stieng	5.4
Alaskan Eskimos	0
North American Indians	0
Asian Indians	3–8
Peruvians	0
Punjabis	13.5

Sources: Keller (1971); Bowman and Walker (1963); Bowman et al. (1971).

is second in the incidence of malaria. The Abaluhya (12 percent) also live in an area of considerable endemic malaria. The Kikuyu live in the Kenya Highlands, where the incidence of malaria is low; they have the lowest prevalence of G-6-PD deficiency (5 percent). The hetero-

Table 6.3
G-6-PD Variants in Peoples of African Origin

Variant	Population	G-6-PD Activity (% of normal)	Mobility (% of normal)
B	All	100	100
Severe enzyme deficiency associated with chronic nonspherocytic hemolytic anemia			
Charleston	African	14	104–107
San Diego	African	20–40	103 (tris)
Lawndale	African	0	104 (tris)
East Harlem	African	10	104 (TEB)
New York	U.S. black	0.6	normal
Atlanta	U.S. black	25	93 (TEB)
El Morro	African	10	100 (TEB, tris)
Matam	African	?	100 (TEB)
Hotel-Dieu	African, Senegal	0.1	107 (tris)
Severe enzyme deficiency (less than 10%)			
Galveston	African	2–5	100 (TEB)
Dakar	African	4–6	100 (pH)
Unnamed	East African	2.5–12.0	120 (same as
Mali	African	1–5	A+)
			100 (pH)
Moderate to mild enzyme deficiency			
A− (common)	Black	8–20	110 (TEB, tris)
Chibuto	Black, Bantu	20	108 (TEB)
Columbus	Black	36	100 (tris)
Washington, D.C.	Black	16–33	95 (tris)
Cape Town	Cape Coloured	53–80	55–65 (TEB)
Kuanyama	African	73	90 (TEB)
Martinique	African	65	100 (pH)
Ilesha	West Africa	25	75 (TEB)
Very mild or no enzyme deficiency			
Inhambane	African-Bantu	100	112 (TEB)
Steilacom	Black	100	>110 (TEB)
A+	Black	80–100	110 (TEB, tris)
Lourenzo	Africa-Bantu	100	106 (TEB, pH)
Marques			
King County	Black	100	105 (TEB)
Baltimore	Black	75	90 (tris)
Austin			
Ijebu-Ode	Black	100	85 (TEB)
Tacoma	Black	100	94 (TEB)

(*continued*)

Table 6.3
(*Continued*)

Variant	Population	G-6-PD Activity (% of normal)	Mobility (% of normal)
B	All	100	100
	Austin		
Madrona	Black	70–80	80 (pH)
Ibadan/Austin	Black	72	80 (tris)
Ita-Bale	Black	100	65 (tris)
Adame	West Africa	100	96 (TEB)
Abeokuta	West Africa	100	91 (TEB)
Lanlate	West Africa	100	91 (TEB)
Ekiti	West Africa	100	72 (TEB)

Abbreviations: pH = phosphate buffer, pH 7.0; TEB = tris-ethylenediamine tetraacetic acid (EDTA)-borate; tris = tris-hydrochloric acid (HCL).

Source: Yoshida et al. (1971); Beutler and Yoshida (1973); Yoshida and Beutler (1978); Yoshida (1978); Yoshida and Beutler (1982).

zygous GdA−/GdB females but not the hemizygous Gd A− males appear to be relatively resistant against falciparum malaria (Bienzle et al., 1972). It may also be that individuals who carry the Gd A+ variant are also more resistant to malaria than those who are homozygous for the Gd B+ variant (Bienzle et al., 1972), but are less resistant than the A− variant.

The mechanism of this resistance is unclear. Presumably, some kind of alteration in the deficient cells inhibits or does not support parasite growth. Parasite multiplication was reduced when deficient cells were exposed to oxidative stress (Friedman, 1979; Golenser et

Table 6.4
Gd Gene Frequencies in Subdivisions of African Populations

"Anthropological" Subdivision	Range of Gene Frequencies	
	Gd A−	Gd A+
Sudanic	0.22–0.28	0.10–0.22
Bantu	1–18	0.01–0.23
Hamitic	—	0
Nilotic-Hamitic	—	0.02–0.18
Pygmies	0.004–0.05	0.11
San	0.02–0.03	0.01–0.02
Khoikhoi	0	0
All black Africans	0.02–0.28	up to 0.30

Source: Luzzatto (1974).

Table 6.5
Heterogeneity of "Common" Variants of G-6-PD Deficiency Associated with Malaria

Variant	Red Blood Cell Activity (% of normal)	Population Source	Malaria Very Endemic?
A−	8–20	Africa	Yes
Mali	5	West Africa	Yes
Mediterranean	0–7	Mediterranean area	Until recently
Athens	20–25	Greece	Until recently
Mahidol	5–32	Thailand	Yes
Taiwan-Hakka	2–9	Hakka-China	No
Canton	4–24	South China	Yes
Union	0–3	Philippines	Yes
Indonesia	0–5	Indonesia	Yes
Markham	1–10	New Guinea	Yes

Source: Luzzatto (1974).

al., 1983). It has been demonstrated that parasite invasion and intracellular development were reduced under conditions in which intracellular GSH was oxidized to GSSG and membrane intrachain and interchain disulfides were produced (Miller et al., 1984). These findings suggest that an altered thiol status in the erythrocytes deficient in G-6-PD might be responsible for the selective advantage they seem to have in the presence of malaria. (See Chapter 8 for a more detailed discussion of the malaria hypothesis.)

G-6-PD GENE MUTATIONS

The diverse clinical manifestations of G-6-PD deficiency have been partially explained by the finding of point mutations in the *G-6-PD* gene (Vulliamy et al., 1988). Seven mutant G-6-PD alleles were cloned and sequenced. Vulliamy and colleagues confirmed the asparagine → aspartic acid amino acid replacement that had been noted by Yoshida (1967), which has now been located at position 142. Although Vulliamy and co-workers give a position number of 126, they assumed that the two different numbers represented the same amino acid because of a difference in the N-terminal portion (fifty-two versus thirty-six amino acids) of the two published protein sequences. In G-6-PD Mediterranean, a single amino acid replacement was found (serine → phenylalanine). This substitution was presumed to be responsible for the decreased stability and the reduced catalytic efficiency of the enzyme. Other mutants were described, which showed a predominance of C → T transitions with CG doublets involved in four of the cases.

Accordingly, diverse point mutations probably account for the phenotypic heterogeneity of G-6-PD deficiency.

SICKLE HEMOGLOBIN AND G-6-PD DEFICIENCY

Lewis and Hathorn (1965) proposed that G-6-PD deficiency is more common in persons who have sickle cell anemia (HbSS) than in the general population. Other studies suggest that G-6-PD deficiency is also more common in people who have sickle cell trait (HbAS) (Lewis, 1967; Gelpi, 1967; Piomelli et al., 1972.) To account for the increased prevalence of G-6-PD deficiency in individuals with hemoglobin S Lewis and Hathorn suggested that this enzyme deficiency has a beneficial effect on the course of sickle cell anemia. Nevertheless, investigations of G-6-PD deficiency in conjunction with HbSS have not demonstrated any favorable effect on the course of the disease, according to Konetey-Ahulu (1972). Bernstein, Bowman, and Kaptue-Noche (1980) studied the possibility of a protective effect of the simultaneous inheritance of G-6-PD deficiency and HbS in a population of black blood donors at the University of Chicago and in populations in Cameroon, Equatorial Africa. The number of men with both HbAS and G-6-PD deficiency was significantly greater than expected ($P <$ 0.05) among 371 Cameroonian males. The examination of 668 male blood donors in Chicago revealed that the number of men with both HbAS and G-6-PD deficiency also exceeded the expected number, but the finding was not statistically significant.

HEMOLYTIC REACTIONS

The mechanisms of the hemolytic reactions associated with infection, diabetic acidosis, and neonatal jaundice are unknown. It does appear, however, that neonatal jaundice is much more likely to occur in premature infants than in full-term infants (Oski, 1965).

It is most important to note that even though primaquine and other hemolytic drugs may not cause severe hemolysis in Africans and their descendants, hemolytic drugs may initiate hemolysis so severe that transfusions may be needed in non-African peoples with G-6-PD deficiency (Ziai et al., 1967). Earlier one of the authors (J. E. B.) noted from many personal observations that in favism a G-6-PD deficiency could be diagnosed even though the hemoglobin was as low as 2 grams per deciliter. The enzyme defect in non-Africans in the Middle

East, at least, was not confined to older erythrocytes, as in Africans and their descendants.

Neonatal jaundice in patients with G-6-PD deficiency has been the subject of many reports. For example, Olowe and Ransome-Kuti (1969) observed that a major cause of neonatal morbidity and mortality in Lagos, Nigeria, was severe neonatal jaundice in babies who have G-6-PD deficiency. The authors astutely noted that the jaundice was more severe in outpatient than in inpatient infants, and looked for an exogenous cause. The investigators tested mentholated powder, which was commonly used in many clinics and homes to dress umbilical cords and found that this was the offending agent. It was recommended that the use of menthol- or camphor-containing products be discontinued. One of the authors of the paper (Ransome-Kuti) became the minister of health in Nigeria, so it is likely that the recommendation of the investigators was instituted.

Gibbs, Gray, and Lowry (1979) found G-6-PD deficiency in sixteen (70 percent) of a group of twenty-three neonates in Jamaica who had unexplained moderate or severe jaundice. This finding exceeded the 9.4 percent rate observed and the 22 percent expected in Jamaican neonates who were not jaundiced. Phenobarbitone and phototherapy reduced the need for exchange transfusion, which was required for eight of the newborns. Two infants developed kernicterus and one died. The investigators suggested that the jaundice was apparently precipitated by unknown factors that had a disproportionate effect on infants with G-6-PD deficiency.

Roux, Karabus, and Hartley (1982) concluded that G-6-PD deficiency contributed significantly to the severity of neonatal jaundice in a group of newborns in South Africa. An investigation of 3,718 newborn infants with jaundice in excess of physiological levels revealed that prematurity, hemolytic disease, hematomas, or infections were found in 1,278 patients. Of the remaining 2,440 newborns, 137 had G-6-PD deficiency and 2,303 had idiopathic hyperbilirubinemia. Exchange transfusion was necessary in 59 (43 percent) of the infants with G-6-PD deficiency and in 426 (18.5 percent) of those with idiopathic hyperbilirubinemia.

In Israel, Ashkenazi et al. (1983) studied 800 infants and found that although early jaundice and the use of phototherapy were somewhat more frequent in the group with G-6-PD deficiency, the results were not significant. The authors, however, recommended that preterm infants and those with low birth weights be routinely screened for G-6-PD deficiency.

TREATMENT

Most individuals with G-6-PD deficiency will probably never have a clinical hemolytic episode during their lifetimes and those who do will usually not require any therapy. It is simply a matter of avoiding the drug that precipitated an episode. On the other hand, hemolysis may be so severe in patients with some variants of G-6-PD deficiency that transfusion may be necessary. Exchange transfusion may be required in neonatal jaundice if the hemolysis is very serious. The primary method of dealing with G-6-PD deficiency is to avoid exposure to the drugs or other agents that are responsible for triggering the hemolytic episodes.

There is a difference of opinion about whether or not it is justified to screen the black population for this condition and to provide preventive counseling to affected individuals or parents. The main reason for such a program is to provide information to these individuals so that they might avoid the complications of G-6-PD deficiency even though the effects are usually self-limited. At the very least, some individuals may be able to avoid hospitalization. On the other hand, widespread population screening for G-6-PD deficiency could lead to some of the abuses and stigmatization that resulted from mass population screening for sickle hemoglobin (see Chapter 15).

7

Hemoglobinopathies and Thalassemias

Hemoglobinopathies and thalassemias are the best studied of all genetic disorders. Since sickle hemoglobin is the most important abnormality found in peoples of African origin, most of the discussion in this chapter concerns this variant. (For more detailed discussions of the world distribution of abnormal hemoglobins and thalassemias, refer to Bowman, 1983; Honig and Adams, 1986; and Winter, 1987.) Although here we will refer to the world distribution of abnormal hemoglobins and thalassemias, we mainly emphasize the anthropological, genetic significance of DNA analysis of α and β-globin gene segments. Whenever possible, we point out the clinical significance of alterations of DNA within and adjacent to these loci.

Human hemoglobins (Honig and Adams, 1986) consist of four interlocking polypeptide chains, each of which has an attached heme group. The polypeptide chains are named alpha (α), beta (β), gamma (γ), delta (δ), epsilon (ϵ), and zeta (ζ). The α-chains are common to the β-, δ-, γ- and ϵ-chains and interlock in the following manner:

Normal hemoglobins after 1 year of age
(percent of total hemoglobin)

$$A = \alpha_2\beta_2 \quad 94\text{–}97 \text{ percent}$$
$$F = \alpha_2\gamma_2 \quad <2 \text{ percent}$$
$$A_2 = \alpha_2\delta_2 \quad 2\text{–}3 \text{ percent}$$

Embryonic hemoglobins

$$\text{Gower } 1 = \epsilon_4$$
$$\text{Gower } 2 = \alpha_2\epsilon_2$$
$$\text{Portland} = \gamma_2\zeta_2$$

Hemoglobin F (Hb F, fetal hemoglobin) is the principal hemoglobin of the fetus from the third to the sixth months of intrauterine life. The proportion of Hb F decreases three months before birth at the same time as the β-chain is produced. The amount of Hb F decreases after birth, and by the end of the first year it is less than 2 percent of hemoglobin.

Schroeder et al. (1968) discovered that fetal hemoglobin is produced by the action of two nonallelic genes, $^{G}\gamma$ and $^{A}\gamma$. These were so named because the γ-chain products differ at position 136 in the globin chain: glycine is found in the $^{G}\gamma$ chain, and alanine is present in the $^{A}\gamma$ chain. The γ-chain is also heterogeneous at position 75. In this region the isoleucyl residue may be replaced by a threonyl residue. Apparently, the Ile-Thr heterogeneity is confined to the $^{A}\gamma$-chain. Accordingly, the $^{A}\gamma^{I}$- and $^{A}\gamma^{T}$-chains are the product of allelic genes (Huisman et al.. 1981).

The β-chains and the δ-chains are coded by one locus each, but the γ-chains have two loci (these loci are described later). The α-chain is composed of 141 amino acids, and the β-, γ-, and δ-chains each have 146 amino acids (Honig and Adams, 1986).

DISTRIBUTION OF ABNORMAL HEMOGLOBINS AND THALASSEMIAS

Sickle hemoglobin is found in high frequency in diverse peoples besides Africans and their descendants; namely, Greeks, Italians (particularly Sicilians), Eti-Turks, Arabs, southern Iranians, and Asian Indians (Bowman and Goldwasser, 1975). It has long been known that the clinical picture of sickle cell anemia differs between Africans and their descendants, on the one hand, and Middle Easterners and Asian Indians, on the other. There is also variation at the α- and β-globin DNA level within and among these populations, which may explain, in part, some of these clinical differences.

Ghana is an important region for the study of the impact of hemoglobinopathies on populations. In this country, of every 1 million children born, about 30,000 will have some form of hemoglobin disorder. More than 20,000 newborns per million have sickle cell disease. About 20 percent of Ghanaians in the southern part of the country and about 10 percent of Ghanaians in the north have the sickle gene. On the other hand, about 10 percent of southern Ghanaians and 20 percent of northern Ghanaians have hemoglobin C trait. About one in three Ghanaians has either sickle cell trait or hemoglobin C trait. Accordingly, one in nine matings is between carriers of abnormal hemoglobins, with a 25 percent chance to produce children with Hb SS, Hb SC, or Hb CC variants (Konotey-Ahulu, 1971).

The distribution of abnormal hemoglobins in Africa is variable.

Table 7.1
Number of Heterozygotes for Hemoglobinopathies

Region	Number
Africa	50×10^6
United States, Caribbean, South America, North Africa, Middle East, India	10×10^6
Subtotal	60×10^6

Source: Modified from report of a WHO working group (Community control of hereditary anemias, 1982).

The prevalence of sickle cell trait in West Africa averages about 20–30 percent, but there are differences. The frequency of sickle cell trait in the Ivory Coast is about 14 percent; it is only 12 percent in Liberia, but the highest frequencies of β^+-thalassemia in Africa are found in Liberia (Livingstone, 1983). The prevalence of sickle hemoglobin in North and South Africa is low. Sickle hemoglobin is even more variable in East Africa. For example, the frequency is 45 percent in the Bamba of Uganda, 18 percent in the Toro, and 0.8–3.0 percent among some Hamitic peoples of Uganda. In some regions of Tanzania, the Hb AS prevalence is 38 percent; in Zambia, the range of Hb AS is from 13–27 percent.

Hemoglobin K is found in high frequency in North Africa, but there are isolated reports from Ghana and Nigeria. There are several types of K hemoglobins.

With respect to population numbers, the most important abnormal hemoglobin next to Hb S is E. This variant is found in high frequency in Southeast Asia (Wasi, 1983), a region that also has very high frequencies of α- and β-thalassemias.

Hemoglobin D variants are found in Punjabis of India and in Iranians (Bowman and Ronaghy, 1967).

A World Health Organization (WHO) working group (Community control of hereditary anemias, 1983) found that hundreds of millions of peoples are heterozygous for the major hereditary anemias and that at least 200,000 homozygotes are born annually. This figure is equally divided between sickle cell disease and the thalassemias. In tropical Africa about 4 percent of matings risk producing an affected child, and an estimated 100,000 infants with sickle cell anemia (Hb SS) are born each year. This contrasts with approximately 1,500 per year in the United States, 700 per year in the Caribbean, and 140 per year in Great Britain. Data for South America were not given.

World figures for heterozygotes for the hemoglobinopathies are shown in Table 7.1. The annual births of patients with major hemoglobinopathies are estimated in Table 7.2.

Table 7.2

**Estimates of Births of Homozygotes or Compound Heterozygotes
for Major Hemoglobinopathies Affecting Blacks**

Region	β-Thalassemia	Hb S/β-Thalassemia	Hb SS & SC	Total
Sub-Saharan Africa	+	+	100,000	100,000
North Africa	850	300	100	1,250
Middle East	1,650	530	3,100	5,280
Europe	2,350	100	100	2,550
North America and Caribbean	>100	+	2,200	2,300
S. America	−	−	+	−

Source: Modified from report of a WHO working group (Community control of hereditary anemias, 1982).

Note: + present (estimates not given), − not described.

GLOBIN GENES

Human globin genes have been isolated by molecular cloning, and their nucleotide sequences have been determined. Each gene consists of regions that code for globin mRNA (exons) interspersed with intervening sequences (IVS or introns). The *exons* are so named because they are transported to the cytoplasm and are thus external to the nucleus; the stretches of DNA are called *introns* because they stay inside of the nucleus. Each globin gene has two intervening sequences that divide the mRNA coding region into three coding blocks (Figure 7.1). The globin gene (which includes both IVS) is first transcribed by RNA polymerase into a large mRNA precursor. Through RNA processing, the transcribed IVS are excised, and the regions of the transcript that contains mRNA are ligated. The mRNA is also modified at its 5′ and 3′ ends. Then, RNA is transported from the nucleus into the cytoplasm, combined with ribosomes, and translated into the amino acid sequence of the globin polypeptide. Globin then combines with heme and other globin chains to form hemoglobin. (For detailed descriptions of these processes see Maniatis, et al., 1980; Antonarakis, Phillips and Kazazian, 1982; Orkin, Antonarakis, and Kazazian, 1983; Orkin and Kazazian, 1984; Kazazian, 1985; Schwartz and Surrey, 1986.)

The α-globin genes are found on chromosome 16 in a 25-kb region. The location of the two ζ, two α, and the φ α (pseudoalpha) genes are shown in Figure 7.2. The β-globin group is found on the short arm of chromosome 11 in a 50-kb region. The arrangement of the genes for the embryonic ε-globin gene, the fetal $^Gγ-$ and $^Aγ-$

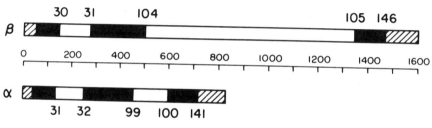

Figure 7.1. The α- and β-globin genes. The coding regions (exons) are black ■ and the intervening sequences (introns) are white □

Source: Maniatis et al., 1980. Reproduced with permission. Copyright © 1980 by Annual Reviews Inc., and T. Maniatis

globin genes, the pseudogene, $\varphi\beta$ and the $\delta-$ and $\beta-$ globin genes are also shown in Figure 7.2.

Hemoglobin G-Philadelphia ($\alpha^{68Asn\rightarrow Lys}$) was reviewed by Rucknagel and Bruzdzinski (1983). This variant occurs in the heterozygous state in about 1 in 5,000 African-Americans, and is the most common of the α-globin mutants with the exception of Hb Hashaorn

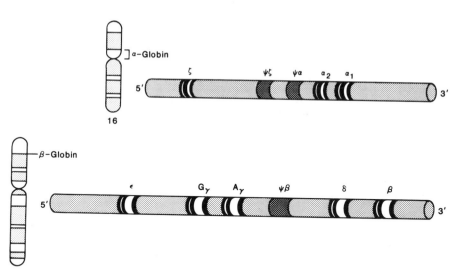

Figure 7.2. The human genome contains both copies and analogues of globin genes. The α-gene on chromosome 16 exists in two copies of α_1 and α_2; adjacent to these genes are three analogues: pseudoalpha ($\psi\alpha$), zeta (ζ, the embryonic α-gene), and pseudozeta ($\psi\zeta$). The β-gene exists in only a single copy on chromosome 11. The adjacent δ-gene analogue produces normal hemoglobin β-chain, but very inefficiently; the pseudobeta ($\psi\beta$) produces nothing. The γ-gene (two copies) is the fetal β gene; the ε (embryonic) gene serves the same function during the early stages of gestation

Source: Kazazian, 1985. Reprinted with permission of H. Kazazian

in Jews. In heterozygotes the amount of Hb G in the blood of heterozygotes is trimodally distributed with modes at 20, 30, and 40 percent
(Baine et al., 1976). Rucknagel and Bruzdzinski pointed out that
about one-half as many α-globin structural variants have been described as have β-globin mutants, and that the α-globin genes expressed in utero are under considerably more selection pressure than
are the β-globin genes. Interestingly, α-G Philadelphia is found on
both ($\alpha\alpha$) and ($-\alpha$) chromosomes, which suggests a connection between the two through recombination or gene conversion (Higgs et
al., 1989).

SICKLE CELL DISEASE

Sickle cell disease includes sickle cell anemia, a designation that by
custom is restricted to patients who are SS (homozygous S), and patients who have sickle hemoglobin plus one other abnormality of β-
globin or defect in β-globin production, the most common of which
are Hb SC disease and Hb S/β-thalassemia. Under conditions of reduced oxygen tension, the red blood cells become distorted into various forms, some of which resemble sickles, from which the name of
the disorder is derived. When this occurs within the small blood vessels the red cells form "logjams," and this leads to a partial or complete blockage of blood supply. Any part of the body or any organ
may be affected, but frequently the heart, lungs, kidneys, spleen,
pelvic bones, and brain are damaged (Bowman and Goldwasser,
1975).

The character and severity of sickle cell anemia differs for various
age and population groups. The major features of sickle cell anemia
are outlined in Figure 7.3. It should be emphasized, however, that all
people with sickle cell anemia do not necessarily have all of these signs
and symptoms. Some people may be free of serious illness, some have
only a few of these signs and symptoms, and others experience many
of these clinical features. The life expectancy of individuals with sickle
cell anemia is also quite variable. Some die early, but others lead
productive lives and reach an advanced age.

The clinical picture of sickle cell anemia also differs among populations. Gelpi (1982), Kamel and Moafy (1983), Perrine, Gelpi, and
Perrine (1983), and El-Hazmi (1983) described sickle cell disease in
the Middle East, and Brittenham (1983) discussed this disorder in
India. In general, the clinical course of sickle cell anemia is milder in
populations in these regions than it is in Africans and their descendants, but there are exceptions. The reports of the distribution and

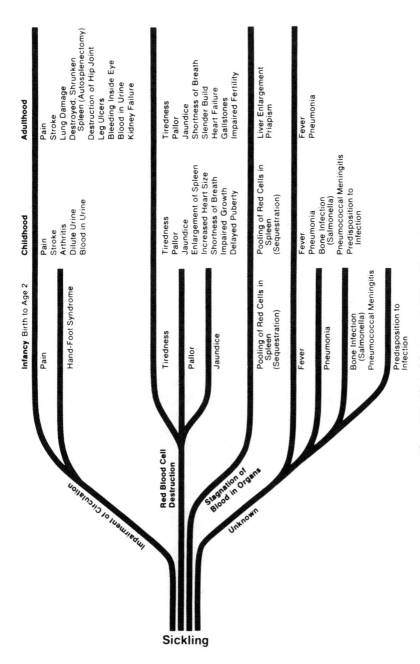

Figure 7.3. Clinical features of sickle cell anemia

Source: Bowman and Goldwasser, 1975

197

clinical picture of sickle cell disease in the Middle East are mainly from Saudi Arabia, the United Arab Emirates, Iran, Turkey, Kuwait, and Israel (Arabs).

Gelpi (1982) described the clinical course of sickle cell disease over a period of twenty years in several hundred Shi'ite Moslems from eastern Saudi Arabia who live in one or the other large oases near the Persian (Arabian) Gulf. (In the Middle East, the gulf is referred to as the Persian Gulf in Iran and as the Arabian Gulf by Arab countries.) These Arab populations are geographically and culturally isolated from the Sunni Moslem population that dominates Saudi Arabia. (Shi'ites also live in Saudi Arabia, mostly in the east, and number about 1.5 million out of a total population of about 6 million.) The slave trade from Africa to Saudi Arabia has already been mentioned in Chapter 1, and Gelpi indicated that the Shi'ite Moslem population has a high frequency of the cDe and Fy a-b- blood group markers. This population also has G-6-PD deficiency, usually of the Mediterranean type (Perrine, Gelpi, and Perrine, 1983), and the frequency of Hb AC is low (0.29 percent). But Hb AC also occurs in very low frequency in East Africa.

Gelpi (1982) reported that patients usually came to the attention of the physicians because of unexplained arthralgias, anemia, or splenomegaly. Leg ulcerations were absent; pregnancy complications were unusual; and among newborns of women who have sickle cell anemia, infant mortality was no different from that of infants of unaffected mothers. Other features of sickle cell anemia in this population included the absence of sepsis in infants and early childhood, freedom from cholelithiasis, the rarity of the hand-foot syndrome in infants, and a very low frequency of aseptic necrosis of the femoral head in adolescents. The mean hemoglobin was 10.6 grams per deciliter for males and 9.6 grams per deciliter for females. And most important: the mortality rate in patients with sickle cell anemia was not significantly different from that of an unaffected cohort group of Saudi Arabs.

El-Hazmi (1983) described the distribution of sickle cell disease throughout Saudi Arabia and indicated that the principal foci are not only in Al-Hafouf and Al-Qateef in the Eastern Province but also in Tehamat-Aseer in the Southwestern Province and Khaiber in the Northwestern Province (Figure 7.4; Table 7.3). The mild form of sickle cell disease was also prevalent in the Sunni Arab populations of Tehamat-Aseer and Khaiber. Further, El-Hazmi reported that the number of individuals who were homozygous for Hb S was higher than the expected number, which was similar to a deviation from Hardy-Weinberg equilibrium reported by Gelpi (1982). Both El-

Figure 7.4. Map of the Arabian peninsula. - - -, boundaries of the provinces, -.-.-. = approximate boundaries of Saudi Arabia; a, Riyadh; b, Al-Dammam; c, Al-Hafouf; d, Al-Quateef; e, Abha; f, Tamnia; g, Tehamat-Aseer; h, Al-Medina; i, Quba; j, Khaiber; k, Tabuk; m, Sakaka; n, Domet Al-Jandel (northern province); o, Al-Aflaj; p, Al-Ain; q, Khusaibah (Al-Qaseem)

*Source:*El-Hazmi, 1983. Reprinted by permission of Elsevier Science Publishing Co., Inc., New York, NY. Copyright © 1983

Hazmi and Gelpi attributed this finding to the survival of Saudi patients with sickle cell anemia to adult life. El-Hazmi (1983) reported an expected number of five and an observed number of seventy-three Hb SS homozygotes in the population from Al-Hafouf. Actually, the expected number of Hb SS homozygotes is sixty-four rather than five (Table 7.3). Even so, it is evident that El-Hazmi is correct in his conclusion. In a black population in Africa or the Americas, the observed and the expected numbers of people with sickle cell anemia would not be as close as that found in populations in Saudi Arabia. Hence, in Saudi Arabia the mortality of individuals with sickle cell anemia has probably been minimal before adulthood.

Miller et al. (1987) examined the molecular basis of the elevated production of hemoglobin F by looking for mutations in the promoter

Table 7.3
Actual and Expected Numbers of Homozygotes for Hemoglobin S in Four Regions
in Saudi Arabia

	Tehamat-Aseer N = 1,361	Khaiber N = 576	Al-Hafouf N = 1,591	Al-Qateef N = 510
AS	213 (0.16)	138	495 (0.31)	128 (0.25)
Observed SS	12	12	73	12
Expected SS	14	12	64	10

Source: Modified and recalculated from El-Hazmi (1983).
Note: The figures in parentheses show the frequency of AS in that population.

regions of the two hemoglobin F γ-globin genes ($^G\gamma$ and $^A\gamma$). The DNA sequences 450 base pairs (bp) upstream of both the $^G\gamma$ and $^A\gamma$ globin genes were normal except for a single base cytosine-to-thymidine (C → T) substitution at −158 bp 5′ to the cap (preinitiation) site of the $^G\gamma$-globin gene of the high hemoglobin F chromosome. This substitution was found in nearly 100 percent of individuals who have sickle cell disease or sickle cell trait, but it was also observed in 22 percent of normal Saudis. Homozygosity for this mutation had no effect on hemoglobin F production in the normal Saudi population.

Kamel and Moafy (1983) found that in Abu Dhabi, United Arab Emirates, mild sickle cell anemia was also present in Sunni Arabs, but there was heterogeneity in the clinical picture of this disorder. The level of hemoglobin was quite variable; some patients had a number of painful crises; infections were common in the more severely affected group; and other patients needed transfusions at least once a year. The sample size of the patients with sickle cell anemia was small (eleven), but it was sufficient to indicate the variability of the clinical picture of sickle cell anemia on the Arabian peninsula.

Miller et al. (1986) investigated high fetal hemoglobin production in sickle cell anemia in the Eastern Province of Saudi Arabia. Circulating fetal hemoglobin levels of 16 ± 7.4 percent were found in individuals with mild sickle cell anemia, but levels of only 1.09 ± 0.97 percent in patients' parents with sickle cell trait. To determine whether or not the patients with sickle cell disease inherited an increased capacity to synthesize fetal hemoglobin, a radioimmunoassay of fetal and adult hemoglobin was performed on erythroid progenitor (BFU-E) derived erythrocytes from patients with sickle cell disease and their parents. Mean fetal hemoglobin content per BFU-E-derived erythroblast from patients with sickle cell disease was 6.2 ± 2.4 picogram per cell or 30.4 ± 8.6 percent fetal hemoglobin. Despite virtually normal Hb F in the Hb AS parents of patients with sickle cell anemia, the mean fetal hemoglobin production per BFU-E-derived erythroblast was elevated to 3.42 ± 1.79 pg/cell or 16.1 ± 6.4 percent fetal hemoglobin, and the magnitude of fetal hemoglobin found in the parents correlated with that of the patients. The authors concluded that the high fetal hemoglobin is genetically determined, but is expressed only during erythropoesis. Further evidence of genetic determination was provided by analysis of DNA polymorphisms within the β-globin gene cluster. A distinctive 5' globin haplotype was found (+ + − + +) on at least one chromosome 11 in all of the persons with high Hb F SS and Hb AS. The relationship of the haplotype to Hb F production is yet to be determined.

The clinical picture of sickle cell disease in India is variable (Brittenham, 1983). Sickle cell anemia in the southern aboriginal groups is mild, but the sickle cell anemia that is found in a few caste groups in central and western India was apparently somewhat severe.

Alpha-Thalassemia and Sickle Cell Disease

Although some nondeletion defects have been found in α-thalassemias, the major cause of α-thalassemia is gene deletion (Antonarakis et al., 1982). Three types of abnormalities are usually found: a chromosome in which α-globin gene sequences are absent, and a chromosome with a deletion of only five nucleotides at the junction of a coding region and intervening sequences (splice junction) that affects processing of mRNA. The chromosome in which a single α-globin gene is absent is found in about 20–25 percent of U.S. blacks. It was found that DNA from normal individuals (those with four α genes) digested with *Bam*HI and hybridized with an α-globin gene probe yields a 14-kb fragment. If an individual is heterozygous for a normal chromosome 16 ($αα$) and a chromosome 16 that contains one α-globin gene ($α-$), both the normal 14-kb and 10-kb fragments are seen on autoradiography. In Hb H disease, there is only one functional α-globin gene ($--/α-$); thus only a 10-kb fragment is found. If a chromosome 16 contains three α-globin genes, an 18-kb fragment is found instead of the normal 14-kb fragment (Figure 7.5). It is most important to note that in this figure the deletions of the α-alleles in blacks differ from that usually observed in, say, Southeast Asians or Chinese: ($α-/α-$) in blacks and ($--/αα$) in Southeast Asians or Chinese. This is the reason why ($--/--$) or deletion of four α-globin genes is uncommon in blacks. Consequently, hydrops fetalis caused by α-thalassemia is rare in blacks but common in Southeast Asians and Chinese.

Embury et al. (1982) studied the effects of α-thalassemia on the severity of sickle cell disease in 147 patients. There were twenty-five

Alpha Gene Patterns

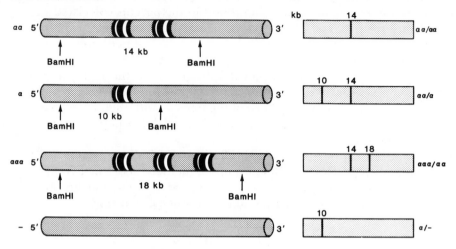

Figure 7.5. The location of restriction site using *Bam*HI produces patterns depending on the number of α-genes present. Two copies on each chromosome is normal. One chromosome may carry a deletion (αα/α−) or a duplication αααα/αα, or the gene(s) may be absent from one chromosome; pattern α/− is of total/partial deletion heterozygote
Source: Kazazian, 1985. Reprinted with permission of H. Kazazian

subjects with the normal four α-globin genes; eighteen with three, and four with two α-globin genes. The mean hemoglobin, hematocrit, and absolute reticulocyte counts (S.D.) were 7.90 ± 0.9 grams per deciliter, 22.9 ± 2.9 percent, and $501,000 \pm 126,000$ per cubic millimeter, respectively, in the group with four α-globin genes; 9.8 ± 1.6 grams per deciliter, 29.0 ± 5.0 percent, and $361,000 \pm 51,000$ per cubic millimeter in the group with three α-globin genes; and 9.2 ± 1.0 grams per deciliter, 27.5 ± 3.0 percent, and $100,000 \pm 15,000$ per cubic millimeters in the group with two α-globin genes. Deletion of α-globin genes was also accompanied by a decreased mean corpuscular hemoglobin concentration (MCHC) in postreticulocyte erythrocytes and by increased hemoglobin F levels. The decreased intraerythrocytic hemoglobin S concentration and elevated hemoglobin F levels that was associated with α-thalassemia apparently decreased the degree of hemolytic anemia in sickle cell disease. The presence of α-thalassemia correlated directly with the hemoglobin F level and inversely with the MCHC of postreticulocyte erythrocytes.

In a Jamaican population, Higgs et al. (1982) compared the clinical and hematological features of forty-four patients who had sickle cell disease and homozygous α-thalassemia-2 (α−/α−) with those of normal (αα/αα) controls. Patients with homozygous α-thalassemia-2 had significantly higher erythrocyte counts, hemoglobin, and hemoglobin A_2, as well as significantly lower Hb F, mean corpuscular hemoglobin, mean corpuscular hemoglobin concentration, mean corpuscular volume, reticulocyte counts, irreversibly sickled cells, and serum total bilirubin levels than those with a normal α-globin gene number. Heterozygotes ($\alpha-/\alpha\alpha$) had intermediate values. In the group with homozygous α-thalassemia-2, fewer patients had episodes of acute chest syndrome and chronic leg ulceration, and more of them had splenomegaly as compared with individuals in the other two subgroups. The authors interpreted these findings as confirming previous suggestions that α-thalassemia inhibits in vivo sickling in Hb SS and may be an important genetic determinant of its severity.

De Ceulaer et al. (1983) showed that α-thalassemia reduces the rate of hemolysis in patients with sickle cell anemia. Their investigation demonstrated a significantly prolonged red cell survival in patients with sickle cell disease and homozygous α-thalassemia-2 ($\alpha-/\alpha-$) as compared with age-matched and sex-matched controls with a normal α-globin gene component ($\alpha\alpha/\alpha\alpha$). Hematologic indices showed that patients with the $\alpha-/\alpha-$ genotype had significantly higher levels of Hb A_2, lower Hb F levels, reticulocyte counts, mean cell volume, and irreversibly sickled cells than did the normal ($\alpha\alpha/\alpha\alpha$) controls. Red cell survival was significantly prolonged in subjects with the $\alpha-/\alpha-$ genotype (13.8 days compared with 9.4 days in the $\alpha\alpha/\alpha\alpha$ subjects). It was emphasized that the mechanism whereby α-thalassemia reduces the degree of hemolysis is debatable. The factors considered were those described by Noguchi and Schechter (1981) and by Embury et al. (1982), namely, a reduction in the intracellular concentration of Hb S by a reduction in MCHC (which is lowered in those with the $\alpha-/\alpha-$ genotype), and the lower MCV, which may allow easier negotiation of small vessels in the capillary bed.

Steinberg et al. (1983c) studied the effects of α-thalassemia and microcytosis upon the hematological and vaso-occlusive severity of sickle cell anemia. The incidence of painful episodes, acute chest syndrome, aseptic bone necrosis, and leg ulcers in three patient groups with sickle cell disease was examined. There were 2,197 patients over age two who were stratified according to MCV; 183 patients were selected on the basis of an elevated Hb A_2, and on whom globin biosynthesis studies were done; and 125 patients were found who had α-globin genotypes assigned by restriction endonuclease gene mapping. When patients were arranged by MCV, there was a reciprocal relationship between Hb A_2 levels and MCV, a finding that singled out patients with α-thalassemia in the low MCV groups. Interestingly, neither microcytosis, β^0, nor α-thalassemia appeared to provide any protection from vaso-occlusive complications, and the frequency of aseptic necrosis was increased in patients with microcytosis over age twenty and in groups with α-thalassemia. It was postulated—contrary to some reports—that the effects of reduced MCV and MCHC, when accompanied by a reduction in hemolysis and rise in hemoglobin concentration, as in Hb SS α-thalassemia, may cause a rise in blood viscosity sufficient to impair blood flow and negate any amelioration of vaso-occlusive complications in Hb SS. Steinberg et al. (1983b) also studied the interaction between Hb Sβ^0-thalassemia and α-thalassemia. The clinical and laboratory characterization of patients with Hb Sβ^0-thalassemia combined with α-thalassemia was defined by erythrocytic indices, Hb A_2 and Hb F levels, globin biosynthesis, and by α-globin gene mapping. Patients with Hb Sβ^0 + α-thalassemia resembled those with Hb Sβ^0-thalassemia—except for balanced globin synthesis ratios and a low Hb F level. There was no difference in frequency of painful crises, leg ulceration, aseptic necrosis of bone and acute chest syndrome in patients with Hb Sβ^0 + α-thalassemia, sickle cell anemia (Hb SS), Hb SS α-thalassemia and Hb Sβ^0-thalassemia.

The effect of α-thalassemia in Hb SC disease was studied by Steinberg et al. (1983a). They ascertained the prevalence of α-thalassemia in fifty-three adults with Hb SC disease and correlated α-globin gene deletion with the hematological and clinical findings. There was no relationship between α-globin genotype and hematocrit, pain crises, bone lesions, proliferative retinopathy, or clinical severity score. The authors also stated that they were unable to document a clinical benefit of α-thalassemia on Hb SC disease and suggested that any beneficial or deleterious effect of α-thalassemia on Hb SC might be difficult to detect because of the intrinsic mildness of Hb SC when compared to Hb SS. They further stated that the number of patients examined was relatively small and that additional studies should be done.

The studies of Steinberg et al. (1983c) confirm in part the observations of Powars et al. (1980), who found no relationship between red cell indices and vaso-occlusive episodes such as painful crises, aseptic necrosis, and acute chest syndrome in 214 patients with Hb SS with an average age of 19.2 years.

Evidence that α-thalassemia had a different distribution in subjects with and without sickle hemoglobin and that α-thalassemia is related to prolonged survival in sickle cell anemia was provided by Mears et al. (1983). In twenty-nine West Africans with Hb AA, the $(-\alpha)$ chromosome frequency was 0.16; in the seven subjects with Hb AS, it was 0.18; and in the fifteen patients with Hb SS, it was 0.33. In twenty-five U.S. blacks with Hb AA, the $(-\alpha)$ chromosome frequency was 0.12; in ten subjects with Hb AS, 0.20; and in twenty-eight patients with Hb SS it was 0.30. In the twenty-five subjects with sickle cell anemia from Equatorial Africa, the $(-\alpha)$ chromosome frequency was 0.22. The $(-\alpha)$ chromosome frequency is much lower than that of the other groups, but subjects with Hb AA and Hb AS were not studied. The samples from Africa and the United States were pooled, and the ages of sixty-six of sixty-eight patients with sickle cell anemia were recorded. Of nineteen subjects between the ages of one and ten years, the $(-\alpha)$ chromosome frequency was 0.18; however, for the twenty-six subjects in the second decade, the $(-\alpha)$ chromosome frequency was 0.36. Significant age differences were observed only in the pooled sample. The trend is evident in all three populations, but sample sizes were too small to approach statistical significance. Age stratification of subjects with Hb AS was not mentioned.

Fabry et al. (1984) correlated the frequency of α-globin gene deletion with age in 100 Hb SS patients who ranged in age from birth to sixty-one years. Between birth and ten years, patients who had two and three α-globin genes constituted 17 percent of the population; between ten and twenty years of age the percentage of these genotypes was 38 percent; and in the twenty and over age group the percentage of these genotypes rose to 49 percent.

Steinberg and Embury (1986) reviewed α-thalassemia in blacks in the United States, looking at the genetic and clinical aspects, and the interactions with the sickle hemoglobin gene. These authors showed that even though the prevalence of α-thalassemia varies, the heterozygote frequency is about 0.30, and about 2 percent of blacks are homozygous for this condition. The most prevalent deletion in α-thalassemia in blacks involves about 3.7 kb of DNA $(-\alpha^{3.7})$, rightward deletion. The leftward deletion $(-\alpha^{4.2})$ is seen mainly in Asians but may be seen in blacks. In blacks, α-thalassemias are rarely clinically significant, but it is important to differentiate α-thalassemias from iron deficiency to avoid unwarranted treatment with iron and also to distinguish the condition from heterozygous β-thalassemia. Although persons with the $\alpha-/\alpha\alpha$ genotype are not usually anemic and have a normal mean corpuscular volume (MCV), the $-\alpha/-\alpha$ genotype is associated with microcytic erythrocytes and occasional mild anemia. After reviewing the disparate clinical effects of α-thalassemia on sickle cell anemic mentioned earlier, the authors concluded that the fundamental mechanisms by which α-thalassemia benefits sickle cells is related to greater membrane redundancy and the protection it provides against sickling-induced membrane stretching, cation loss, and cell dehydration. The lower MCHC of better hydrated α-thalassemic cells retards polymerization of Hb S. Even so, the authors did point out that the higher hemoglobins in patients with α-thalassemia and sickle cell anemia may lead to increased intravascular complications.

Stevens et al. (1986) found that α-thalassemia modified the hematologic expression of Hb SS by increasing total hemoglobin and A_2, decreased Hb F, mean cell volume, reticulocytes, irreversibly sickled cells, and bilirubin levels.

The molecular basis of α-thalassemia and its interaction with sickle hemoglobin was investigated by Kulozik et al. (1988). Studies of the α-globin genotype of 282 Asian Indians showed an overall α-globin gene frequency of 0.29, generally caused by the $-\alpha^{3.7}$ and $-\alpha^{4.2}$ deletions. Patients with sickle cell disease and α-thalassemia had higher hemoglobin levels, erythrocyte counts, and hemoglobin A_2 levels, and lower reticulocyte counts, mean corpuscular volume, mean corpuscular hemoglobin and Hb F levels than those with the normal genotype. Further, a higher prevalence of α-thalassemia was found in patients with sickle cell anemia over ten years of age, which suggests differential survival of patients who have sickle cell disease and α-thalassemia.

Differential survival of patients with both α-thalassemia and sickle cell disease, as previously shown by Mears et al. (1983) and by Kulozik and co-workers (1988) in Asiatic Indians, was not found in Jamaica (Higgs et al., 1982), Senegal (Pagnier et al., 1984b), or Nigeria (Falusi et al., 1987). Thus, the end is not in sight for an understanding of the full implications of the interaction of α-thalassemia and sickle cell anemia. (See Chapter 8 for further discussion of this interaction.)

GLOBIN GENE POLYMORPHISMS

Kan and Dozy (1978) described a DNA polymorphism that is linked to the β-globin gene in African-Americans. Using the restriction enzyme HpaI, the β^A-globin gene was found on a 7.6-kb fragment and rarely on a 7.0-kb segment in persons without sickle hemoglobin. In 60–70 percent of individuals with sickle hemoglobin, the sickle gene was linked with a 13.0-kb fragment (Figure 7.6).

Pagnier et al. (1983) described this polymorphism in Senegal, Ivory Coast, Burkina Faso (formerly, Upper Volta), Togo, Peoples Republic of Benin, Cameroon, Central African Republic, and Gabon. In Senegal, nine subjects with Hb AA, nineteen with Hb SS, and seven with Hb AS were examined for HpaI linkage. Out of the forty-five sickle genes, three were on 13-kb segments (7 percent); and four of twenty-five β^A globin genes (16 percent) were on 13-kb fragments. In the Ivory Coast, slightly higher frequencies of association of the 13-kb

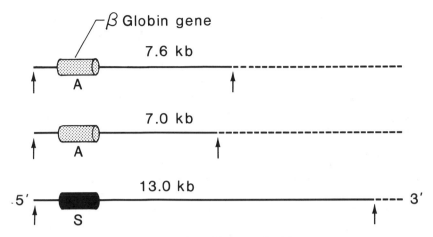

Figure 7.6. β-globin DNA fragments cleaved by *Hpa*I. β^A-globin DNA associated with 7.6 and 7.0 fragments are depicted in gray and β^S-globin DNA associated with the 13.0 globin DNA fragment is shown in black. The *arrows* indicate the cleavage sites
Source: Kan and Dozy, 1980 *(redrawn).* Reprinted with the permission of Y. W. Kan and *Science.* Copyright © 1980 by the American Association for the Advancement of Science

fragment were found with both β^A and β^S-globin genes. In Burkina Faso, subjects with the following genotypes were examined: Hb AA (five subjects); Hb SS (one); Hb AS (five); Hb SC (three); Hb CC (two). All of the sickle genes (five) were linked with the 13.0-kb *Hpa*I segment. Of the ten β^A-globin genes, one (10 percent) was on the *Hpa*I fragment. It was believed that the genotypes of the Hb AS and Hb AC subjects was consistent with close linkage of the β^A-globin gene to the 7.6-kb *Hpa*I segment. Furthermore, seven of the β^C globin genes that were assignable were on the 13-kb *Hpa*I segment. Thus the β^C-globin gene (in this part of Africa) is associated with the *Hpa*I 13.0-kb fragment in a frequency that approaches 100 percent. This study confirmed the work of others (Kan and Dozy, 1980; Mears et al., 1981). In the Peoples Republic of Benin one subject had Hb AA, fourteen had Hb SS, one had Hb CC, and six had Hb SC. All thirty-four β^S- and all eight β^C-globin genes were on the 13-kb *Hpa*I fragment, and the two β^A genes were on the 7.6-kb segment. These results confirmed that of a previous study by the same authors of the same ethnic group as that in the Peoples Republic of Benin (Mears et al., 1980) in which the sickle gene was always linked to the 13.0-kb *Hpa*I fragment. In Togo, however, the β^A globin gene was associated with the 13.0-kb fragment 18 percent of the time (Mears et al., 1980). In Cameroon, seven Hb AA and fourteen Hb SS subjects were examined. Of the twenty-eight *Hb S* genes, twenty-one (75 percent) were linked to the 13.0-kb *Hpa*I fragment, but no β^A globin genes were so linked. In the Central African Republic, two Hb AA, thirteen Hb SS, three Hb AS, and one Hb SC subjects were examined. Two of the twenty-eight (7 percent) of the sickle genes were linked to the 13.0-kb HpaI fragment, and all six β^A-globin genes were on the 7.6 kb HpaI fragment. In Gabon, two persons with Hb AA and eight with Hb SS were investigated. Six of the sixteen β^S-globin genes (38 percent) were on the 13.0-kb HpaI segment, but none of the four β^A-globin genes were linked to this fragment.

Nagel (1984) summarized these studies and those of others (Kan and Dozy, 1980; Mears et al., 1981; Pagnier et al., 1984a) (see Figure 7.7). Close linkage between the 13-kilobase HpaI fragment was found in Central West Africa, as previously noted by Kan and Dozy (1980). There was also a strong linkage between the 7.6-kb segment and the β^A-globin gene in Bantu-speaking Africa. Interestingly, there was also close linkage between the 7.6-kb HpaI and the β^S globin segment in parts of Atlantic West Africa; and the North African populations of Morocco and Algeria had 100 percent linkage between the 13.0-kb HpaI fragment and the β^S globin gene.

Using the restriction enzyme HpaI, Kan and Dozy (1980) noted that the usual 7.0- or 7.6-kb fragments associated with Hb A were replaced by a 13.0-kb fragment in 80 percent of U.S. blacks that they studied from the San Francisco area who had the β^S-globin gene. The frequency of this association with β^S-globin gene varied from region to region. This meant that in a variable minority of U.S. blacks the β^S-globin gene was associated with 7.0- or 7.6-kb fragments.

The RFLP HpaI-β globin gene linkage has been studied in several African populations to determine the tightness of the linkage of the β^S-globin gene to the 13.0-kb HpaI fragment (Mears et al.,1981). In North Africa, in a predominantly Arab population, these loci appear to be tightly linked because linkage was found in all of forty-two assignable HbS genes. Quite a different situation is found in West Africa, where 17–18 percent of the β^A-globin genes have been found linked to the 13.0-kb HpaI fragment. In Togo, on the other hand, the β^S-globin and β^C genes are tightly linked to the 13.0-kb HpaI fragment. In contrast, 72 percent of the β^S-globin genes in individuals from the Ivory Coast are found on the 7.6-kb fragment. These studies were then consistent with the hypotheses that β^S-globin genes arose on at least two different chromosomes in two geographically close but ethnically separate regions of West Africa with subsequent spread to North, Equatorial, and East Africa. Kan (1984) further delineated the association of the HpaI 13.0-kb fragment in various populations (Figure 7.7).

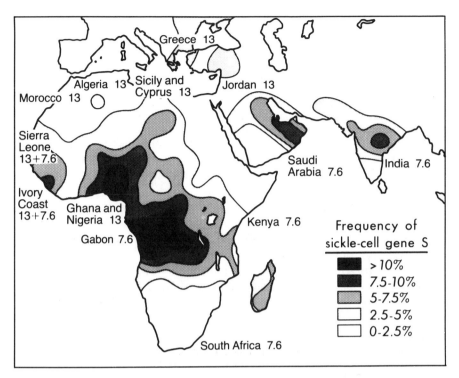

Figure 7.7. Association of the 7.6 and 13.0 kb *Hpa*I DNA fragments with the sickle gene in populations in Africa, Europe, the Middle East, and Asia
Source: Kan, 1984. Reproduced from the original with the permission of Y. W. Kan

A number of other polymorphisms in the β-globin gene cluster were reviewed by Antonarakis et al. (1983) and are shown in Figure 7.8. Each polymorphic site can be either present (+) or absent (−). Seven of the twelve polymorphisms are in flanking DNA; three are found in intervening sequences (numbers 3, 4, and 9 in Figure 7.9); one is in a pseudogene; and one is in the coding region of the β-globin gene (number 8). Sequencing data on numbers 3, 4, 7, 8, and 9 show single nucleotide substitutions. Nine of these occur in all populations. *Taq*I, and *Hinf*I (numbers 1 and 7) are polymorphic in blacks, but not in other populations.

Orkin and Kazazian (1984) reported seventeen common polymorphisms (Table 7.4) in the β-globin gene cluster, which was an expansion of the investigations from a previous report of twelve polymorphisms in this region by Antonarakis et al. (1983). Of these polymorphisms, fourteen were described in all populations studied: these were called public. Three were found in one ethnic group and

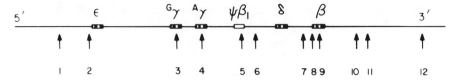

Figure 7.8. The location of twelve polymorphic restriction sites in the β-globin gene cluster depicted on a physical map of the region. The sites are from 5′ to 3′ : ε: embryonic gene; $^{G}\gamma$, $^{A}\gamma$, : fetal genes; $\psi\beta_1$: pseudogene; δ, β: adult genes. 1, *Taq*I; 2, *Hinc*II; 3, *Hind*III; 4, *Hind*III; 5, *Hinc*II; 6, *Hinc*II; 7, *Hinf*I; 8, *Hgl*AI; 9, *Ava*II; 10, *Hpa*I; 11, *Bam*HI; 12, *Rsa*I polymorphic sites, respectively *Source:* Antonarakis et al., 1983. Reprinted by permission of Elsevier Science Publishing Co., Inc., New York, NY. Copyright © 1983

were termed private; twelve were in flanking DNA; three were in introns; and one was in the first exon of the β-globin gene. The frequencies and the considerable differences of these polymorphisms in various populations are outlined in Table 7.4.

Nagel (1984) summarized the experiments of Pagnier et al. (1984a, see Figure 7.9), in which the frequency of the haplotypes of eleven polymorphic sites in the β-globin gene cluster associated with the β^S

Table 7.4
Frequency of DNA Polymorphic Sites in the β-Globin Gene Cluster

	Greece		Italy		Blacks		India		S.E. Asia	
Sites	β^A	β^T	β^A	β^T	β^A	β^S	β^A	β^T	β^A	β^E
(1) *Taq*I*	1.00	1.00	1.00	1.00	.88	.41	1.00	1.00	1.00	1.00
(2) *Hinc*II	.46	.85	.76	.54	.10	.02	.78	.75	.72	.20
(3) *Hind*III	.52	.14	.26	.48	.41	.35	.30	.26	.27	.73
(4) *Hind*III	.30	.07	.06	.37	.16	.05	.06	.09	.04	.00
(5) *Pvu*II	.27	.16					.62	.04		
(6) *Hinc*II	.17	.07	.20	.11	.15	.04	.17	.10	.19	.73
(7) *Hinc*II	.48	.12	.28	.31	.76	.81	.27	.17	.27	.73
(8) *Rsa*I	.37		.77		.50		.79			
(9) *Taq*I	.68		.23		.53		.27			
(10) *Hinf*I	.97	.92	.95	.92	.70	.10	1.00	.86	.98	1.00
(11) *Rsa*	No data listed									
(12) *Hgi*AI	.80	.90	.86	.73	.96	.96	.82	.38	.44	.73
(13) *Ava*II	.80	.90	.86	.73	.96	.96	.78	.38	.44	.73
(14) *Hpa*I	1.00	1.00	1.00	1.00	.93	.35	1.00	1.00		1.00
(15) *Hind*III	.72				.63		.56			
(16) *Bam*HI	.70	.78	.74	.82	.90	1.00	.82	.84	.74	.73
(17) *Rsa*I	.37	.21	.18	.17	.00	.00	.18	.08		

Source: Modified from Orkin and Kazazian (1984).

Note: Superscripts A, T, S, and E = normal β-globin, β-thalassemia, sickle β, and β-E, respectively.

*Numbers in parentheses refer to restriction sites from 5′ to 3′ positions of the cluster.

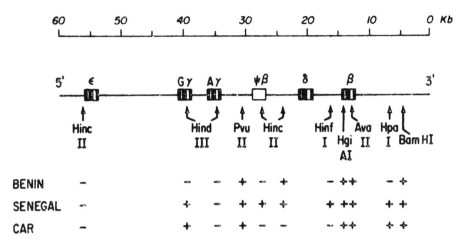

Figure 7.9. The sites and endonuclease enzymes used to determine eleven DNA polymorphisms in the β-gene cluster. All of the twenty Hb S-bearing chromosomes studied in Benin corresponded to the specific haplotype depicted. In the Central African Republic (CAR), a different specific haplotype was found among 86 percent of the Hb S-containing chromosomes studied; of the rest, one chromosome was of the Benin type and another was atypical. Of the forty-six Hb S-containing chromosomes studied in Senegal, 82 percent exhibited the specific haplotype depicted in the figure; of the rest, 14 percent were of the Benin type and 7 percent corresponded to two atypical haplotypes. All of the twenty Hb S chromosomes studied in Algeria were of the Benin type
Source: Nagel, 1984. Reprinted with the permission of the author and Springer-Verlag, Heidelberg

globin gene are shown. Three different haplotypes were linked to the βˢ globin gene in three different regions of Africa (Peoples Republic of Benin, Senegal, and Central African Republic). Nagel then queried whether there are other major haplotypes associated with Hb S in other parts of Africa. From the studies of Wainscoat et al. (1983) and Antonarakis et al. (1984), Nagel concluded that there are no other major haplotypes in Africa that are significantly associated with Hb S.

Wainscoat et al. (1983) and Antonarakis et al. (1984) studied 244 chromosomes from patients with Hb SS in Jamaica. In the sample studied by Wainscoat and colleagues, seven polymorphic sites were determined for 152 chromosomes. Nagel (1984) pointed out that of the 152 chromosomes, 84 percent corresponded to the three haplotypes that were associated with three geographic regions of Africa. The remaining haplotypes occurred with very low frequency: they were postulated to have arisen by independent mutation, crossing over, conversion, or mutation from the major haplotypes. In Antonarakis and colleagues' data, 90 percent of the haplotypes corresponded to the three major haplotypes. Nagel next compared our historical knowledge about the sources of slave imports into Jamaica from Curtin's estimates (Curtin, 1975) with the origin of slaves as

Table 7.5

Origin of Slave Imports to Jamaica (1655–1807): Comparison of Historical Data
with Percentages Calculated on the Basis of Hb S-linked Haplotypes

African Ports of Origin	Slave Imports (%)	Origin by Haplotypes (%)
Atlantic West Africa (Senegambia, Sierra Leone, Windward Coast)	15	10
Central West Africa (Gold Coast, Bight of Benin, Bight of Biafra)	68	72
Bantu Africa (Most of Gabon Congo, Angola, Namibia)	17	17

Source: Modified from Nagel (1984). The slave imports are estimates from Curtin (1975).

calculated by haplotypes (Table 7.5). The origins of the slaves corresponded with the estimates of Curtin.

Nagel (1984) noted that in Cuba (Curtin, 1975), from 1817 to 1843, 3.4 percent of the slaves originated from Atlantic West Africa, 42 percent from Central West Africa, and 55 percent from Bantu-speaking Africa. Nagel cited an earlier study of seventy Cubans with Hb SS and found that 50 percent of the β^s chromosomes are linked to the *Hpa*I fragment and 50 percent to the 7.6-kb restriction fragment, which is in agreement with the importation figures. Nagel further predicted that the 50 percent *Hpa*I positive chromosomes will be largely of the Bantu haplotype as well as a small source from the Senegalese haplotype. Nagel then made projections of the probable composition of the haplotype frequency in African-Americans with the β^s allele. Nagel postulated that because of the bias of the South Carolina slave market, which preferred slaves of Senegalese origin and avoided slaves from the Bight of Biafra, the proportion of haplotypes would be expected to deviate from those of the British trade of the eighteenth century; there should be a higher proportion of individuals from Senegambia and a lesser frequency of people from Central West Africa.

Nagel (1984) recalculated the data of Antonarakis et al. (1984) on seventy-six chromosomes in seventy-six persons with Hb SS from Baltimore, Maryland (Table 7.6). Individuals with Hb S in the Baltimore sample were probably descendants of slaves imported by the Virginia trade. The haplotype figures for the Atlantic West Africans were lower than the historical estimates of imports. Nagel postulated that since the proportion of slaves from different regions of Africa varied during the 150 years of the British slave trade, and the importation of slaves from Senegambia decreased in the last 20 years of the trade, the slaves that were imported at the end of the slave trade may have

Table 7.6
Origin of Slave Imports to North America: Comparison of Historical Data
with Percentages Calculated on the Basis of Hb S-linked Haplotypes

African Ports of Origin	Slave Imports (%)		Origin by Haplotypes (%)
	Virginia	S. Carolina	
Atlantic West Africa			
(Senegambia, Sierra Leone, Wind-ward Coast)	27	43	15
Central West Africa			
(Gold Coast, Bight of Benin, Bight of Biafra)	54	17	62
Bantu Africa			
(Most of Gabon, Congo, Angola, Namibia)	20	40	18

Source: Modified from Nagel (1984).

contributed more to the gene pool of present-day African-Americans. Other reasons were also given, but Nagel wisely concluded that better data must be gathered to obtain more accurate African origins of African-Americans. Of course, all of these estimates assume that there has been very little movement of present-day African populations since the days of the slave trade. Additional samples from descendants of Africans outside of the United States and surveys in Africa are needed if we want to be certain of the virtual exclusive association of these three haplotypes with the β^S locus, and wish to confirm the historical origins of the descendants of slaves from Africa.

Nagel (1985) studied the hematologic characteristics of patients with sickle cell anemia from Atlantic West Africa and central West Africa. Senegalese (Atlantic West Africa) patients have higher levels of hemoglobin F, a preponderance of $^G\gamma$-chains in hemoglobin F, a lower proportion of very dense red cells, and a lower percentage of irreversibly sickled cells than from those from central West Africa (the Peoples Republic of Benin). These data were interpreted to indicate that the γ-chain composition, and the hemoglobin F level are haplotype-linked and that the decrease in the percentage of dense cells and irreversibly sickled cells is secondary to the elevation of the hemoglobin F level.

Nagel et al. (1987) studied the hematologic and genetic characteristics of the third haplotype, the Bantu, in studies of sixty-four unrelated patients from the Central African Republic. Of these 128 chromosomes, 116 β globin genes were linked to the Bantu haplotype, and 12 chromosomes were atypical. In fifty-four individuals in whom the Bantu cluster haplotype was found, twenty-six were $\alpha a/\alpha a$ (48 percent), 22 were $-\alpha/\alpha a$ (41 percent), and six were $-\alpha a/-\alpha a$ (11 percent). The gene frequency of the $-\alpha$ haplotype was 0.315, higher than that found in Hb AA individuals of the Central African Republic. Accordingly, 62 percent of these patients with Hb SS have α-thalassemia.

Nagel and colleagues (1987) then analyzed the hematologic and genetic background of patients with Hb SS with the Bantu haplotype and compared them with patients with Hb SS who had the Benin and Senegal haplotypes. The mean and the mode of the percentage of Hb F is high and like that of the Hb SS patients with Senegal haplotypes; however, the distribution of the values was different in that no patients with Hb SS had Hb F levels less than 5 percent, but 11 percent of those with the Bantu haplotype had Hb F levels between 0 and 5 percent. On the other hand, the patients with the Benin haplotype had Hb F distribution with a mode between 2.6 and 5 percent. Individuals with the Bantu haplotype also had a low percentage of dense cells, a finding similar to that found in patients with the Senegal haplotype. An important feature that delineates the Senegal from the Bantu haplotype is the expression of $^G\gamma$. The Senegalese have a high $^G\gamma$ in adults, which is an unusual finding among normal individuals who switch from the newborn 70 percent $^G\gamma$ to 30 percent $^A\gamma$ ratio to the adult 40 percent $^G\gamma$ to 60 percent $^A\gamma$ ratio during the first four months of life. Undoubtedly, according to Nagel and co-workers, there were at least three independent origins of the sickle gene in Africa, and the *Hb S* gene may be accompanied by a different set of linked epistatic genes in each of the three haplotypes.

Seltzer et al. (1988) found that individuals with sickle cell anemia and Hb F levels greater than 25 percent differed widely in the clinical severity of their disease, and that differences in cellular distribution of Hb F did not entirely account for the clinical picture. Neither the presence of the C → T substitution in a position −158 to the $^G\gamma$ cap site nor specific haplotype combinations explained the unusually high Hb F levels. We have seen that Nagel (1985) found that the Benin haplotype was associated with lower levels of Hb F in individuals with sickle cell disease than that found in patients with the Benin haplotype. Seltzer and colleagues however, found that in five families in which Hb SS was found with high levels of Hb F, five patients were homozygous for the Benin haplotype; one had 19 percent Hb F, and the rest ranged from 28 to 42 percent. Homozygotes for the Senegal haplotype were not found; however, of the five others with Hb SS, four were heterozygous for the Benin/Senegal haplotype and had Hb

F percentages ranging from 19 to 32 percent; and one was heterozygous for the Central African Republic/Benin haplotype with a Hb F
of 31 percent. The Seltzer group postulated that there may be novel
hereditary patterns for nondeletional hereditary persistance of fetal
hemoglobin whereby some children with Hb SS may be homozygous
or doubly heterozygous for putative nondeletional hereditary persistence of fetal hemoglobin mutations.

Kulozik et al. (1986) undertook a geographical survey of β^S-globin
gene haplotypes and described evidence for an independent Asian
origin of the sickle mutation from that found in Africa. The haplotypes of 152 β^S chromosomes were determined in six different population groups: Orissa (tribal, India); Orissa (nontribal, India); Poona
(India); eastern oases (Saudi Arabia); Riyadh (Saudi Arabia); and
Ibadan, Nigeria. One particular haplotype was found in 77 percent of
individuals from Riyadh in the southwest portion of the Arabian peninsula and in 97 percent of a sample of individuals from Nigeria. A
different haplotype was found in 90 percent of a sample from eastern
Saudi Arabia, and in western India (50 percent), eastern India, both
in tribal (88 percent) and in nontribal (100 percent) groups. Chromosomes of individuals from Nigeria and from the southwest section of
the Arabian peninsula have the haplotype $(---++-+)$ previously
found in West African, Jamaican, and U. S. American blacks, but
those from the west and east coasts of India showed a different
haplotype not found in Africa $(++-++++-)$. It was also pointed
out that the distribution of the Asian haplotype corresponded to the
geographical distribution of the mild clinical phenotype of Hb SS.
The most likely interpretation of these data is that there was a multicentric origin of the β^S-globin gene, making it probable that this mutation had an independent origin in Asia as well as in parts of Africa.

Ramsay and Jenkins (1988a) studied β-globin gene–associated restriction length polymorphisms in southern-African Bantu-speaking
black and Kalahari !Kung San populations. It was hoped that the
investigations could provide answers to the question as to how a family
of closely related languages, the Bantu, could diffuse over a vast region in the short time span of 2,000 years, and also to assess the
affinities of the populations of Southern Africa. The subjects were
randomly selected from the San, Southern African Bantu-speaking
black populations, and from eleven patients with sickle cell anemia,
and one with sickle cell trait.

Eleven polymorphic restriction endonuclease sites were tested. The
frequencies of the presence of these sites are shown in Table 7.7. The
β^S-associated haplotypes were determined using eleven polymorphic
restriction endonuclease sites following the scheme of Pagnier et al.

Table 7.7

β^S-Associated Haplotypes in Various African or African-Derived Population

Population and Region	Total	Haplotypes			
		Senegal	Benin	Bantu	Rare
West Africa					
Senegal	56	46	8	0	2
Central West Africa					
Benin	20	0	20	0	0
Algeria	20	0	20	0	0
Nigeria	34	0	33	0	1
Bantu-speaking Africa					
Central African Republic	28	0	2	24	2
Southern Africa	23	0	0	20	3
U.S. blacks	76	5	45	17	9
Jamaican blacks	94	6	63	15	10
Riyadh, Saudi Arabia	22	0	17	0	5

Source: Ramsay and Jenkins (1987).

(1984a). The data, along with β^S-associated haplotype information on other African populations is summarized in Table 7.7.

The β-globin haplotypes were also determined in the San and Southern African blacks (see Table 7.8). Ramsay and Jenkins (1988a) used this comparison to show that in both the northern and southern extremities of Bantu-speaking Africa the β^S allele is found in the same haplotype. These investigators postulated, accordingly, that the sickle mutation arose only once in the Bantu-speakers before the migrations of the Bantu about 2,000 years ago.

Structural analysis of the 5' flanking region of the β-globin gene in African patients with sickle cell anemia provided additional evidence for at least three origins of the sickle mutation in Africa (Chebloune et al., 1988). Evidence that haplotype analysis of the β-globin cluster shows two regions of DNA characterized by nonrandom association of restriction site polymorphisms, as described earlier in Antonarakis et al. (1983), was first presented. One of these regions extends over 38 kilobases 5' to the δ gene and the other, 3' to the δ gene, which includes the β gene and extends 20 kb downstream. Interestingly, as Chebloune et al. pointed out, this region is probably a hot spot for recombination in that the recombination rate in this region was discovered, by Chakravarti et al. (1984a), to have a recombination rate three to thirty times higher than expected for a region of this size.

Table 7.8

Frequencies of the Various *β*-Globin Gene Cluster Polymorphic Restriction-Endonuclease Sites in Southern African Populations

Restriction Endonuclease Site	Population	
	Bantu-Speaking Mean ± S.E.	!Kung San Mean ± S.E.
Hinc-5′ *ε*	0.167 ± 0.054 (48)	0.059 ± 0.033 (51)
*Hind*III-*Gγ*	0.438 ± 0.005 (48)	0.680 ± 0.066 (50)
*Taq*I-inter *γ*	0.773 ± 0.086 (24)	0.739 ± 0.065 (46)
*Hind*III-*Aγ*	0.146 ± 0.051 (48)	0.080 ± 0.038 (50)
*Hinc*II-*β1*	0.106 ± 0.045 (47)	0.275 ± 0.063 (51)
*Hinc*II-3′*β1*	0.872 ± 0.049 (47)	0.980 ± 0.020 (50)
*Hinf*I-5′*β*	0.833 ± 0.108 (12)	0.813 ± 0.097 (16)
*Ava*II-*β*	0.933 ± 0.037 (45)	0.760 ± 0.060 (50)
*Hpa*I-3′*β*	0.762 ± 0.119 (42)	1.000
*Hind*III-3′*β*	0.563 ± 0.072 (48)	0.765 ± 0.060 (51)
*Bam*HI-3′*β*	0.826 ± 0.056 (46)	0.788 ± 0.057 (52)

Source: Ramsay and Jenkins (1987).
Note: Numbers in parentheses denote numbers of chromosomes tested.

Chebloune et al. (1988) further summarized features of this 9-kb region before presenting their own findings. There are segments of tandemly repeated sequences $(ATTTT)_n$ $(AT)_x T_y$, that are found 1.4- and 0.5-kb 5′ to the cap site, respectively. The ATTTT arrrangement had been found in four, five, and six copies. Three structures of $(AT)_x T_y$ had been found in human DNA. It was postulated that these repetitions may participate in high-frequency recombination events. In contrast to the normal $β^A$ chromosome, the 5′ and 3′ sub-haplotypes are nonrandomly associated. As has been mentioned, the geographical specificity of the three major haplotypes associated with $β^S$-globin DNA supports the thesis of three origins of the mutation, but this hypothesis may be disputed. Three reasons for arguing against it were outlined: (1) the distance between the restriction polymorphism sites that define the haplotypes is sufficiently long to allow easy recombination; (2) the framework for normal *β*-globin genes is identical in the three major *β* haplotypes; (3) no differences other than the *Hpa*I polymorphism were found in the 3′ subhaplotype between Benin chromosomes on the one hand and Central African and Senegal haplotypes on the other. Further, Senegal and Central African haplotypes can be derived one from the other by a single crossover event, which could suggest only two instead of three origins. And most of the minor haplotypes that are found in African-Americans and Jamaican blacks may be explained by one or two recombination events.

Chebloune et al. (1988) used their technique of S1 nuclease mapping of genomic DNA suitable for examining polymorphisms derived from $(ATTTT)_n$ and $(AT)_x T_y$ repeats. The S1 nuclease genomic mapping was performed on the DNA of patients with sickle cell anemia who originated from the Senegal, Benin, and the Central African Republic region of Africa. Three additional structures of the $(AT)_x T_y$ repeat were detected, which coincided with the geographical origin of the subjects. One β^S chromosome of each region was analyzed by sequencing. Ten more nucleotide positions were variable. These findings were interpreted as further evidence for three origins of the β^S gene in Africa.

Ramsay and Jenkins (1988b) studied the α-globin cluster haplotypes in the Kalahari San and Southern African Bantu-speaking blacks. Significantly, there was a low level of variation in the San-associated haplotypes and a high level of variation in the South African blacks. Nineteen different haplotypes were found among the thirty-six haplotypes studied in the black population: however, only seven different haplotypes were found among the thirty-seven haplotypes in the San. Five were common to both populations.

Livingstone (1989) ran computer simulations of β^S-globin DNA haplotype diffusion and challenged the thesis that separate mutations in β^S-globin DNA occurred in Africa, the Middle East, and India. Livingstone instead postulated a Middle Eastern origin of hemoglobin S and that the worldwide distribution is the result of the diffusion of a single mutant throughout Africa and the Middle East by Arab or Muslim expansion. He argued that the β^S-globin gene is a recent mutant that has spread rapidly because of the enormous selective advantage in environments with endemic falciparum malaria. Because the hemoglobin AS heterozygotes have the highest fitness of any β-globin heterozygote, the β^S-gene has a much faster rate of diffusion than other mutants. Livingstone asserted that the association of β^S with different haplotypes in different areas is not evidence of separate mutation if within the given time period it could have spread throughout the population in which it is found, and if other forces, such as recombination and interallelic conversion could have changed the haplotype association of the β^S-gene during its diffusion.

THE ORIGIN OF THE β^C-GLOBIN GENE IN BLACKS

Boehm et al. (1985) reported a single origin of the β^C-globin gene in blacks. This study involved an examination of twenty-five chromosomes bearing this gene at eight polymorphic restriction sites.

Twenty-two of the twenty-five chromosomes were identical at all sites and had a haplotype seen only infrequently among β^A-bearing chromosomes in U.S. blacks. Two different haplotypes were seen among the three exceptional chromosomes. The β^C-globin gene is not as widely distributed as is the β^S-globin gene, which may explain, in part, the limited β-globin haplotypes associated with this gene.

MOLECULAR DEFECTS OF THE BETA-THALASSEMIAS

Orkin, Antonarakis, and Kazazian (1983); Orkin and Kazazian (1984); and Kazazian and Boehm (1988) published excellent reviews of the molecular aspects of the β-thalassemias. The β-thalassemias are found in Mediterraneans, North Africans, and Middle Easterners, Asian Indians, Chinese, Southeast Asians, and black Africans (Kazazian and Boehm, 1988). These authors also estimated that about 3 percent of the world's populations (150 million people) carry a β -thalassemia gene, and that about twenty alleles account for 90 percent of β-thalassemia genes worldwide, and fifty-four known alleles account for almost all of the thalassemia genes.

Until the development of techniques to characterize polymorphisms in DNA, the β-thalassemias were classified into β^+ and β^0 groups. The β^+-thalassemias were more commonly found in blacks and the β^0-thalassemias were described more often in non-African populations. In the β^+-thalassemias there is some production of β-globin, and in the β^0-thalassemias, there is no production of β-globin. Most of the thalassemias show alterations within or adjacent to the β-globin gene. Because the thalassemias are so complex, and the changes within and without β-globin DNA are so variable, we will emphasize the β-thalassemia variants that are more commonly found in Africans and their descendants. We will refer to thalassemias that occur in other populations when they are needed for comparison purposes.

Before we begin, a clinical note is in order. Since the thalassemias found in Africans are usually characterized by the production of some β^A-globin and the β-thalassemias that are found in non-Africans are not, it is no surprise that the β-thalassemias of Africans are milder that those found in non-Africans. Patients who are homozygous for β^+-thalassemia commonly have a mild clinical picture with low transfusion requirements, in contrast to severely affected non-Africans with this disorder. But, since there is considerable heterogeneity in the thalassemias, many patients who are presumed to be homozygotes are compound heterozygotes.

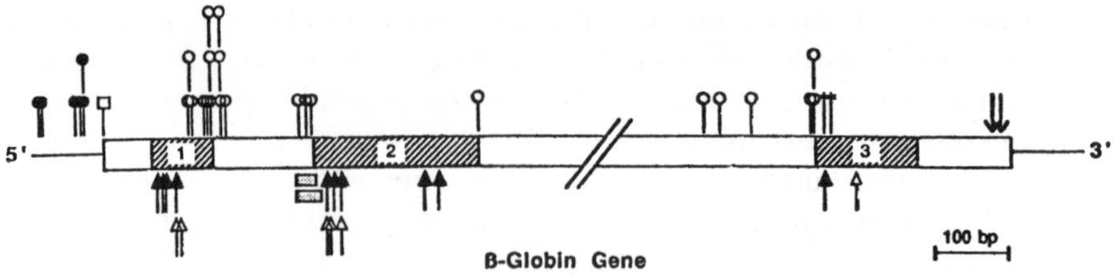

Figure 7.10. Point mutations in β-thalassemia. The *β*-globin gene is shown with *numbered hatched areas* representing the coding regions of exons. *Boxed open areas* between the exons are introns, and *boxed open areas* at the 5′ and 3′ ends of the *genes* are untranslated regions that appear in the mRNA. The various types of mutation are depicted by different symbols. For example, twenty-two of the fifty-one mutations affect RNA splicing and are shown as ⚲ . ⚲ , transcription; ⚲ cap site; ↓, RNA cleavage; ↑, frameshift; ↑ , nonsense codon; † unstable globin; ▭, small deletion
Source: Kazazian and Boehm, 1988. Reprinted with the permission of Grune & Stratton and H. H. Kazazian

Orkin et al. (1983) pointed out that although there is a linkage disequilibrium for some polymorphic restriction sites and thalassemia genes, haplotypes alone cannot delineate normal from thalassemic chromosomes. Apparently, polymorphisms in the globin gene cluster and the gene frameworks were established before most of the β-thalassemia mutations. Orkin and Kazazian (1984) preferred to divide the thalassemias into simple β-thalassemia—a condition in which only β-globin synthesis is affected—and complex β-thalassemia, in which the production of β-like globins is compromised. That classification will be followed; this discussion is based on the reviews of Orkin, Antonarakis, and Kazazian (1983), Orkin and Kazazian (1984), and Kazazian and Boehm (1988).

Simple β-Thalassemias

Orkin and Kazazian (1984), Kazazian (1985), and Kazazian and Boehm (1988) identified at least fifty-one point mutations and three deletions that produce "simple" β-thalassemia. Deletions that involve other genes in the *β*-globin gene cluster in addition to the *β*-gene result in more complex mutations (Kazazian and Boehm, 1988). The sites of the *β*-thalassemia mutations in the gene and the effect on gene expression are shown in Figure 7.10 and Table 7.9. We modified this table to emphasize the mutants that occur in blacks. The transcription mutants are particularly important for our discussion. The −88 mutant that is found in blacks and the −87 mutant that is but rarely observed in Mediterraneans have mild clinical effects. These would therefore belong in the category of β^+-thalassemia. Additionally, the mutants within the TATA box (Table 7.9) have been reported in three

Table 7.9
Point Mutations in β-Thalassemia in Blacks

Mutant Class	Type	Origin
I. Nonfunctional mRNA		
Nonsense mutants		None reported
Frameshift mutants + 1 codons	0	U.S. black
II. RNA-processing mutants		
Splice junction changes		
IVS-2 3'-end (A-G)	0	U.S. black
IVS-2 3'-end (A-C)	0	U.S. black
Consensus changes		None reported
Internal IVS changes		None reported
Codon region substitutions affecting processing		
Codon 24 (T-A)	+	U.S. black
III. Transcriptional mutants		
88 C-T	+	U.S. black, Asian Indian
29 A-G	+	U.S. black, Chinese
IV. RNA cleavage + polyadenylation mutants		
AATAAA-AACAAA	+	U.S. black

Source: Modified from Kazazian and Boehm (1988).
Note: IVS, intervening sequences (introns).

β-thalassemia genes in persons of widely dissimilar ethnic backgrounds. In this group, the base changes in positions -28 and -29 were found in Kurdish, black, and Chinese patients. These genes produce about 20–30 percent of β-RNA that is usually generated by a normal gene.

Kazazian and Boehm (1988) analyzed the population specificity of the β-thalassemia alleles. In the Mediterranean group, there were fifteen alleles and six accounted for 92 percent of the genes. In the Chinese/Southeast Asian populations there were nine alleles and four accounted for 91 percent of the genes. In the Asian Indian population, there were ten alleles and five accounted for 90 percent of the genes. In the black population, there were nine alleles. Finally, in the North African/Middle Eastern populations, there were twelve alleles plus the Mediterranean alleles and one Asian Indian allele.

Complex β-Thalassemias

The complex β-thalassemia syndromes are characterized by disturbances in the expression of other β-like genes (Orkin and Kazazian, 1984; Kazazian and Boehm, 1988). The sites of the mutations are also shown in Figure 7.11. In this group the typical hereditary persistence of fetal hemoglobin (HPFH), is commonly seen in blacks and is associated with a deletion of the β– and δ-globin genes. Although these are technically thalassemias, homozygous HPFH is generally without clinical effect. The double heterozygote S/HPFH is also asymptomatic.

CLINICAL AND GENETIC HETEROGENEITY IN AFRICAN-AMERICANS WITH HOMOZYGOUS BETA-THALASSEMIA

Huisman's group studied nineteen black patients with homozygous β-thalassemia from the southeastern United States (Gonzalez-Redondo et al., 1988). These authors first indicated that β-thalassemia is relatively rare in blacks, and that of the different types, the A → G substitution at position −29 in the TATA box is most common, followed by the T → C replacement in the cleavage-polyadenylation signal and the C → T substitution at nt −88. Other forms include the T → A substitution at codon 24, the A → G substitution at position 829 in the acceptor splice of the β IVS-II, the A → C substitution at the same location, the G → A replacement at position 1 in the donor site of the IVS-II, and the frameshift at codons 106 and 107 through an insertion of G. Gonzalez-Redondo and colleagues also assert that in the southeastern United States the incidence of β-thalassemia (homozygous) is between 0.3 and 0.5 percent.

Of the thirty-eight chromosomes tested, twenty-one (55 percent) had a C → T substitution at nt −88; three (8 percent) had the substitution at codon 24. One each of several other abnormalities were detected. One of these, which was new, had an A → T mutation in codon 61 (AAG → TAG, which resulted in the creation of a stop codon, and thus β⁰-thalassemia.

Hereditary Persistence of Fetal Hemoglobin and Gamma-Chain Variants

Hereditary persistence of fetal hemoglobin (HPFH) is a condition characterized by the continuation of fetal hemoglobin (Hb F) synthesis without major hematologic abnormality (Weatherall, 1983). There are two general forms of HPFH: a pancellular form in which Hb F is uniformly distributed within the erythrocytes, and heterocellular HPFH in which there is heterogeneous distribution of Hb F. It should be emphasized, however, that in the pancellular form, there is slight variation of Hb F from cell to cell. (This important point is rarely mentioned in the description of pancellular HPFH.) The most common form of HPFH in blacks is apparently a deletion of the $^G\gamma$

Table 7.10
Types of Hereditary Persistence of Fetal Hemoglobin in Blacks

	Black $^G\gamma^A\gamma(\delta\beta)^0$	Black $^G\gamma(\delta\beta)^+$	Black $^G\gamma(\delta\beta)^+$ (Kenya)	Black $^G\gamma(\delta\beta)^0$
Homozygotes				
Hb F (%)	100	—	—	—
Gly 136	0.5–0.6	—	—	—
Hb A$_2$ (percent)	0	—	—	—
Heterozygotes				
Hb F (%)	17–36	15–20	4.5–10.5 (Kenya 5.5–27)	15–25
Gly 136	0.3–0.5	1.0	1.0	1.0
Hb A$_2$ (%)	1.2–2.7	1.6–2.2	1.0–2.1	1.5–2.5
Distribution of Hb F		All Pancellular		
Molecular defect	Deletion δ + β	?	βγ fusion gene, deletion of parts of $^A\gamma$ + β genes	?
Interaction with Hb S	Hb S + F	Hb S + F + A	Hb S + Kenya + F	Hb S + F

Source: Modified from Weatherall (1983).

and $^A\gamma$ genes, with persistence of $^G\gamma$ and $^A\gamma$ synthesis. Even though this type of HPFH may be viewed as a form of δβ-thalassemia, it is compensated (Weatherall, 1983). The several forms of HPFH are shown in Table 7.10.

Generally, HPFH may be divided into deletion and nondeletion HPFH. In deletion HPFH, large genomic deletions remove the δ- and β-chains, and there is usually overproduction of both $^G\gamma$- and $^A\gamma$-chains. In the nondeletion form of HPFH, overproduction of Hb F is primarily $^G\gamma$ or $^A\gamma$. The most common deletion type, HPFH-1, is found in U.S. blacks and is characterized by a 5′ breakpoint 4 kilobases 5′ to the δ-gene. Another type, HPFH-2, which is found in Ghanaian blacks, has both 5′ and 3′ endpoints shifted about 5 kb in the 5′ direction; however, the total amount of DNA lost is within 1 kb of being similar to HPFH-1. Homozygotes for both deletion forms of HPFH have no abnormality of phenotype other than a high level of HPFH (Collins et al., 1987).

Collins et al. (1987) indicated that the most common forms of hereditary persistence of fetal hemoglobin include large deletions that remove the δ and β genes but leave the fetal $^G\gamma$ and $^A\gamma$ intact. Using pulsed-field gel electrophoresis and the enzyme *Sfi*I these authors prepared a large-scale restriction map of the β-globin cluster in normal and HPFH DNA. The deletions in HPFH-1 and in HPFH-2 were approximately 105 kb in length. Collins and co-workers postulated that the mechanism for the deletion was probably the loss of a complete chromatin loop.

Collins et al. (1984a, 1984b) described the characteristics of the β-globin gene cluster in $^G\gamma\beta^+$ HPFH individuals who do not have deletions of the β-globin cluster. In heterozygotes for $^G\gamma\beta^+$ HPFH, about 15–25 percent of Hb F is formed. These authors confirmed that there were no rearrangements in 40-kilobase fragments that contained the $^G\gamma$, $^A\gamma$, and δ, and β genes from the $^G\gamma\beta^+$ HPFH chromosomes, but they discovered a point mutation 202 base pairs (bp) 5' to the $^G\gamma$ gene (Figure 7.11). Collins and co-workers (1984b) queried whether the point mutation was responsible for the phenotype, or whether it was a variant that happened to be linked to the mutant gene—but with no functional significance. In a population survey of 170 non-HPFH alleles in blacks in which there were 61 β^A globin genes and 109 β^S globin genes, none had the mutation 202 bp 5' from the $^G\gamma$ gene. They concluded that the −202 mutation is the cause of the $^G\gamma$ HPFH, but they could not exclude the possibility that it is a polymorphism.

The gene products of the γ-chain genes are usually found in a $^G\gamma$ to $^A\gamma$ ratio of 70 : 30. Approximately thirty γ-chain variants have been described, but most of them have not been observed in blacks (Huisman, 1983). Huisman emphasized that the distribution of γ-chain variants is limited by the locations of laboratories that are actively involved in the structural analysis of these hemoglobins. Thus, unlike many other hemoglobins and thalassemias, the delineation of the world distribution of γ-chain variants may not be very informative at this time. Nevertheless, Huisman's elegant studies indicate that a variant, $^A\gamma^T$, described next, could be an important anthropological marker.

A γ-chain variant that was initially named F Sardinia, or 75(E19) Ile → Thr, has been found in high incidence in various populations and has been renamed the $^A\gamma^T$-chain. Huisman (1983) collaborated with scientists from many countries to obtain f($^A\gamma^T$) values (f = the frequency of the $^A\gamma^T$ gene). They obtained 2,588 umbilical cord blood samples, and samples from 283 patients with Hb SS, 23 patients with Hb Sβ-thalassemia, and 53 patients who were homozygous for β-thalassemia. There was a high frequency of $^A\gamma^T$ in Yugoslavian and in Italian newborns and in white newborns from the southeastern United States. The results were quite different in a black population in Ghana. None of the 82 Hb SS patients studied in Ghana produced Hb F with the $^A\gamma^T$-chain, which suggested that this mutant may be

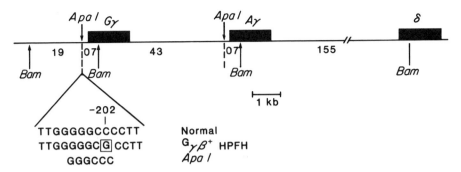

Figure 7.11. Point mutation at position −202 of the $^G\gamma$ gene in $^G\gamma$B+ HPFH. The mutation abolishes a normal *ApaI* restriction site at this location. The location of other *ApaI* and *BamHI* sites and the sizes of fragments (in kilobases) expected from a double digest with both of these enzymes are shown

Source: Collins et al., 1984. Reprinted with the permission of Grune & Stratton and F. S. Collins

absent in Ghanaians. But when β-thalassemia was present, the $^A\gamma^T$-chain was found in three of ten patients with Hb S-β-thalassemia in Ghana. The world distribution of this variant is summarized in Figure 7.12. These studies suggest that most of the $^A\gamma^T$ variants observed in black populations may be the result of European admixture.

Odenheimer et al. (1987) studied the relationship between fetal hemoglobin and disease severity in children with sickle cell anemia. A sample of 140 children with sickle cell anemia was evaluated to determine the relationship between hematologic variables, percent Hb F, percent Hb A_2, percent Hb, and mean cell volume. Fetal hemoglobin was found to be a strong predictor of a child's hemoglobin and transfusion status. A decrease of 4.76 percent of HbF was associated with 3.58-fold greater odds of being hospitalized both before initial assessment and at a follow-up examination, compared with not being hospitalized at either evaluation stage. A decrease of 4.76 percent in Hb F was associated with 5.56-fold greater odds of having a transfusion both before an initial assessment and at follow-up evaluation, compared with not having a transfusion at either evaluation. Patients who were both hospitalized and given a transfusion at initial assessment and at a follow-up examination, had a mean percent Hb F of 7.59 percent, and patients who were not hospitalized or given transfusions at either evaluation had a mean percent Hb F of 13.61 percent. Fetal hemoglobin was not a predictor of pain crises, and none of the other hematological parameters were predictors of disease severity. The authors concluded that a single percent Hb F measurement may be useful in predicting the clinical course of children with sickle cell anemia.

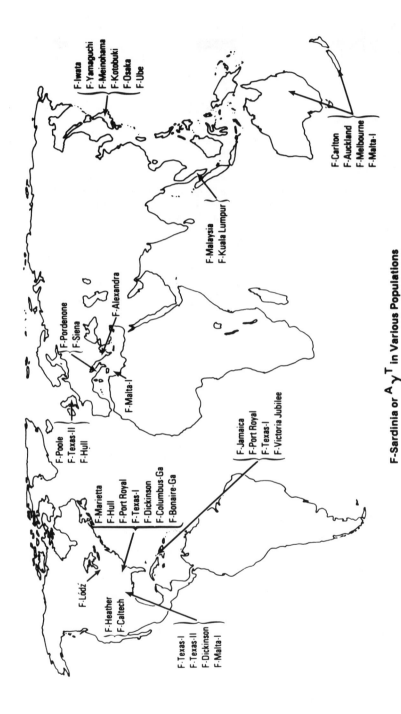

Figure 7.12 The distribution of γ-chain variants

Source: Huisman, 1983. Reprinted by permission of Elsevier Science Publishing Co., Inc., New York, NY. Copyright © 1983

F-Sardinia or $^A\gamma^T$ in Various Populations

Table 7.11
Restriction Enzyme Identification of Some Abnormal Hemoglobins

Abnormal Hb	AA Substitution	DNA Change	Enzyme
Hb S	$\beta^{6\,glu—val}$	CCTGAG—CCTGTG	*Dde*I or *Mst*II
Hb O Arab	$\beta^{121\,gly—lys}$	GAATTC—AAATTC	*Eco*RI
Hb D Punjab	$\beta^{121\,glu—gln}$	GAATTC—CAATTC	*Eco*RI
Hb J Broussais	$\alpha^{90\,lys—asn}$	AAGCTT—AATCTT	*Hind*III
Hb F Hull	$\gamma^{121\,glu—lys}$	GAATTC—AAATTC	*Eco*RI

Source: Rowley et al. (1985).

PRENATAL DIAGNOSIS OF HEMOGLOBINOPATHIES AND THALASSEMIAS

Restriction enzyme analysis of DNA has made possible the antenatal diagnosis of Hb S, other variants of hemoglobinopathies, and some α- and β-thalassemias. Here, most of our attention is directed to Hb S. We have previously mentioned the DNA polymorphism that Kan and Dozy (1978) described, using the restriction enzyme *Hpa*I. With this enzyme, the 13.0-kb fragment was invariably associated with the β^S allele, but since this association was inconstant, prenatal ascertainment was not always possible. Fortunately, other restriction enzymes are specific for the β^S allele and for some other hemoglobins.

Chang and Kan (1982) and Orkin et al. (1982) noted that the sequence CCTGAGG in the region 5 to 7 in the normal DNA sequence is recognized by the enzyme *Mst*II. Cleavage of human β -globin DNA with *Mst*II generates multiple DNA fragments, which includes two fragments from region 5 to 7. One fragment is 200 base pairs and the other is 1,150 base pairs in length. Sickle β-globin contains the sequence CCTGTGG, which is not recognized by *Mst*II. Accordingly, cleavage of sickle β-globin DNA with *Mst*II generates only one DNA fragment, 1,350 bp in length. These different cleavage patterns can detect Hb AA, Hb AS, and Hb SS DNA. The restriction enzyme *Dde*I produces similar results, but the fragment enzymes are smaller than those that are cleaved by *Mst*II and are thus harder to detect. Several abnormal hemoglobins can now be specifically identified by restriction enzyme analysis (see Table 7.11).

Saiki et al. (1985) described a method for increasing the sensitivity of restriction analysis by more than two orders of magnitude. This reduces the laboratory time required for DNA analysis to less than one day. The technique uses the polymerase chain reaction to amplify specific DNA sequences enzymatically before oligonucleotide restriction analysis. Less than 20 milligrams of genomic DNA are needed (Embury et al., 1987).

Chehab and Kan (1988) developed a polymerase color complementation assay, which involves a rapid nonradioactive method to detect gene deletions, chromosomal rearrangements, and etiologic agents. After the polymerase chain reaction is applied, the amplified DNA is separated by gel electrophoresis and the gel is stained with ethidium bromide; then the presence or absence of an amplified band of the target sequence is diagnostic. The technique has been utilized to detect four base pair deletions in α-thalassemia, β-globin gene deletions in homozygous β-thalassemia, the t(14;18) chromosomal translocation in some lymphomas, and the (9;22) translocation in chronic granulocytic leukemia, and the delineation of Hb AA, Hb AS, and Hb SS. This very important technique should revolutionize the diagnosis of a number of genetic disorders without the use of radioisotopes.

8

The Malaria Hypothesis

Malaria is the result of complex interrelationships of many variables: species and strain of plasmodium, immunity in humans and in the mosquito, habits of humans and the mosquito, environmental conditions, such as temperature, rainfall, relative humidity, topography, soil, flora and fauna, mosquito control measures, and genetic differences of the intermediate hosts: humans (Bowman, 1964b). Moreover, stable malaria should be distinguished from unstable malaria. In stable malaria, the transmission is high and not subject to marked fluctuations over the years. The collective immunity of the population is high and epidemics are unlikely. In unstable malaria, transmission varies and is subject to seasonal or other fluctuations to such an extent that the collective immunity of the population is variable and frequently low.

HISTORICAL CONSIDERATIONS

Malaria has been known since antiquity to be one of the great scourges of humans. It was recognized in the ancient civilizations of Assyria, Egypt, and China. In the 1960s, it was believed that malaria could be eradicated, but it is still a menace. There may be at least 300 million patients each year with this disease, and approximately 1 million malaria-related deaths occur annually (Wyler, 1983). The present-day distribution of malaria is shown in Figure 8.1.

The first clinical description of malaria was given by Hippocrates (born c. 460 B.C.). He noted its cyclic character and its association with swamps. Later, the Romans recognized the relationship of malaria with swampy areas and avoided building military camps in such regions. Malaria was widespread in Europe during the Middle Ages and until at least the nineteenth century (Sigerist, 1965). It is generally accepted that malaria was unknown in pre-Columbian America and that the disease was first introduced in the Americas after 1492. But as

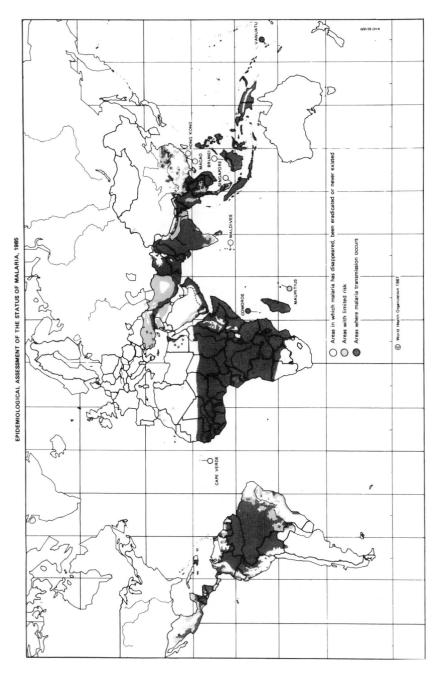

EPIDEMIOLOGICAL ASSESSMENT OF THE STATUS OF MALARIA, 1985

○ Areas in which malaria has disappeared, been eradicated or never existed

◐ Areas with limited risk

● Areas where malaria transmission occurs

© World Health Organization 1987

Figure 8.1. Epidemiological assessment of the status of malaria, 1987.

Source: Copyright © World Health Organization 1989

we have already noted in the first chapter, malaria could have been introduced in the Americas much earlier. Even so, we shall soon see that there is other evidence for the relatively "recent" introduction of malaria to the Americas.

Sigerist (1965) analyzed malaria and ancient Greek history and postulated that the Greeks who surrendered to the Roman legions did so because they had been ravaged by malaria. Sigerist examined the writings of ancient and medieval authorities and came to the conclusion that there is only the slightest evidence that malaria existed on the mainland of Greece in very early times; that the disease was probably common in about 500 B.C. in Magna Graecia and on the coast of Asia Minor; and that it invaded the mainland in the second half of the fifth century B.C. There was probably a serious outbreak in Attica during the Peloponnesian War. Malaria would have then been endemic throughout Greece in about 400 B.C. Sigerist (1965) believed that there can be no doubt that malaria must have affected the Greeks, but he questioned whether the effects were far-reaching.

Malaria was very influential in the history of another region, Roman Compagna, the land that surrounds Rome. Sigerist (1965) noted that early sources suggested that the Compagna was sometimes a wasteland, and at other times it was a flourishing region, teeming with life. During the course of twenty-five centuries, the Compagna was inhabited in four periods: (1) in the pre-Roman era; (2) when Imperial Rome was at the height of its power; (3) in the early Middle Ages (eighth and ninth centuries); and (4) in modern times from the fifteenth to the seventeenth centuries. Between these periods the Compagna was a place of desolation, with ruined and abandoned hamlets and a few wild sheep grazing around the city. Emperors and popes made great attempts to repopulate the region, but the efforts were in vain. Why? Sigerist stated that it has been assumed that the many wars that afflicted the city of Rome were responsible for the devastation of the Compagna. Sigerist believed the notion that malaria was responsible for the vicissitudes of the Compagna. Other historical studies showed that in certain regions malaria occurs in periodic cycles, driving all life away. And then, after centuries, for unknown reasons, the disease gradually recedes and life returns.

The effect of malaria on the early colonization of West Africa was considerable. This disease was one of the main factors in confining Europeans to coastal areas during the early periods of exploration and the slave trade. The later introduction of quinine in treating malaria eventually led to the colonization of regions of Africa that were previously uninhabitable for Europeans.

Malaria played a major role in the colonization of the Americas, the subsequent development of the West Indian sugar industry, and the consequent need for slaves from Africa. One of the best analyses of malaria in the Americas was a study by Effertz (1909). Effertz analyzed the differential clinical effects of malaria on American Indians, Europeans, and Africans. He observed that the impact of malaria on Amerindians was devastating, that the effect of malaria on Europeans was less harmful, and that the African slaves were only slightly, if at all, affected. Effertz surmised that since malaria took such tremendous toll on Amerindians, the disease was new to the Americas. Further, the moderate effect of malaria on Europeans was the result of residual immunity from European malaria. But if the Europeans were only mildly affected in the Americas, why was malaria in Africa so harmful to them? We can only speculate that there might have been strain differences in falciparum malaria. After the discovery of quinine, Europeans took this drug in large amounts, but, for some reason, the Amerindians had a horror of quinine and were unaware of its medicinal properties. Effertz also thought it inconceivable that the Amerindians would not have developed some treatment for a disease that had been present in their part of the world for thousands of years. (For further corroboration that malaria was not indigenous to the Americas before 1492, see Chapter 1's discussion of the distribution of mammalian malaria parasites.)

Malaria was not eradicated in the United States until just before World War II. Most, if not all, cases of malaria now observed in the United States are imported; there have been occasional reports of transmission from people who have recently visited malarious regions.

FITNESS AND HETEROZYGOTE ADVANTAGE

The Hardy-Weinberg equilibrium can be disturbed by selection, migration, genetic drift, and mutation. Selection disturbs this equilibrium by increasing or decreasing fitness. There have been various definitions of fitness. Fitness has been expressed as:

1. The total number of offspring (excluding stillbirths and abortions).
2. The number of offspring who reach reproductive age.
3. The number of offspring who reach the mean age at which the parents reproduced.
4. The number of offspring who complete their reproductive life (Clarke 1959).

Heterozygous advantage may be calculated—say, for sickle hemoglobin—by the following method for the determination of the coefficient of selection (s) (Emery, 1976).
Let us assume that the gene frequency of the sickle gene (q) is 0.200, then,

$$s = \frac{q}{1 - q}$$

$$s = 0.25$$

Accordingly, the heterozygote (Hb AS) has a 25 percent advantage over the homozygotes (Hb AA and Hb SS).

MALARIA HYPOTHESIS PREDICTIONS

Haldane (1949) first suggested that disadvantageous disorders like thalassemia are maintained in high frequency in some human populations because the heterozygote is relatively resistant to the lethal effect of malaria, and Allison (1954) introduced the concept that people who have sickle cell trait are relatively resistant to falciparum malaria. (We emphasize the word *relatively*, because some physicians are under the misconception that persons who have sickle cell trait cannot develop falciparum malaria.) The distribution of hemoglobins S, C, E, α- and β-thalassemias, and G-6-PD deficiency are associated with the endemicity of *Plasmodium falciparum*. In short—and with few exceptions—the malaria hypothesis predicts that wherever these variants are found, falciparum malaria is or has been endemic.

The strongest association of red cell variation with malaria is the important finding that the absence of the Duffy blood antigens Fya and Fyb protects the red cell from invasion by *Plasmodium vivax* (Livingstone, 1983). This discovery explains, in part, the present-day absence of *P. vivax* in most of West Africa, and the resistance of West Africans and their descendants to vivax malaria. Other forms of human malaria, *Plasmodium malariae* and *Plasmodium ovale*, have not been associated with hemoglobins, thalassemias, enzyme deficiencies, or blood groups.

The malaria hypothesis has been most attractive to population geneticists, because sickle hemoglobin in particular occurs with high frequency in many groups, and in the past, at least, there was considerable mortality in patients with sickle cell anemia. Thus, it could easily be shown that the sickle gene must be maintained by factors other than mutation—otherwise the high frequencies could not be maintained. But if the heterozygote has an advantage over either homozygote (Livingstone, 1958)—that is, if the individual with sickle cell trait is relatively resistant to the lethal effects of falciparum malaria—the gene could be maintained.

Seven predictions have been used to test the malaria hypothesis:

1. In regions where falciparum malaria is or has been endemic, high frequencies of Hb S and G-6-PD deficiency should be found, and there should be a positive correlation between present or past endemicity of falciparum malaria and the frequency of G-6-PD deficiency or Hb S.

This prediction of the malaria hypothesis has been widely reported as substantiated (Allison, 1954; Motulsky, 1964; Siniscalco et al., 1966). Generally, in regions where falciparum malaria is or has been endemic, high frequencies of G-6-PD deficiency have been recorded. The absence of G-6-PD deficiency and Hb S in Amerindians (excluding populations where there is African admixture) probably has no bearing on the malaria hypothesis if malaria is of comparatively recent origin. But this assertion was brought into question in the first chapter of this book. Malaria was also endemic in Europe before the continent was cleared of marshes. Hence, the absence of G-6-PD deficiency in many northern Europeans, and some southern Europeans, could affect the validity of the malaria hypothesis.

Many studies claim a correlation of past or present endemicity of falciparum malaria with the frequency of G-6-PD deficiency (Allison, 1954; Motulsky, 1964; Siniscalco et al., 1966). The relevance of these investigations to the malaria hypothesis could be open to question, for there may be too many variables to allow valid conclusions. Most human populations have a history of migration and breeding with other groups to such an extent that the habitat of the population has been altered many times, and the extent of admixture may defy resolution even over periods as short as one or two generations. The endemicity of falciparum malaria today may have no relationship to that of a few years ago or 500 or 1,000 years ago.

A geographical correlation that would favor the malaria hypothesis should demonstrate a high frequency of G-6-PD deficiency in regions that are hyperendemic for falciparum malaria, a low-to-absent prevalence in hypoendemic or nonmalarious regions, and comparable allele frequencies for other genetic markers in the populations. These criteria have been fulfilled only in the series of surveys directed by Siniscalco et al. (1966). Even so, the genetic markers other than G-6-PD were confined to ABO, MN, and Rh blood groups, which are frequently invariant in small geographic regions.

Many populations have an absence or very low frequency of G-6-

PD deficiency in regions that are or were endemic for falciparum malaria. Some examples are the Zoroastrians and the Armenians of Iran, the Armenians of Lebanon, the Samaritans of Israel, the Ethiopians, several groups in New Guinea, and some Europeans (Livingstone, 1967; Bowman and Walker, 1963; Bowman, 1964b; Sheba, 1963; Sheba and Adam, 1962; Kidson and Gorman, 1962; Taleb et al., 1964). Although these data do not necessarily disprove the malaria hypothesis, they do not support it.

Studies of the association of G-6-PD deficiency and of Hb E in South Vietnam were inconclusive (Bowman et al., 1971). Low frequencies of Hb E and G-6-PD deficiency occurred in a Sedang population in a region where the endemicity of malaria was low; however, in the An Phu region where the malaria parasite rate was also low, the Cham had a high prevalence of G-6-PD deficiency and of Hb E. But such correlations are valid only if (1) there has been a static environment in the region for many hundreds of years; (2) there has been no significant migration or admixture; and (3) there has been fairly constant population size over many centuries. If these criteria are impossible to meet, then the malaria hypothesis is untestable by this method of investigation.

2. If the malaria hypothesis is valid, then Hb S and G-6-PD deficiency should not be found in frequencies higher than those obtained by mutation pressure alone in regions where falciparum malaria has not existed.

Wherever Hb S has been found, falciparum malaria is present or existed in the past. The same does not hold for G-6-PD deficiency. Malaria is unknown in Micronesia. Kidson and Gajdusek (1962) found that 9 percent of 58 males on the island of Angaur had G-6-PD deficiency; on Koror, the figures were 8 percent of 24 males, and on Ifalik, 6 percent of 34 males had G-6-PD deficiency. The G-6-PD deficiency was not found among 117 males on Ulitki. Livingstone (1967) suggested that G-6-PD deficiency on these islands may be the result of ancient gene flow from the Philippines. But since it is now known that G-6-PD deficiency is quite heterogeneous, these populations should be restudied and compared with those in the Philippines. When the investigations were performed, the multiple variants of G-6-PD were not foreseen. If the variants of G-6-PD on the Micronesian islands are unlike those on the Philippines, the G-6-PD–malaria hypothesis could be seriously questioned.

3. If a population with a high prevalence of Hb S or G-6-PD deficiency migrates from a falciparum malarious environment to a nonmalarious environment, the frequency of Hb S or G-6-PD deficiency in that population should decrease.

The prevalence of G-6-PD deficiency and of Hb S in West Africans is higher than the frequency of these variants among blacks in the United States. Motulsky (1964) attributed these differences to be the result of European gene flow into the black population of the United States and by migration from a malarious to a nonmalarious environment. In fact, Allison (1955) stated that given a population with 22 percent sickle cell trait and a one-third admixture with a population without sickle cell trait, a trait frequency of 15.4 percent would be expected, but the prevalence of sickle cell trait rarely exceeds 8 percent in blacks in the United States. Allison asserted that this fall would have taken place in about twelve generations, or over 300 to 350 years, which fits the period of slave importations into the United States. But the African slaves brought their malaria with them; and falciparum malaria was only eradicated from the southern part of the United States after World War II. Sickle hemoglobin gene frequencies in West Africa are not uniform today; and there is no reason to suspect that they were not variable during the days of the slave trade. In fact, Hb AS frequencies differ in blacks in the United States today, from about 7 percent to as high as 10–15 percent (Pollitzer, 1958; Huisman, 1980; Bowman, 1983). The highest frequencies are found in blacks who live around Charleston, South Carolina. But in this region, the European admixture in blacks is about 5 percent (Pollitzer, 1958). Furthermore, the high frequencies of both G-6-PD deficiency and Hb S in blacks in Nova Scotia is not consonant with Allison's or Motulsky's thesis, unless genetic drift, or inbreeding was (or is) the explanation.

4. If the malaria hypothesis is valid, Hb S and G-6-PD deficiency should result in a suboptimal environment for the growth of *P. falciparum*.

The development of techniques for the culture of erythrocytic *Plasmodium falciparum* has made it possible to directly investigate the development and course of infection of *P. falciparum* in sickle erythrocytes. Friedman (1978) found that in an 18 percent oxygen atmosphere there was no difference in growth and multiplication of *P. falciparum* in Hb AA, AS, and SS erythrocytes. Cultures of erythrocytes in an atmosphere of less than 1–5 percent oxygen demonstrated that the sickling of Hb SS cells killed most or all falciparum malaria parasites; parasites in Hb AS erythrocytes were mainly killed at the large ring cell stage.

Pasvol, Weatherall, and Wilson (1978) compared the rates of invasion and growth of *P. falciparum* in cells with and without sickle hemoglobin in aerobic conditions and under circumstances of reduced oxygen tension. The growth of parasites was retarded or inhibited in Hb AS and Hb SS cells as compared with Hb AA cells. Moreover, at least

83 percent of the singly infected cells were not sickled, which indicated that sickling was not necessary for growth retardation of *P. falciparum*.

Friedman (1978) and Pasvol, Weatherall, and Wilson (1978) commented on the work of Luzzatto (1974) and Luzzatto et al. (1970), who found preferential sickling of *P. falciparum* infected erythrocytes when they subjected blood from a person who had Hb AS cells to complete deoxygenation and postulated that resistance is due to sickling and phagocytosis of infected red cells in the spleen. Friedman (1978) repeated these experiments and found that schizont-infected cells do not sickle; the schizonts in Hb AS cells survived low oxygen culture. Friedman's studies showed that 90 percent of young parasites were killed in a population of Hb AS cells in which only 60 percent were sickled, which suggested that the parasite contributes to conditions that induce sickling.

Pasvol, Weatherall, and Wilson (1978) believed that it is unlikely that a parasite can bring about sufficient reduction in oxygen tension in Hb AS cells to cause them to sickle and remain sickled in peripheral blood. Rather, limited growth of *P. falciparum* in Hb AS cells in areas of reduced oxygen tension appeared to be a more likely basis for resistance.

Roth et al. (1978) also studied the sickling rates of Hb AS cells infected in vitro with *P. falciparum*. These experiments supported the work of Luzzatto. The rate of sickling of cells that were infected with *P. falciparum* and of noninfected cells was examined in the absence of oxygen and at several different concentrations of oxyhemoglobin that may be obtained in vivo. Erythrocytes that contained trophozoites and schizonts sickled less readily than uninfected cells by light microscopy; however, electron micrographs showed polymerized deoxyhemoglobin S with a high frequency.

McGhee and Trager (1950) and Fulton and Grant (1956) discovered that *P. lophurae* and *P. knowlesi* require GSH for growth in vitro, and 50 percent of the GSH of red cells contributes to the cysteine requirements of the parasite. Motulsky (1964) extrapolated these findings to apply to *P. falciparum*. Luzzatto (1969) has indeed confirmed that the growth of *P. falciparum* is poor in erythrocytes deficient in G-6-PD.

In fact, Bienzle et al. (1972) found a greater resistance to falciparum malaria in females heterozygous for enzyme deficiency and in males with the nondeficient glucose-6-phosphate dehydrogenase (Gd) A variant. These authors studied 700 boys and girls from a rural area with holoendemic falciparum malaria. In the boys, there was no evidence that enzyme-deficient subjects had a greater resistance to falciparum malaria than did normal subjects. Interestingly, however, boys with the G-6-PD variant A had significantly lower parasite counts than did those with the B variant of the enzyme. Homozygous girls also had no protection. On the other hand, the girls with the *GdA-/GdB* genotype had significantly lower parasite counts than any other group of males or females. These authors concluded that the high frequency of *GdA* − in the population is probably maintained by the advantage against malaria selection of hemizygous males, and perhaps homozygous females, and that the high frequency of the *GdA* -gene maintained principally by the advantage against falciparum malaria of heterozygous females. This thesis does not address why the so-called suboptimal environment of hemizygous erythrocytes deficient in G-6-PD in males is more conducive to faliciparum malaria than the minimally suboptimal environment of heterozygous female erythrocytes in G-6-PD.

Finally, Yoshida and Roth (1987) extended the studies of Hempelman and Wilson (1981) and Usanga and Luzzatto (1985) and confirmed that *P. falciparum* contains G-6-PD. Southern blot hybridization indicated that the parasite genome contained nucleotide sequences that were hybridizable with the human G-6-PD cDNA. The experiments were interpreted as an indication that *P. falciparum* is capable of adapting to erythrocytes deficient in G-6-PD by producing its own G-6-PD.

5. Nonimmune subjects with Hb S and G-6-PD deficiency should be more resistant to experimental falciparum malaria than those who do not have these variants.

Powell and Brewer (1965) studied the duration of the prepatent period and the early clinical course of falciparum malaria in sixteen nonimmune black males in the United States. Eight of the men had G-6-PD deficiency, and eight did not. There was no retardation of erythrocytic development or of early proliferation of asexual erythrocytic forms of *P. falciparum* in the subjects who were G-6-PD deficient. These investigators concluded that their studies did not disclose that G-6-PD deficiency confers a biological advantage against falciparum malaria. Powell and Brewer also were careful to state that their work did not invalidate the malaria hypothesis, because the nature of their experiments did not shed light on the critical issue of very high levels of parasitemia attended by substantial risks of mortality.

6. Falciparum malaria parasite rates and counts should be significantly lower in erythrocytes of subjects with Hb S and G-6-PD deficiency than in subjects without these markers.

Studies of *P. falciparum* rates in people with G-6-PD deficiency versus rates in unaffected children have shown no significant differences between these groups (Bowman and Frischer, 1964; Kruatrachue et al., 1962; Bernstein, Bowman, and Kaptue-Noche, 1980). Allison and Clyde (1961) found significant differences between G-6-PD-deficient and non-G-6-PD-deficient subjects with parasite counts greater than 1,000 per microliter. Wilson (1961), however, disputed that the differences provided any evidence of protection from the lethal effects of falciparum malaria. Bernstein, Bowman, and Kaptue-Noche examined 1,183 blood samples from Cameroonians and discovered that sickle cell trait and G-6-PD frequencies were heterogeneous among villages as well as within geographic regions and ethnic groups. The mean *P. falciparum* parasite counts were positively correlated with the frequency of Hb AS for children six years of age and under, but there was no correlation for mean parasite counts and the prevalence of G-6-PD deficiency. The mean parasite counts of Hb AS and Hb AA children did not differ. Interestingly, the mean age of persons with sickle cell trait was found to be significantly greater than the mean age of Hb AA individuals.

7. The severity of falciparum malaria should be greater in subjects who are non-G-6-PD deficient than in people who are G-6-PD deficient and in subjects with Hb AA as compared to those with Hb AS.

People with G-6-PD deficiency do develop falciparum malaria. But what about severity? Gilles et al. (1967) studied the relationship between G-6-PD deficiency and falciparum malaria in 55 boys under four years of age. All had parasite counts of 100,000 per microliter or higher. These were compared with 102 boys who attended the clinic for other reasons. There was no difference in prevalence of G-6-PD deficiency between the two groups. But an important prediction of the malaria hypothesis would be that the prevalence of G-6-PD deficiency in the severe malaria category would be significantly lower than the frequency of G-6-PD deficiency in a control population without malaria.

Martin et al. (1979) found that Nigerian children with convulsions and *P. falciparum* parasitemia above 100,000 per microliter did not have a decreased frequency of G-6-PD deficiency, and that a reevaluation of earlier studies led to the conclusion that *clinical* evidence of protection against falciparum malaria in people with G-6-PD deficiency is baseless.

Martin (1980) proposed a mechanism whereby G-6-PD deficiency, oxidant stress, and malaria interact. He postulated that individuals who do not have G-6-PD deficiency are not protected against malaria and also are not susceptible to oxidant stress, but persons who are homozygous or hemizygous affected are protected from lethal malaria but are also selected against by oxidant stress. On the other hand,

240 Genetic Variation and Disorders in Peoples of African Origin

the female heterozygote has both the *normal* and the *G-6-PD-deficient* allele and she is protected from the negative selection of favism and falciparum malaria. Thus there would be a balanced polymorphism. Bernstein and Bowman (1980), however, pointed out that the polymorphism will be balanced only if in those who are unaffected the susceptibility to malaria is greater than the protection against oxidant stress alone, and in those who are homozygous and hemizygous affected the susceptibility to oxidant stress is greater than the protection against malaria. That is, the fitness of the homozygotes and of the hemizygotes must be less than that of the heterozygotes. If either condition is not met, the polymorphism will be transient, with the eventual loss of one allele and the fixation of the other. Bernstein and Bowman also stated that in Africa the G-6-PD polymorphism is a three-allele polymorphism; the *GdA +* and *GdA − alleles* have a similar geographic distribution. But there is no reasonable explanation for the maintenance of the *GdA +* allele, other than the ubiquitous malaria association.

Livingstone (1971) reviewed evidence that there is indeed increased mortality in persons with Hb AA cells as compared to those with sickle cell trait. In every investigation mortality from cerebral malaria was virtually nonexistent in individuals with sickle cell trait.

The prevalence of hemoglobins and their interaction with malaria was studied in Garki, Kano State, Nigeria (Fleming et al., 1979). Sickle cell trait was found in 24 percent of newborns and 29 percent of those over age five. The authors concluded that the *Hb S* gene frequency is apparently maintained by a fitness in heterozygotes of 21 percent over normal homozygotes, and that increased fertility and a high mutation rate did not appear to make a significant contribution.

The reported interaction between falciparum malaria, G-6-PD deficiency, Hb S, and β-thalassemia is now joined by suggestions that α-thalassemia may also be protective (Oppenheimer et al., 1984; Ramsay and Jenkins, 1984; Willkox, 1984; Flint et al., 1986; Higgs et al., 1989). But, as we saw in Chapter 7, sickle hemoglobin and β-thalassemia also interact with α-thalassemia. Thus genic interaction—which once used to be in the province of the mathematical geneticists—is now a potentially useful tool for medical geneticists and genetic counselors, and, perhaps, eventually for therapy.

The most extensive studies that attempted to show how high frequencies of α-thalassemia are the result of natural selection by falciparum malaria were provided by Flint et al. (1986), through analysis of the association between α-thalassemia and falciparum malaria in the southwest Pacific. Populations in this region offered the advantage that malaria endemicity varied with altitude in Papua New

Guinea and with latitude in Melanesia. This situation allowed an investigation of the prevalence of α-thalassemia and the endemicity of malaria. Additionally, Hb S and Hb E, which are also subject to selection, are also absent in Melanesia, as is β-thalassemia. It was found that α+-thalassemia gene (the single α gene deletion) frequencies correlated strongly with malaria, whether malaria varied with latitude or altitude. The authors also excluded genetic drift or founder effects to account for the α-thalassemia gene frequencies.

Higgs et al. (1989) indicated, however, that in vitro studies have not shown an abnormality of invasion or growth of *P. falciparum* in red cells of α+-thalassemia heterozygotes unless the cells were maintained under conditions of unusual oxidant stress. Further, α-thalassemia is in high frequency in Polynesians (0.09–0.12 perent), Micronesians (0.02 percent), and Australian aborigines (0.01–0.13), populations that have not been exposed to endemic falciparum malaria.

Bowman and Bloom (1985) studied the possible relationship of genic interaction and falciparum malaria. They began by outlining certain interactions and associations between and among some polymorphisms. First, as noted in Chapter 7, α-thalassemia may ameliorate morbidity and mortality in patients with sickle cell disease. Second, the clinical picture of the Hb SC double heterozygote is milder than that found in people with Hb SS. Third, α-thalassemia in conjunction with homozygous β[0] thalassemia produces a milder clinical picture than that of β[0]-thalassemia without α-thalassemia (Rostatelli et al., 1984). Next, patients with ααα/αα and β[0]-thalassemia are reported to have a more severe clinical picture than patients with the normal four copies of the gene (Galanello et al., 1983). Is there a similar effect of ααα/αα on the clinical picture of sickle cell anemia? In the fifth place, populations with Hbs S, C, or E also have α- and β-thalassemias. Finally, α- and β-thalassemias may be present in some populations without hemoglobins S, C, or E. Bowman and Bloom suggested that although β-thalassemia, Hb S, Hb C, and Hb E may be maintained by the selective effect of falciparum malaria, allelic effects and principally the epistatic effect of unlinked α-thalassemia may also contribute to the most disadvantageous of the hemoglobin variants, β[S]- and β[0]-thalassemia. In short, genic interaction may complement natural (environmental) selection in the maintenance of β[S]- and the β[0]-thalassemias. Wagner and Cavalli-Sforza (1975) have written on the possible role of "hitchhiking" and epistasis on the rate of natural selection and on the establishment of beneficial mutant combinations. The work of Nagylaki (1977) should also be consulted as an excellent contribution to this literature.

Durham (1983) demonstrated that direct assessments can be made of the role of malaria selection in the evolution of S allele frequency differences. Through the use of a surrogate variable for malaria—monthly maximum rainfall—Durham showed that the selective action of the parasite can account for a great deal of the variability of the frequency of the sickling gene that is found among the Kwa, West Atlantic, and Mande language subgroups of West Africa. No consistent pattern of covariation between β^S frequencies and the rainfall surrogate were found among the Voltaic (Gur), Nilo-Sahara, and Afro-Asiatic speakers of West Africa. Further, when the results of the surrogate tests for the Kwa and West Atlantic speakers are superimposed, there was no unexplained difference in the central tendencies of the data. At all levels of the surrogate variable up to about 500 millimeters of rainfall, the Kwa speakers had higher frequencies of β^S than the West Atlantic speakers. Durham offered a number of possible explanations, for example, "systematic bias in T [time], the duration of selection . . . hidden differences in microclimate, residential arrangements . . . human activity patterns not correlated with any choice of surrogate . . . the existence of so-called modifier genotypes, like fetal hemoglobin or α-thalassemia which are capable of reducing the deleterious effects of the homozygosity" (p. 69).

THE ROLE OF POLYGAMY IN MAINTENANCE OF SICKLING IN AFRICA

Konotey-Ahulu (1970) outlined other factors that could account for, in part, the high prevalence of sickle hemoglobin in Africa. He suggested first that adult individuals with sickle cell disease now often live long enough to produce children and freely procreate. Second, some studies have shown that people with sickle cell trait have a slightly higher fertility than do people who are Hb AA (Roberts, Lehmann, and Boyo, 1960). Third, the contribution made by female hyperfertility is much less than that made by a polygamous man.

Konotey-Ahulu then developed his thesis. Suppose N is the number of wives of a man with sickle cell trait, and n is the average number of children per wife. The number of individuals with sickle cell trait donated to the next generation by the polygamous man will be

$$Sp = \frac{Nn}{2} \qquad (8.1)$$

The wife of a monogamous marriage generally has more children than one who shares her husband with others. Suppose the wife of the monogamous man has twice as many children as the wife of a polygamous man, then the number of children with sickle cell trait donated through the one wife of a monogamous man with sickle cell trait is

$$Sm = \frac{1 \times 2n}{2} = n \qquad (8.2)$$

The numerical difference that polygamy would make in the prevalence of sickle cell trait if we imagine the same person with sickle cell trait in the two foregoing circumstances is the difference between equations (1) and (2) (Konotey-Ahulu, 1970).

$$Sp - Sm = \frac{Nn}{2} \qquad (8.3)$$

From equation (3) we can see that $Sp - Sm = 0$ under two conditions; (1) when $n = 0$; or when married men with sickle cell trait have no children, and (2) when $N = 2$, when men with sickle cell trait limit themselves to two wives. The first condition is rare, and the second condition does occur. However, according to Konotey-Ahulu, the men's polygamy will make no difference to the rate of sickle hemoglobin in the population only if their wives each have half the number of children that wives in a monogamous situation would have. Such a situation is rare, because the two wives of a bigamous husband usually produce more children than the one wife in a monogamous situation.

Konotey-Ahulu (1980) asserted that the adult African male, unlike the average European man, can have a biological fitness exceeding that of his wife. This author developed what he called the male procreative superiority index (MPSI) by dividing the total number of a man's children by the average number of children born to each wife. The countrywide mean MPSI for 3,095 fathers contacted throughout Ghana for 3,095 fathers was 2.03, which indicated that a Ghanaian father had on the average twice as many children as the mother. (This figure assumes, of course, that there is no illegitimacy.)

Konotey-Ahulu (1970) suggested that the role of polygamy may not be limited to the *HbS* gene, but could obtain for other high-frequency hemoglobins or heritable characters. Konotey-Ahulu pointed out that King Solomon was supposed to have had 400 wives and 300 concubines. If King Solomon had had an abnormal gene, his contribution to posterity would have been considerable.

And since we all have at least five to eight recessive genes . . .

9

Lactose Intolerance
and Malabsorption

There are several names for the inability to digest lactose properly. They include lactose intolerance, lactose malabsorption, lactase deficiency, adult lactase deficiency, and hypolactasia.

In most mammals, the enzyme lactase is present in the intestines in infancy and diminishes in the adult. Should the adult condition be termed a deficiency or part of the normal course of lactase development and regression? Johnson, Cole, and Ahern (1981) pointed out that high lactase activity in adults is found in only a minority of ethnic groups and, perhaps, in some species of nonhuman primates. Accordingly, why not consider the persistence of lactase activity in the adult as abnormal? These dilemmas serve as a reminder that the terms *abnormal* and *normal* are often relative.

LACTOSE

Milk is the sole source of the nonabsorbable lactose (Sahi, 1978). Lactose is hydrolyzed to the absorbable monosaccharides, glucose and galactose, by lactase, an enzyme that is found in the brush borders of the epithelial cells of the small intestine. Normal lactase activity in the jejunum and proximal ileum has been defined as greater than 2 units per gram of wet weight mucosa (Bayless, 1976). Bayless also indicated that lactase rises to peak levels at birth in a full-term infant, in contrast to sucrase and maltase, which attain adult levels before birth.

Even though the activity of lactase diminishes in some adults, it does not disappear (Sahi, 1978); but the activity is so low that the benefits of ingested lactose are impaired. If lactose digestion is compromised, the lactose absorbs water into the intestine; lactose is then split in the colon to lactic acid, hydrogen, and carbon dioxide (CO_2)—which in turn causes the symptoms of lactose malabsorption: meteorism, borborygmus, abdominal pain and distention, and loose stools and diarrhea (Sahi, 1978). An eight-ounce glass of milk contains 12 grams of lactose. An individual who has minimal levels of lactase will be able to drink about 1–2 liters of milk, or up to 96 grams of lactose, without maldigestion (Bayless, 1976). This important observation will be explored later.

DIAGNOSTIC TESTS

Several tests measure lactose intolerance, namely, the simple lactose tolerance test (or its variants), the lactose tolerance with ethanol test, the $^{14}CO_2$ breath test, and the hydrogen breath test (Johnson, Cole, and Ahern, 1981). We will briefly describe two that are most commonly used: the lactose tolerance test and the hydrogen breath test.

Lactose Tolerance Test

The lactose tolerance test is probably the best objective procedure for measuring lactase activity (Bayless, 1976). If lactase levels are normal, lactose will be split, and glucose and galactose will be absorbed. After the ingestion of about 1–2 liters of milk, or the equivalent in a lactose load of 2 grams per kilo up to 100 grams per kilo, blood glucose rises by greater than 20–25 milligrams per 100 milliliters in an individual who is a lactose absorber. Bayless believes that this index is more dependable than the detection of the milder gastrointestinal symptoms (such as bloating, cramps, flatulence, or loose stools), and more reliable than waiting for the development of diarrhea, which may not occur in all people who have lactose intolerance. Bayless also mentioned a more expedient method that does not entail laboratory tests: one can merely give an individual two glasses of cold milk—which produces more symptoms than warm milk—to drink on an empty stomach; and then observe any symptoms.

Hydrogen Breath Test

Hydrogen gas is produced by the fermentation of carbohydrate by bacteria within the lumen of the gastrointestinal tract. Hydrogen is exogenously formed, and is passed as flatus or it diffuses into the blood and is exhaled (Ostrander et al., 1983).

Rosado and Solomons (1983) studied the increase in the hydrogen ion concentration in human breath samples after the administration of lactose in the dietary range. To establish standards for lactose absorbers or malabsorbers, 360 milliliters of whole cow's milk was given to forty-one subjects. Breath samples were taken and tested at thirty-minute intervals for five hours. An increase of hydrogen concentration of greater than or equal to 20 microliters per liter above basal values at any of the ten intervals was diagnostic of malabsorption. Increases of greater than or equal to 18, or greater than or equal to 15 microliters per liter were 85 percent as specific in classifying the same person. A simplified procedure was devised for field studies and entailed the collection of four samples at zero, two, three, and four hours. The results were analyzed, using greater than or equal to 20 microliters per liter as the cut-off value. These findings showed 80 percent sensitivity and 100 percent specificity when they were compared with the eleven sample tests. Ostrander et al. (1983) believed that although the interpretation of the results of breath hydrogen analysis in neonates and young infants remains problematic, hydrogen breath analysis may be the best available method for estimating semiquantitatively the degree of lactose absorption.

GENETIC CONSIDERATIONS

In a genetic study of lactose digestion in Nigerian families, the ability to digest lactose was ascertained in several Nigerian ethnic groups (Ransome-Kuti et al., 1975). The data consisted mainly of family pedigrees. In families in which both parents had lactose malabsorption, all progeny had lactose malabsorption. If one parent—usually of Northern European origin, or of the Fulani group—could absorb lactose, the offspring consisted of some or all who could digest lactose. These investigators studied ten families in which one parent was a lactose absorber and the other parent was a lactose malabsorber. From these matings, there were eighteen lactose absorbers and eleven lactose malabsorbers. The authors concluded that the ability to digest lactose is transmitted as an autosomal dominant trait.

Lisker, Gonzalez, and Daltabuit (1975) studied lactase deficiency in sixty-one families with 177 children over six years of age. Their findings were, however, compatible with an autosomal recessive mode of inheritance of lactose intolerance.

Newcomer et al. (1977) investigated the pattern of inheritance of lactase deficiency in 104 Amerindian and 2 white subjects in nineteen families. In three families in which both parents had normal lactase activity, 40 percent of the offspring had deficient lactase activity; and in three families with one parent who had normal lactase activity and one parent with deficient lactase activity, 65 percent of the children had lactase deficiency. In seven families in which both parents were lactase deficient, 93 percent of the offspring had lactase deficiency. It was concluded that the distribution of lactase deficiency suggests an autosomal recessive mode of inheritance of lactase deficiency.

Metneki et al. (1984) studied the breath of 102 pairs of adult twins in Budapest, using the hydrogen breath test. In this sample, 52 twins were monozygous (MZ) and 50 pairs were dizygous (DZ) twins of identical sex. All MZ twins were concordant for lactase phenotypes. The DZ distribution corresponded with Hardy-Weinberg expectation—which was derived from the frequency of hypolactasia in MZ and DZ twins, and in the Budapest population.

Sahi (1978) considered the question of whether hypolactasia is manifest because the enzyme lactase is adaptive; that is, is lactase activity maintained because of the ingestion of milk, and does lactase diminish in the absence of milk? Sahi believed that hypolactasia was the basic defect; consequently, those who have low lactase do not consume milk, because they know that they will have symptoms. Thus, the cause of hypolactasia is not the result of the absence of induction of the enzyme by its substrate, lactose.

Johnson, Cole, and Ahern (1981) believed that a mutant dominant allele is the basis of lactase intolerance and that a natural "wild type" allele results in the cessation of lactase production, once the period of infancy is over. (In other words, normal lactose absorption is dominant; lactose intolerance is recessive.) This dominant gene increased in frequency through selection in groups with a long history of milk drinking. Johnson, Cole, and Ahern also suggested that if a dominant gene prevents malabsorption and increases fitness, this can explain why populations with a long history of cattle herding (and use of milk products) are low in malabsorption and intolerance, while other groups without such a history are high in both. Additionally, given the brief period that cattle have been domesticated (about 6,000 years), the best way to explain these differences is to postulate that the absence of malabsorption/tolerance is the result of a mutant dominant allele that has higher fitness than the normal (lactase deficient) recessive allele.

An argument against the induction hypothesis to explain the marked ethnic differences in adult lactose malabsorption was provided by Johnson et al. (1978). The prevalence of adult lactose malabsorption and the pattern of milk ingestion were studied among 109 Amerindian groups of the Great Basin of the United States and the Southwest. In this population, 100 reported themselves as full-blooded Amerindians; 3 had Mexican admixture; and 6 had European ancestors. Lactose malabsorption was found in 92 percent of the full-blooded Amerindians and in 50 percent who had European ancestors. Most of the Amerindians had consumed large quantities of milk during and after childhood, but were malabsorbers as adults.

Jackson and Latham (1979) reported on lactose malabsorption in Masai children of East Africa. (The Masai regularly drink large quantities of milk and experience no symptoms of lactose intolerance.) Yet, as measured by the lactose tolerance test, 62 percent of Masai were malabsorbers. Further, the milk used by the Masai is in two forms, fresh and fermented. The lactose content of the fermented product has not been determined, but a fairly high level of lactose (about 3 grams per deciliter) remains in matured fermented milk. The Masai children drank more than 750 milliliters of milk per day, some fresh and some soured. The proportion varied and could not be determined (Jackson and Latham, 1979). In short, one would not expect that a milk-drinking population would have a high frequency of lactose intolerace.

The authors stated that the finding of lactose malabsorption in a nomadic cattle-raising and milk-drinking group is interesting and contrary to the observations of some anthropologists and other investigators. It is also is inimical to the induction hypothesis of the persistence of lactase in milk drinkers.

On the other hand, the association of cattle herding and lactose intolerance, and speculation thereof, must consider the type of milk product. Segal et al. (1983) ascertained lactase deficiency in different black South African populations. Even though two of the groups were cattle herders and milk drinkers (Zulu and Xhosa), lactase deficiency was found in 78 percent of these populations. The authors suggested that the apparent anomaly occurs because the Zulu and the Xhosa consume a fermented buttermilk, which has a low lactase content, instead of fresh milk (but see comments on the Masai). They also suggest that South African blacks came from a West and South African zone of nonmilking and that they took up dairying recently. But, as we have indicated in Chapter 1, some historical evidence places some South African black groups in South Africa many centuries before the advent of the Europeans. And, as we have seen, the Masai also have a high frequency of lactase deficiency and are milk drinkers. Additionally, cattle raising is observed focally in many regions of West Africa today and could have been imported into South Africa long ago.

Johnson, Cole, and Ahern (1981) cautioned, however, that it is equally likely that there could be an environmental explanation for population differences. They point out that two African groups, one of which herds cattle and drinks milk (the Fulani) may differ from the

Ibo, who do not. About 20 percent of Fulani have lactose intolerance, but about 100 percent of Ibo are intolerant. Palestinian Arabs have a high rate of malabsorption, but Saudi Arab Bedouins have a low rate of malabsorption. Johnson and co-workers doubted that such neighboring groups form sufficient separate gene pools across hundreds of generations to produce differences of this magnitude. But this reasoning is questionable. The Fulani and Ibo are dissimilar populations, as any superficial observer would note.

Johnson, Cole, and Ahern (1981) then reviewed lactose absorption in persons of mixed ethnic backgrounds in order to test the dominant gene thesis. In an examination of families in which one parent is from a group with a low rate of malabsorption and the other is from a group with a high frequency of malabsorption—and when both parents manifest malabsorption—the children almost always have malabsorption. And when one parent manifests malabsorption, and the other does not, the offspring experience malabsorption more than 50 percent of the time. Other cross-ethnic studies were cited to examine a dominant allele hypothesis. If only 10 percent of Europeans manifest malabsorption, and if malabsorption is the result of a single autosomal recessive allele, about 47 percent of Europeans should be homozygous for the normal allele, 43 percent should be heterozygous, and 10 percent should be homozygous for the malabsorption recessive allele. Orientals and Asians show about 90 percent malabsorption: thus, 0.3 percent should be homozygous dominant, 9.7 percent heterozygous, and 90 percent homozygous recessive. If the Asian and European data are combined, Europeans who mate with Asians should have offspring who are homozygous dominant or heterozygous about 71 percent of the time; slightly under 30 percent of offspring would have malabsorption.

The findings of Johnson et al. (1980) and Johnson, Cole, and Ahern (1981) did not support the usual genetic explanation of lactose tolerance from population and family studies. Subjects from various ethnic groups in Hawaii and homeland Chinese, Japanese, and Koreans reported on milk drinking in childhood and adulthood and on symptomatology after the ingestion of milk. Lactose intolerance was infrequent in all populations. The investigators claimed that differences between Hawaiians of Asian ancestry and Asians born abroad in mean discomfort scores—the amount of milk needed to produce symptoms—and a high proportion of symptoms among Hapa-Haoles (individuals of European-Oriental ancestry) do not support a genetic explanation. Survey data from thirty-eight families— which consisted of both parents and one offspring—did not support the thesis that lactose intolerance results from autosomal dominant inheritance and did not substantiate the notion that Europeans are more frequently tolerant of lactose than are Orientals.

Johnson, Cole, and Ahern (1981) maintained that lactose intolerance and lactose malabsorption are different but correlated phenotypes. A high proportion of people who are malabsorbers are lactose tolerant, and a small proportion of absorbers are intolerant. These authors do not believe that absorption or tolerance is controlled by a single gene pair; instead, they assert that absorption is controlled by a single gene in families of Europeans, and that tolerance—but not absorption—is controlled by dietary factors. Further, if lactase deficiency is defined as rise in blood glucose of less than 20 milligrams per 100 milliliters after a loading dose of 50 milligrams of lactose, people who do not have such a rise are categorized as malabsorbers. But although symptoms are correlated with malabsorption, the association is inconclusive. McCraken (1971) found only a 0.59 correlation between a report of symptoms by history and test results, and a 0.54 correlation between clinical test results and symptoms after the clinical test. Evidently, the relationship between malabsorption, as ascertained by tests, and intolerance, as measured by distress after lactase ingestion, is quite imperfect.

Flatz and Rotthauwe (1977) recorded that among 349 blacks in the United States who were over age fifteen, 76 were lactose absorbers, and 273 were not. If U.S. blacks have about 30 percent of their genes from Europeans, and if 90 percent of Europeans absorb lactose, and if African ancestors average 10 percent absorbers of lactose, over 40 percent of blacks should be absorbers, but only 22 percent of blacks were observed to be lactose absorbers. It does appear, however, that there are far too many variables that extend over generations: for example, the 30 percent admixture does not take into consideration the range; blacks in the United States vary in the extent of European admixture; and the tests are imprecise.

Johnson et al. (1981) presented other evidence that opposed the thesis that malabsorption is inherited as an autosomal recessive, in other words, they asserted that lactose absorption is dominant. These authors stated that the single gene theory apparently assumes that a person is either lactose tolerant, or lactose intolerant. (We believe that such a statement presumes that there is no variability in the manifestation of gene action, which is fallacious.) Nevertheless, Johnson and co-workers reiterate that malabsorption and intolerance are far from perfectly correlated. But this could be the fault of the tests, and not the mechanism of inheritance.

LACTOSE MALABSORPTION AND LACTOSE INTOLERANCE

But even so, we agree with Johnson's thesis that social policy should be guided by data on lactose intolerance, not malabsorption. It is a misconception that intolerance is an inevitable association of malabsorption. People who test as intolerant frequently have no symptoms caused by ingestion of lactose. Tolerance is also related to dose. More than one-half of people who have intolerance develop symptoms only after drinking 0.5–1.0 liter of milk at one sitting. Furthermore, people who are unaccustomed to lactose, and who initially have symptoms after drinking, eventually adapt so that they can drink milk without problems (Johnson et al., 1981). We now examine some of the evidence for these assertions.

Paige et al. (1975a) found significant improvement in absorption of 90 percent hydrolyzed milk in people with low lactase activity. These workers suggested that lactose-hydrolyzed milk may serve as an alternative to food planners who want to provide milk to populations who have a high risk of lactose intolerance. On the other hand, since it has been shown that even individuals who are lactose intolerant can drink moderate quantities of milk, Paige et al. (1975a) studied the relationship of milk consumption to blood glucose in lactose-intolerant persons. Of eighty-nine black elementary school children who were tested, 54 percent had a flat lactose tolerance curve. The maximal blood glucose rise was 12.3 milligrams per deciliter in the lactose malabsorbers who were defined as milk drinkers. It was concluded that some children who are malabsorbers of lactose have sufficient levels of lactase to hydrolyze moderate amounts of milk.

Later, Paige et al. (1977) further demonstrated that lactose malabsorbers could tolerate milk. A lactose tolerance test was performed on 116 black children in the United States between thirteen and fifty-nine months of age. Of the 116 children, 34 (28 percent) had lactose malabsorption, and diarrhea and cramps were noted singly or in combination in 18 percent of this group. Interestingly, 12 percent of the lactose absorbers showed serious signs of intolerance. Eighty-seven percent of those who had malabsorption, and 92 percent who were absorbers, drank 240 milliters or more of milk per day.

Garza and Scrimshaw (1976) studied the frequency of lactose intolerance by the standard lactose tolerance test compared with the prevalence of intolerance to graded amounts of milk in sixty-nine black and thirty white children. Of the black children, no child was intolerant to 240 milliliters (one glass) of milk, but 11 percent of 4-to-5-year-old children, 50 percent of 6-to-7-year-old children, and 72 percent of 8-to-9-year-old children were lactose intolerant. It was also stated that symptom response to greater amounts of milk did not indicate that the frequency of lactose intolerance might be a reason for limiting milk programs for children.

In the same investigation, no significant differences were observed between the milk intake of lactose-tolerant and lactose-intolerant black children 6 to 10 years of age. But 8-to-9-year-old black children did drink significantly less milk than did 8-to-9-year-old white children. And most important, the blood glucose response at zero, twenty, and forty-five minutes after the ingestion of 2 galactose per kilogram of body weight (maximum, 50 milligrams) was unreliable as an indicator of symptomatic response to lactose. (Recall that the Masai are able to drink large quantities of milk without symptoms, even though a large proportion of them are lactose malabsorbers.)

Kwon, Rorick, and Scrimshaw (1980) used a double-blind test procedure to evaluate the relationship between lactose malabsorption and lactose intolerance in eighty-seven healthy teenagers of different ethnic origins, 14 to 19 years of age. Capillary blood glucose analysis after an oral dose of 50 grams of lactose identified forty-five of the youths as lactose malabsorbers. A lactose-free (LF) chocolate dairy drink and one that contained 4.5 percent lactose (LC) were given on consecutive mornings. No significant differences were found in the prevalence of symptoms in malabsorbers and absorbers after drinking 240 milliliters of the LF or LC preparation. But twelve lactose absorbers reported symptoms that were not the result of lactose ingestion after 240 milliliters of LF, both LF and LC, or after 240 milliliters—but not 480 milliliters—of LC. Apparently, none of the malabsorbers experienced symptoms from the lactose in 240 milliliters of LC, but 16 percent of malabsorbers reacted to lactose in 480 milliliters of LC.

The Committee on Nutrition, American Academy of Pediatrics (1978), emphasized that intolerance to the consumption of milk is rarely seen in preadolescents. It was suggested that the symptoms observed in lactose-intolerant subjects may even be unrelated to lactose or may be mild enough to be of little practical import.

Finally the absorption of nutrients in lactase deficiency was investigated by Dobongnie et al. (1979). It was concluded that the nutritional consequences of malabsorption are probably minimal.

ADVANTAGE OF PERSISTENCE OF LACTASE IN ADULTS

There has been considerable speculation about the selective advantage of the persistence of lactase (Flatz and Rotthauwe, 1977). As we have seen, there is a low absorption of vitamin D in Europeans, because of diminished ultraviolet light in the winter (Chapter 2). Lactose enhances the absorption of calcium, which prevents rickets and osteomalacia. Women with lactose tolerance would have fewer pelvic deformities, and thus produce more children than would women with hypolactasia.

POPULATION DISTRIBUTION

Simoons (1981) stated that all populations that originate in zones of nonmilking in Africa, Southeast and East Asia, New Guinea, and the Americas have been found to have a high frequency of lactose malabsorption, and all groups who have a long tradition of consuming milk in its lactose-rich form in large quantities have a low prevalence of lactose malabsorption. The lowest frequencies of hypolactasia are in the Scandinavian region with concentric genetic clines of lactose malabsorption extending to the south and the east (Socha et al., 1984).

The Nilotic peoples are an exception to the finding that peoples who drink large quantities of milk have a low prevalence of lactose intolerance. There is a question, however, as to whether the differences in the frequency of lactose malabsorption could have evolved in the historical period since humans first started practicing dairying and using milk. But Simoons (1981) quoted earlier studies of the 1970s to the effect that frequencies of malabsorption from 1 to 3 percent are what would be required. Simoons also states that it is unlikely (and we agree) that the high prevalence of lactose absorption originated in one region and then became disseminated. It is more likely that there was a single selective advantage that prevailed everywhere. (Notably, in the early days of speculation about sickle hemoglobin distribution, attempts were made to find an original source of the mutant, but it is now agreed that this gene and most others, for that matter, had multiple foci of origin.)

The distribution of lactose intolerance in various populations is shown in Table 9.1. Age is an important variable in population studies of lactose malabsorption. Sahi, Launiala, and Lartinen (1983) studied lactose malabsorption in a fixed cohort of Finnish subjects, ages seventeen to twenty-five. This was the third investigation of this group of subjects. Ten years previously the prevalence of hypolactasia had been 6.2 percent; five years earlier, the frequency was 9.3 percent; and in the current study the prevalence was 10.3 percent. Since the incidence of hypolactasia increases with age, most of the studies on very young

Table 9.1
Lactose Malabsorption in Various Populations

Region or Population	Frequency (%)	Source
North Africa		
Egypt	80	Gabr et al. (1977)
Egypt	73	Hussein et al. (1982)
Egypt	80	Simoons (1981)
Mahgreb	78	" "
West Africa		
Senegal	29	Arnold et al. (1980)
Ghana (Children)	73	Simoons (1981)
Nigeria		
Fulani	20	Johnson et al. (1981)
Fuloni	22	Simoons (1981)
Ibo	100	Johnson et al. (1981)
Yoruba	98	Simoons (1981)
South	75–100	" "
Central Africa		
Cameroon and Zaire	98	" "
Zaire (Shi)	96	" "
Zaire (Batusi)	0	" "
Zambia	90	" "
East Africa		
Ethiopia	90	" "
Kenya: Masai	62	Jackson and Latham (1979)
Batusi	16	Bayless (1976)
Batusi	12	Simoons (1981)
Batusi (Rwanda)	8	Arnold et al. (1980)
Uganda	90	Bayless (1976)
Rwanda Twi Pygmies	77	Simoons (1981)
Tanzania (children)	92	" "
Kenya (children)	73	" "
South Africa		
Zulu	89	O'Keefe and Adam (1983)
Zulu, Xhosa	78	Segal et al. (1983)
Durban blacks	93	O'Keefe and Adam (1983)
Rehoboth Basters	65	Simoons (1981)
Middle East		
Lebanon	78	Nasrallah (1979)
Lebanon	78	Simoons (1981)
Jordan	64	Snook et al. (1976)
Jordan	78	Hijazi et al. (1983)
Bedouin	24	" " "

Table 9.1
(*Continued*)

Region or Population	Frequency (%)	Source
Middle East		
Syria	99	Simoons (1981)
Saudi Arabia		
Bedouin	14	" "
Urban	13	" "
Israel: Jews	61	Bayless (1976)
Iran	68	Sadre and Karbasi (1979)
Yemen		
Jews	44	Simoons (1981)
Arabs	25	" "
Asia		
Afghanistan		
Tajiks	82	" "
Pashtuns	79	" "
Pashi-i	87	" "
Uzbeck	100	" "
Hazara	80	" "
Urban	76	" "
Pakistan	60	Ahmad and Flatz (1984)
Pakistan	55	Abbas and Ahmad (1983)
India		
North (Aryan)	27	Tandon et al. (1981)
Madhya Pradesh, Eastern Uttar Pradesh, Bihar, Orissa	35	Simoons (1981)
Bombay	58	" "
Hyderabad	61	" "
South	100	" "
South	67	Tandon et al. (1981)
From Trinidad	67	Bartholomew and Pong (1976)
Bangladesh	80	Brown et al. (1979)
Sri Lanka		
Ceylonese	73	Senewiratne et al. (1977)
Ceylonese	78	" " "
Ceylonese	66–79	" " "
Sinhalese	73	Simoons (1981)
Tamils	71	" "
Thailand	97	Anh et al. (1977)
Vietnam	100	" " "
China		
Han (North)	92	Wang et al. (1984)
Inner Mongolia	88	" " "
Kazakhs	76	" " "
Overseas	87	Simoons (1981)

(*continued*)

Table 9.1

(*Continued*)

Region or Population	Frequency (%)	Source
Asia		
Japan	90	Nose et al. (1979)
Pacific		
Australia: Aborigines	84	Brand et al. (1983)
New Guinea	100	Cook (1979)
Northern Europe		
Sweden	1	Simoons (1981)
Sweden	8	" "
Denmark	2	Bayless (1976)
Finland	19	" "
Finland	17	Simoons (1981)
Finland	10	Sahi (1978)
Laplanders	25–60	Simoons (1981)
Britain	5	Ferguson et al. (1984)
Switzerland	8	" " "
Germany (overall)	15	Flatz et al. (1982)
Northwest	6–9	" " "
West and South	13–14	" " "
Southwest	23	" " "
East	22	" " "
Austria		
West	15	Rosenkrantz et al. (1982)
East	25	" " "
Eastern Europe		
Hungary	37	Czeizei et al. (1983)
Hungary (Upper)	40–56	" " "
Poland	38	Socha et al. (1984)
U.S.S.R.		
Estonia	28	Simoons (1981)
Leningrad	15	" "
Southern Europe		
Italy	38	Bianchi-Parro et al. (1983)
North	51	Burgio et al. (1984)
South		
Naples	78	Rinaldi et al. (1984)
Sicily	71	Burgio et al. (1984)
Greece	75	Ladas et al. (1982)
Cyprus: Greeks	88–100	Bayless (1976)
North America		
Eskimos	88	" "
Canada: Indians	60	Essestad-Saydad and Haworth (1977)

Table 9.1
(*Continued*)

Region or Population	Frequency (%)	Source
North America		
United States		
Indians	66	Bayless (1976)
	92	Johnson et al. (1978)
Blacks	70	Bayless (1976)
Whites	8	" "
Mexican-Americans	53	Woteki et al. (1977)
	47	Sowers and Winterfelt (1975)
Mexico	66	Lisker et al. (1978)
South America		
Peru	66	Bayless (1976)
Brazil		
Whites	45	Seva-Pereira and Beiguelman (1982)
Blacks	85	" " "
Orientals	100	" " "

children have been omitted in the table, but when figures on children were all that was available, they are given and are so marked.

CATARACTS AND LACTOSE ABSORPTION

Rinaldi et al. (1984) found that 22 percent of a group of adult Neopolitans had persistent high lactase activity and that 16 percent were absorbers as indicated by measurement of breath hydrogen concentration and a rise in blood glucose after ingestion of oral lactose. In adults in the same region of Italy with idiopathic or presenile cataract, 49 percent were lactose absorbers as shown by the breath hydrogen test and 55 percent as shown by a rise in blood glucose. These important investigations indicate that adults who are able to absorb galactose from a lactose-containing diet are especially susceptible to senile or to presenile cataracts. Because cataracts are more common in regions of high ultraviolet light intensity, and because populations in these regions usually have a high frequency of hypolactasia, lactose intolerance in these regions could be advantageous.

CONGENITAL LACTASE DEFICIENCY

Savilahti, Launiala, and Kuitunen (1983) stated that congenital lactase deficiency is one of twenty rare autosomal recessive disorders that are excessively common in Finland. The authors described sixteen infants with this disorder. The initial sign was watery diarrhea after the first feeding of breast milk, or, at the latest, by the age of ten days. Lactose malabsorption was diagnosed at a mean age of thirty-six days, at which time the infants were dehydrated, and fifteen of them weighed less than they did at birth. While the infants were drinking breast milk, their feces contained 20–80 grams per liter of lactose. Peroral lactose tolerance tests showed the greatest rise in blood glucose to be 0.8 millimoles per liter. The jejunum showed slight to moderate villous atrophy in four early specimens, but in later samples the villous height was normal. The mean height of the epithelial cells of the jejunum was decreased. The lactase activities in jejunal biopsy specimens were lower than they are in most patients with acquired lactase deficiency, but with some overlap; the sucrase and maltase activities were within normal limits. The children achieved normal growth after they were given a lactose-free diet.

Freiburghaus et al. (1976) studied the protein patterns of brush border fragments in congenital lactase deficiency and in adult lactase deficiency. Brush border membrane proteins of the mucosa of the small bowel were separated on polyacrylamide gels from intestinal biopsy samples from four children with congenital lactase deficiency and from two adults with lactase deficiency. In three subjects with the congenital form, the protein band that corresponded to brush border lactase was reduced. No differences in gel bands were found when the patterns were compared to those of the two adults with lactase deficiency. The lactase pattern was not detected in one subject with congenital lactase deficiency. The authors concluded that the mechanism that leads to low lactase activity in the congenital and the adult forms of lactase deficiency is similar. We suggest that this conclusion could be premature: there are numerous examples of enzymes with similar protein patterns.

ACQUIRED LACTASE DEFICIENCY

There are many causes of acquired lactase deficiency. They include premature birth, celiac disease, acute gastroenteritis, kwashiorkor, treatment with neomycin, para-aminosalicylic acid, *Giardia lamblia*, small bowel resection, and gastrectomy (Bayless, 1976).

Another form of lactase deficiency occurred in jaundiced infants who were given phototherapy (Bakken, 1967). The infants developed diarrhea, but when they were given a lactose-free diet, their stools became normal. The effect was reversed when breast milk was given while the babies were still jaundiced, but being treated with light. The author indicated that increased unconjugated bilirubin in the intestines of jaundiced infants during light treatment inhibited intestinal border lactase. Lactase was reactivated when the infants' jaundice faded.

In contrast, Ebbesen, Edelsten, and Hertel (1980a) studied sixty newborns of normal birth weight who had uncomplicated hyper-bilirubinemia. They were fed human milk from which lactose was removed and to which either sucrose or lactose was added. Thirty infants had ordinary phototherapy and thirty infants had intensive phototherapy. Fifteen in each group were given sucrose milk and fifteen were given lactose milk. There were no significant differences in infants who had sucrose milk and in those who had lactose milk. There was also no difference between those who were treated with ordinary nonintensive phototherapy. The authors concluded that lactose malabsorption is not the usual cause of the reduced gut transit time during phototherapy. In another study in which infants were given mother's milk, the findings were similar (Ebbesen, Edelsten, and Hertel, 1980b).

10

Twins and Other Multiple Births

The incidence of dizygotic (DZ) twins and of triplets is highest in African peoples. For example, Nylander (1971a) showed that the twinning incidence in Western Nigeria is 45–46 per 1,000 maternities, which is four times that in Europeans. The dizygotic rate among Japanese is about one-third that usually found in Europeans (Bulmer, 1960). The monozygotic (MZ) twinning rate, however, is constant among most populations.

Twinning and multiple births are significant factors in perinatal and infant mortality, probably because the mean birth weight of multiple births is lower than that of singleton births. There is also a higher frequency of congenital malformation in twins when compared to single births (Layde et al., 1980; Bonney et al., 1978). Thus it might be useful to identify factors that contribute to multiple births in the hope that ways may be found to decrease their incidence in populations that are already handicapped by poor nutrition and inadequate health care.

IDENTIFICATION OF ZYGOSITY AND TWINNING RATES

The zygosity of a large population of twins may be determined by the Weinberg difference method (Myrianthopoulos, 1970). This technique assumes that the sex ratio is 1 : 1 (which it is not, at birth) and that the sex of dizygotic twins occurs by chance. Accordingly, the number of all DZ twins is twice that of unlike sexed twins. The number of MZ twins is calculated by subtracting the number of DZ twins from all twins. But this method only provides an approximation of DZ and MZ frequencies.

Other more direct methods may be used. Zygosity can be estimated by a comparison of blood groups, red cell enzymes, HLA, serum protein factors, finger and palm prints, and restriction length polymorphisms, and other markers. Although analysis of the placenta may provide a clue to zygosity, this method is imprecise. Myrianthopoulos (1970) found that in 159 MZ pairs of twins for whom placental information was available, 30 percent were diamnionic-dichorionic. In these samples, zygosity estimates would have been impossible without genetic markers. On the other hand, zygosity of over one-half of the twin pairs could have been ascertained on the basis of sex and placental type. A monochorionic placenta is accepted as evidence for monozygosity (Nylander and Corney, 1977). These authors reviewed the literature and discovered that in both white and black populations, twins with this type of placentation were alike in sex and in all genetic markers that were tested.

In some studies the twinning rates are corrected for maternal age and in other investigations the rates are not adjusted for this factor. This is important because the age of the mother influences the rate of twinning (Myrianthopoulos, 1970). In a study of whites and blacks in the United States, the curve for all twins showed a sharp increase with age up to age thirty to thirty-four, then the curve declines, and then shows another rise after the age thirty-five to thirty-nine. Twinning rates were examined by zygosity. The MZ twinning rate was stable until age thirty-nine and then sharply increased. The MZ twinning rate was slightly higher in blacks than in whites except in the older age groups. The DZ twinning increased with age up to age thirty to thirty-four in blacks and age thirty-five to thirty-nine in whites, and then declined precipitously.

OTHER CAUSES OF TWINNING

Bulmer (1960) surveyed the literature and discovered that there were conflicting findings about the inheritance of twinning. For example, Bulmer reported that a study undertaken in 1909 found that dizygotic twinning was inherited, that monozygotic twinning was not, and that the hereditary influence was limited to the mother's side. On the other hand, a 1934 study (Bulmer, 1960) concluded that there was essentially no difference between the hereditary influence on the mother's and the father's side. Bulmer studied 838 twin pairs in Lancashire, Great Britain. Of the mothers, 33 (4.0 percent) were themselves twins, compared with 14 (1.7 percent) of the fathers. Since the frequency of twin births is about 2.5 percent, but the frequency in adults must be less because of the higher mortality of twins, Bulmer concluded that more of the mothers than expected are themselves twins, but this was not so among the fathers. On the other hand, Nylander (1971a) observed that in a population in Ibadan, Nigeria, higher rates of DZ twinning were seen in women who had previously given birth to twins, and women whose husbands have had twins, but not in women who were themselves twins, or whose husbands have had twins by other wives.

PARITY AND MATERNAL AGE

Selvin (1971) investigated the number of pregnancies before the birth of twins during 1957–59. The sample was 4,407 white twin pairs from 358,617 maternities and constituted one twin pair for 80.4 single births. Mothers of dizygotic twins had more pregnancies before the birth of twins than did mothers of monozygotic twins. Nylander (1971) discovered that a twinning rate of 85 per 1,000 live births was associated with four or more births in Ibadan but a rate of only 20 per 1,000 live births was seen in Aberdeen, Scotland, with the same parity. Bonney et al. (1978) also found a parity effect in their study of 13,000 births in Ghana.

Maternal age has been previously mentioned as a factor to be considered in the comparison of twinning rates in populations. Nylander (1971a) compared DZ twinning in Nigeria with that of Aberdeen, Scotland. There was a striking maternal age effect with a peak frequency of 90 per 1,000 births in Ibadan at age thirty to thirty-four compared to a peak frequency of 20 per 1,000 births in the Aberdeen population at age thirty-five to thirty-nine.

DEMOGRAPHIC AND PHYSICAL VARIABLES

In the comparison of twinning in Nigeria and Scotland, Nylander (1971) also found the following variables:

1. The rate of DZ twinning was associated with increased maternal height in Ibadan, but not in Aberdeen.
2. The rate of DZ twinning in Ibadan was inversely related to social class: the lower the social class, the higher the rate of twinning. This association was not observed in Aberdeen.
3. In a comparison of the levels of gonadotropin follicle-stimulating hormone (FSH) in twin-prone and nontwin-prone mothers, mothers of twins had significantly higher FSH levels than mothers of single births in Ibadan. Such differences in Aberdeen were small and inconsistent.

Nylander believed that these findings in Ibadan—especially the observation that parents who have twin siblings are not more likely to have twins—support a thesis that environmental factors may be important in African populations in contrast to non-African populations. He postulated that this environmental factor may be dietary. The more European the diet of the Nigerian (the higher the social class), the lower the DZ twinning rate. Further, since some of the yams in Nigeria have been found to have an estrogen-like substance, there might be other chemicals that stimulate FSH levels, which, in turn, promote multiple ovulation, which leads to DZ twinning.

Myrianthopoulos (1970) studied the relationship of socioeconomic background to the incidence of twinning in the United States. These findings differed from what had been observed in Nigeria. A socioeconomic index utilized scores for education, occupation, and family income on a scale from zero (low) to 10 (high). Although there are obvious difficulties in analyzing socioeconomic indices among the diverse racial groups in the United States, these problems do not appear to apply to this study. In both whites and blacks, there was an increase of twins in the 6.0–9.9 socioeconomic range. Thus, apparently, mothers in the higher socioeconomic categories have more twins. The increase was highly significant for MZ and DZ twins, as well as for twins in which the zygosity was unknown. After adjustments were applied for maternal age, the socioeconomic differences were highly significant in the DZ group and borderline in the blacks. Myrianthopoulos believed that the residual differences in the maternal age and socioeconomic combinations were negligible in all but the black population. To resolve the apparent discrepancy, the black group was analyzed according to main effects and interactions. Both maternal age and socioeconomic effects and interactions between them were evident. It was concluded that the increase in DZ twinning frequency in the higher socioeconomic groups was independent of the age of the mother, and that the increase in the black twins was a product of interaction of maternal age and socioeconomic effects.

An interesting and puzzling phenomenon is an apparent decline in the rate of twinning in most populations that have been studied (James, 1975). The decline is found exclusively in the rate of DZ twinning. No satisfactory explanation has been offered for this decline.

POPULATION DIFFERENCES IN TWINNING

The marked increase of dizygotic twinning in West African populations has been mentioned. But considerable differences exist among

Table 10.1
Rates of Twinning in Africans and Their Descendants

Region	Monozygotic	Dizygotic
Western Nigeria	4.4	49.8
Midwestern Nigeria	4.6	28.7
Eastern Nigeria	5.1	36.6
Harare, Zimbabwe	2.3	26.6
Johannesburg, S. Africa	4.9	22.3
Nqutu, Zululand	0.5	21.1
Kinshasha, Zaire	3.1	18.7
Lumumbashi, Zaire	3.6	13.3
Jamaica	3.8	13.4
Antigua	3.9	11.5
United States	3.9	11.8

Source: Modified from Bulmer (1960) and Nylander (1978).
Note: Results are given as number per thousand births.

Table 10.2
Rates of Twinning in Europeans and Their Descendants

Country	Monozygotic	Dizygotic
Greece	2.9	10.9
Denmark	3.8	10.2
Czechoslovakia	3.4	9.8
England and Wales	3.6	8.9
East Germany	3.3	9.1
Italy	3.7	8.6
Norway	3.8	8.3
Holland	3.7	8.1
Switzerland	3.6	8.1
Sweden	3.2	8.6
West Germany	3.3	8.2
Luxembourg	3.5	7.9
Austria	3.4	7.5
Belgium	3.6	7.3
France	3.7	7.1
Portugal	3.6	6.5
United States	3.8	6.1
Spain	3.2	5.9

Source: Modified from Bulmer (1960) and Levitan and Montague (1977).
Note: Results are given as number per thousand births.

Africans and their descendants, as will be evident from an examination of Table 10.1. Sample sizes may have influenced some of this variation, but it does appear that Nigeria still has the highest frequencies of dizygotic twinning. The lowest frequencies were found in Gambia, where the frequency is 9.9 per 1,000 live births.

Twinning rates in Europe are listed in Table 10.2. The lowest frequencies were in Spain, where dizygotic twinning occurs at a rate of 5.9 per 1,000; the highest frequencies were in Greece, where the rate is 10.9 per 1,000. The figures from Greece approximate those found in blacks in the United States.

Levitan and Montagu (1977) recorded twinning rates from Australia, Israel, and Japan. The percentages for rates in Australia and Israel were similar (Australia, MZ = 3.8; DZ = 7.7; Israel, MZ = 3.8; DZ = 7.3). Two studies in Japan reported the following percentages: MZ = 3.8; DZ = 2.7 (for three cities); and MZ = 4.1; DZ = 2.3 (for the entire country). Thus Japan not only has the lowest twinning rates of all countries that have been surveyed, but this is the only country in which the MZ twinning rate exceeds the DZ rate.

TRIPLETS AND OTHER MULTIPLE BIRTHS

Nylander (1971b) maintained that the expected incidence of triplet and quadruplet births may be calculated in a population in which the rate of twinning is known by Hellin's hypothesis. The hypothesis stated that if the frequency of twinning is n, the frequency of triplets is n2, and the frequency of quadruplets is n3. Nylander pointed out

Table 10.3
Rates of Triplets in Nigeria, Europe, and the United States

Country	Triplets
Western Nigeria	
Ilesha	1.94
Ibadan	1.78
Igbo-Ora	1.62
United States	
Black	0.14
White	0.09
United Kingdom	0.10
Sweden	0.13

Source: Modified from Nylander (1971).

Note: Results are given as number per thousand births.

that some authors have found discrepancies between Hellin's law and observed frequencies. Table 10.3 compares triplet rates per 1,000 maternities in Western Nigeria, Northern Europe, and the United States.

In Western Nigeria the incidence of triplets and higher multiple births was found to be about 1.6 per 1,000 maternities. The observed incidence was lower than that calculated by Hellin's law. The triplet rates were sixteen times the similar rates in Great Britain and the United States. In Table 10.3, it should be noted that the incidence of triplets in blacks was twice that of in whites in the United States. On the other hand, the prevalence of twins (no breakdown into zygosity) and of triplets was similar in blacks in the United States and in Sweden. Apparently, hereditary, environmental factors, and maternal age interact and affect the incidence of multiple births.

Hollingsworth (1965) examined birth weights and survival of African babies (single births). The data of 4,933 Ghanaian newborns were analyzed for weight, gestation time, parity, maternal age, and infant survival. Babies who survived had significantly higher birth weights, longer gestation times, lower birth order, and younger mothers than did nonsurvivors.

The birth weight and survival of Ghanaian twins was studied by Hollingsworth and Duncan (1966). These authors analyzed 286 Ghanaian twin births for birth weight, maternal age, parity, and survival, and then compared the data with European figures. A significant but unexplained shortage of female-female twins was found in the Ghanaian population (sex ratio, 1 : 22) but not in an Italian study and an English study, where there was no shortage of female-female twins. A very high correlation (r = 0.8) between parity and mother's age in the Ghanaian study was attributed to an association between twinning and these variables. It was postulated that the high parity of the Ghanaian women could account for their high twin rate. The mortality of Ghanaian twins was 15 percent, which was similar to the rate for Italian twins and lower than that for some English twins (21 percent). The authors explained this discrepancy by postulating that the mortality differences could be due to the lower proportion of low-birth-weight Ghanaian twins, and that this may occur because twinning is a more natural phenomenon in Ghanaian women than it is in European women.

11

Congenital Malformations

A congenital malformation is a morphologic defect of an organ or a larger region of the body resulting from an intrinsically abnormal developmental process (Spranger et al., 1982). Another definition (Myrianthopoulos and Chung, 1974) includes the element of time: a congenital malformation is a gross physical or anatomic developmental anomaly that was present at birth or that was detected during the first year of life. The element of time is important because most studies have shown that a substantial number of defects (about a third or more) are discovered after the neonatal period during the first year of life (Myrianthopoulos and Chung, 1974). When a pattern of anomalies appears which are known or thought to be pathologically related and are not known to represent a single sequence or a pattern of anomalies derived from the disturbance of a single developmental field, the pattern is referred to as a *syndrome*. This term often implies a single cause, as for example, in "Down syndrome."

The definitions of major and minor congenital malformations vary considerably from one author to another. No single definition was found to be satisfactory for use in the National Collaborative Perinatal Project (NCPP, described later) (Myrianthopoulos and Chung, 1974). Malformations may be considered major when they are life-threatening, require surgery, cause significant handicap, or have an adverse effect on the function or social acceptability of an individual (Marden et al., 1964; Nelson and Forfar, 1969). Minor malformations are of neither medical nor cosmetic consequence and are unlikely to prove a serious hindrance to normal life or to the achievement of a normal life expectancy (Marden et al., 1964; Nelson and Forfar, 1969).

FREQUENCY AND DISTRIBUTION

Congenital malformations are the most common type of genetically determined abnormality occurring in any group of people, and African people are no exception. A comprehensive study of congenital malformations covering fifteen years and involving twelve institutions (NCPP) showed that African-Americans have a frequency of some congenital malformations significantly different from that of whites (Myrianthopoulos and Chung, 1974). In general, the incidence of minor malformations in blacks is higher than that for whites (8.4 percent versus 6.2 percent), primarily because there is a higher frequency of three particular malformations (polydactyly, branchial cleft anomalies, and supernumerary nipples) in U.S. blacks. Table 11.1 summarizes major and minor malformations found more or less commonly in African-Americans when compared with U.S. whites. There is no difference between the races in overall frequency of major malformations (7.2 percent); however, the frequency of multiple malformations in infants versus all those with single malformations is lower in U.S. blacks (15.4 percent) than in U.S. whites (17.7 percent).

These congenital malformations may be the result of single genes of major effect. For example, polydactyly is nine times as frequent in African-Americans as in U.S. whites. A pattern of occurrence in many families is consistent with autosomal dominant inheritance. Most of these anomalies, however, are distributed in families in a pattern consistent with the action of multiple genes interacting with some aspects of the environment. This is probably the case with hip joint malformation, which is five times more frequent in U.S. white populations than in African-Americans. The differences in the frequencies of these conditions in different ethnic groups is termed the *liability* for expression, which Falconer (1967) defined as the sum total of the genetic and environmental factors that influence the likelihood of a person's manifesting a disorder or genetically determined trait. The phenotype of the condition in question will be expressed when the liability exceeds some threshold beyond which the characteristics of the condition are detectable. For the vast majority of conditions, it is impossible to determine the specific risk genes and environmental factors that would enable us to define the threshold for a particular condition. The degree of genetic-environmental interaction varies along a broad range of expression depending on the degree to which genes are primarily responsible for the expression of the trait or condition. It may be difficult to distinguish whether some congenital malformations have a stronger genetic cause or are primarily produced by a teratogenic factor or factors. But environmental factors can be an extremely important contribution to the expression of a given trait even when the genetic contribution or the "heritability" estimate is high.

Heritability is defined as an estimate of the proportion of total vari-

Table 11.1
Comparison of Frequencies of Malformations
in African-Americans and Whites

Malformation	Rate/10,000 Live Births	
	Whites	Blacks
Major		
Anencephaly	9.9	2.4
Meningomyelocele	6.6	6.8
Macrocephaly	10.8	5.2
Abnormal separation of sutures	9.5	24.7
Adduction of contracture of hip	33.5	9.2
Congenital hip dislocation	39.7	7.6
Metatarsus adductus	151.5	255.5
Scoliosis, lordosis, kyphosis	4.6	10.7
Cleft lip	14.5	7.2
Micrognathia	12.8	3.2
Pectus excavatum	36.8	6.4
Hypoplasia of lung	7.0	19.9
Pyloric stenosis	32.3	8.4
Undescended testes, bilateral	32.8	14.7
Urethral meatal stenosis	7.5	19.5
Cavernous hemangioma	101.4	4.5
Minor		
Pilonidal sinus	25.3	15.1
Polydactyly	15.7	138.9
Syndactyly	41.8	13.5
Abnormal fingers or toes	16.1	5.2
Nasolacrimal duct stenosis	17.0	9.6
Low-set ears	13.2	25.9
Branchial cleft anomaly	27.3	163.1
Cleft uvula	14.1	6.0
Cleft gum	9.5	43.8
Delayed teeth eruption	12.8	49.8
Strawberry/port-wine hemangioma	355.2	99.9
Hairy pigmented nevus	32.3	58.5
Supernumerary nipples	9.1	113.8
Café-au-lait spots	34.0	114.2

Source: Modified from Myrianthopoulos and Chung (1974).

ance in the liability for a trait due to the additive genetic variance. It should also be noted that a heritability estimate is valid only in the population in which it is determined. This means that the heritability of a trait may be high in one population and low in another. In evaluating the significance of differences in the frequency of a particular malformation or malformation syndrome, it is crucial to consider the differences in the genetic backgrounds and environments in which the conditions are being expressed.

IDENTIFICATION OF MALFORMATIONS

Most anomalies are identified by physical examination at birth or during the first year of life. In most of the major studies of congenital anomalies, the infants were observed for a year. Christianson et al. (1981) undertook follow-up studies in population of white and black children from six days of age to five years of age. Although the incidence of congenital anomalies was higher at birth among black infants than white infants, there was no difference between the groups in the cumulative incidence to five years. The incidence increased 3.5-fold for blacks and about fivefold for whites between six days of age and five years of age. The incidence of major malformations was 15 percent at five years. Severe anomalies were more frequent among children who weighed less than 2,500 grams at birth. In view of the fact that blacks have a significantly higher frequency of prematurity and of low-birth-weight babies than whites, one might also have expected the frequency of malformations in blacks to be significantly higher, because the incidence of malformations appears to be higher in this group of infants.

MUSCULOSKELETAL MALFORMATIONS

Musculoskeletal malformations are the most frequent major congenital malformation, making up approximately 37 percent of the major malformations and nearly 15 percent of the minor malformations (Myrianthopoulos and Chung, 1974).

Polydactyly

Polydactyly (extra fingers or toes) is a minor malformation. It is the most frequent single malformation, ranging in frequency from 8.8 per 1,000 births in South African blacks (Pompe van Meerder-voort, 1976) to 22.8 per 1,000 births in a Nigerian population (Scott-Emuakpor and Madueke, 1976). The frequency in the U.S. black population is reported to be 13.9 per 1,000 births (Myrianthopoulos and Chung, 1974). This is nine times more frequent than the rate for U.S. whites. All of the persons with polydactyly found in the

Nigerian study had postaxial malformations and most were type B. Postaxial polydactyly is a malformation of the hand in which the extra digit is attached to the under side of the hand. There are two types, A and B. In type A the extra digit is well formed. In type B the extra digit is not well formed and is frequently in the form of a skin tag. Type B is also known as pedunculated post-minimi. The frequency was slightly lower in females (17.9 per 1,000) than in males (27.1 per 1,000) (Scott-Emuakpor and Madueke, 1976). An analysis of the pattern of occurrence in the families was consistent with an autosomal dominant mode of inheritance, with a penetrance of 64.9 percent. The expressivity of this trait appears to be highly variable. Among South African blacks, most polydactylies consisted of a postaxial skin tag.

Talipes Equinovarus

The frequency of talipes equinovarus in U.S. blacks was not significantly different from that in U.S. whites (3.1 per 1,000 versus 3.8 per 1,000); however, this deformity was three times more frequent in a South African black population (3.5 per 1,000) than in South African whites (Pompe van Meerdervoort, 1976). The frequency of this deformity is similar in both U.S. blacks and U.S. whites, approximately 4.0 per 1,000 and 3.0 per 1,000, respectively (Myrianthopoulos and Chung, 1974).

Hip Malformations

Hip malformations offer an example of a striking difference in the frequency of a disorder among populations. Congenital dislocation of the hip occurs much less often in blacks when compared to whites in all populations that have been studied. In the NCPP data, the frequency of this malformation in U.S. blacks (7.6 per 10,000) was only about 20 percent of the rate for U.S. whites (39.7 per 10,000) (Myrianthopoulos and Chung, 1974). Studies in the Bantu (Roper, 1976) and in other African neonates (Skirving and Scadden, 1979) indicate that intrinsic genetic factors are probably more influencial than extrinsic factors in producing the hip dysplasia. A study of obstetrical history, methods of carrying infants, and arthrography was used to evaluate the theory previously advanced that the much lower frequency of congenital and neonatal hip dislocation in the Bantu was because of the "piggyback" method of carrying used by the mothers. These studies do not support this theory. In another investigation, the hips of twenty full-term neonates were examined in detail by Skirving and

Scadden (1979). Measurements included the degree of anteversion of the femoral neck and the acetabulum and the diameter and depth of the acetabulum. These measurements showed that the acetabulum was deeper on the average and varied within a much narrower range in the African neonates than that reported for white neonates. The measurements of anteversion of the acetabulum and the femoral neck were similar to those given for whites, which provides indirect support for the theoretical role of acetabular dysplasia as the intrinsic factor in the etiology of congenital dislocation of the hip. Besides the lower frequency of congenital dislocation of the hip in blacks, congenital adduction or contracture of the hip in U.S. blacks occurs at lower frequency (9.2 per 10,000 births) when compared to U.S. whites (33.5 per 10,000) (Myrianthopoulos and Chung, 1974).

CENTRAL NERVOUS SYSTEM MALFORMATIONS

The major group of malformations affecting the central nervous system in which there appears to be a difference between whites and blacks, at least in some populations, comprises the neural tube defects, anencephaly and spina bifida. Cornell et al. (1983), in studies in South Africa, reported an incidence for such defects of 1 per 300 births in whites, 1 per 1,250 births in coloureds, and 1 per 2,000 births in blacks. Grace (1981) noted a similar distribution of frequencies in these populations. A major analysis of birth defects in the United States revealed no association of nervous system malformations with birth order, maternal age, socioeconomic status, or sex (Spranger et al., 1982). Greenberg et al. (1983) developed a computer-generated mapping procedure to estimate geographic and race-specific birth prevalence rates of open spina bifida based on birth certificate data adjusted for underascertainment. Separate maps were produced for white and black births. In both groups there appeared to be a declining rate of spina bifida from east to west. The highest rates are 8 per 10,000 births for whites in southern Appalachia, and the lowest rates are less than 1 per 10,000 births for blacks in the Rocky Mountain states and the Pacific Northwest.

The overall incidences of neural tube defects (anencephaly plus spina bifida), as reported in the NCPP data (Myrianthopoulos and Chung, 1975), were 15.4 per 10,000 births for whites, 8.5 per 10,000 births for blacks and 22.4 per 10,000 births for a group called "other." The offspring of white mothers had a significantly higher incidence of anencephaly (9.4 per 10,000 births) than those of black mothers (2.0 per 10,000 births). However, there was no difference in the incidence of spina bifida between these two groups (7.3 and 6.9 per 1,000 births respectively). It should be noted that the offspring of the group called "other" (a group consisting mostly of Hispanic individuals) had a significantly higher frequency of spina bifida than either blacks or whites, 14.9 per 10,000 births. Strassburg et al. (1983) ascertained 536 cases of anencephalus from 1973 to 1977 in Los Angeles County, California, and compared them with a 2 percent random sample of all live births in the county. The children of women with Spanish surnames had an elevated risk for anencephalus and to a lesser extent for spina bifida. Blacks were at lowest risk, especially for spina bifida. There was no association between socioeconomic status and either of the defects. Advanced maternal age was a stronger risk factor for spina bifida than for anencephalus, and there was no increased risk for teenage mothers. There was also no paternal age effect. This difference in the relative incidences of anencephaly and spina bifida among black, white, and Hispanic groups suggests that these types of central nervous system malformations may have different etiologies or that there is a significant difference in the susceptibility to each of these defects. Feldman et al. (1982), in a retrospective study of 174,000 consecutive births at six Brooklyn hospitals over eight years, discovered no significant difference between blacks and whites in the combined prevalence of anencephaly, myelomeningocele, and occipital encephalocele. In this study the prevalence of myelomeningocele appeared to be declining in all ethnic groups.

There seems to be no difference in the incidence of hydrocephaly or microcephaly among whites, blacks, or Hispanics, but white mothers had a significantly higher incidence of children with macrocephaly (12.0 per 10,000 births) than black mothers (5.2 per 10,000) (Spranger et al., 1982).

α-Fetoprotein and Neural Tube Defects

α-Fetoprotein (AFP) is a protein produced by the fetus that appears in the serum of pregnant women. When the maternal serum levels of this protein rise prenatally, it may indicate that the fetus is affected by a neural tube defect (Brock et al., 1974). Measuring this protein in the serum of pregnant women has become a method of prenatal screening for neural tube defects when combined with ultrasonography and measurements of AFP in amniotic fluid. There may also be differences between ethnic groups in AFP concentrations. Crandall et al. (1983) demonstrated that there were no significant differences in serum AFP concentrations when corrections for maternal weight were applied; however, black women showed consistently

higher values, averaging about 10 percent higher at each week of
gestation. These corrections affected only those values falling just
above or below the ninty-fifth percentile cutoff. The authors recommended that corrections for maternal weight and race should be applied when values for AFP in maternal serum are being interpreted.

CLEFT LIP, CLEFT PALATE, AND CLEFT GUM

Cleft lip with or without cleft palate has been studied extensively and
the major genetic determinants of CL(P) are discussed briefly in
Chapter 2. Among the environmental factors known to increase the
risk of CL(P) is maternal diabetes (Myrianthopoulos and Chung,
1975). Diabetic mothers studied in the NCPP had a risk 8.7 times that
of nondiabetic mothers (8.7 per 1,000 versus 1 per 1,000 births).
Isolated cleft palate (CP) is considered separately because of the demonstrated genetic heterogeneity (Fogh-Anderson, 1942). Cleft palate
is often associated with fetal death because it is often a part of multiple malformation syndromes. The analysis of these data confirms
other findings that CL(P) was more prevalent in whites (14.5 per
10,000 births) than in blacks (7.2 per 10,000 births). No racial differences were detected in CP. Khoury et al. (1983) examined the maternal factors in CL(P) by comparing the incidence of oral clefts in offspring of white-white, black-black, and white-black couples. After
adjusting for the father's race, they observed that offspring of white
mothers had a higher rate of CL(P) than those of black mothers. This
maternal racial effect did not apply to CP. This difference persisted
after adjusting for the mother's age, parity, and education. After adjusting for the mother's race, offspring of white fathers had the same
rate or proportion of CL(P) or CP as those of black fathers. The study
shows that the difference in the rates of CL(P) between blacks and
whites is due to the effect of mother's race. This study provides evidence for maternal determinants of CL(P) in humans. Although
blacks have a lower incidence of CL(P) when compared to whites, a
study of a small number of oral cleft patients (Siegel, 1979) showed
that blacks had a significantly higher rate of associated birth defects.
No significant differences were found in recurrence rates between the
racial groups. Iregbulem (1982) reviewed 21,624 births at a Nigerian
teaching hospital from 1976 to 1980. There were 8 infants with CL(P)
or CP, an incidence of 1 per 2,703 live births. In a retrospective
analysis of 360 patients with CL(P) and CP, the frequency was 49
percent for CL, 19 percent for CP, and 32 percent for CLP. Right-
and left-sided lip clefts were found in equal proportions, whereas

with incomplete clefts, left-sided lesions tended to predominate. Associated congenital abnormalities occurred in 18 percent of the patients. Cleft gum is described as a midline cleft in the upper gum. The cleft varies in size. Data from the NCPP confirmed earlier findings that this minor anomaly was significantly higher in black infants (Myrianthopoulos and Chung, 1975). These frequencies were 43.8 per 10,000 births for blacks, 9.5 per 10,000 births for whites, and 7.5 per 10,000 births for the group called "other." The etiology of this variant has not yet been delineated.

EAR MALFORMATIONS

External ear malformations are extremely heterogeneous. This is evident in the wide variation in the incidences reported in the literature, ranging from 2.41 to 88.4 per 10,000 births (Melnick and Myrianthopoulos, 1979). These malformations include preauricular sinus, preauricular tags, anotia/microtia, and other pinna malformations. The frequency of branchial cleft sinuses ranged from 0.31 to 5.03 per 10,000 in these same populations. Of the several studies of this anomaly that have been reported since 1950, the most accurate is probably that of McIntosh (1954), performed in New York City. He conducted follow-up examinations at six and twelve months of age. This time factor was of vital importance because it has been found that less than 50 percent of the malformations found among infants were suspected or noted at birth (Myrianthopoulos and Chung, 1974). McIntosh found a frequency of 59.2 per 10,000 births for all ear malformations and noted no cases of microtia. A large proportion of this population was made up of black infants (44.2 percent of the study group was reported as nonwhite), but the data were not reported according to race.

Racial and ethnic differences in the frequency of preauricular sinuses have been known for many years. As early as 1914, Stannus (1914) found that 4.5 percent of 6,491 Bantus had preauricular sinuses. The frequency of this same abnormality has been reported in Oriental populations as 4–6 percent in one study (Congdon et al., 1932) and 10–14 percent in another (Ride, 1935). Selkirk (1935) examined children in Hamilton County, Ohio, and found that 0.9 percent of white and 5.2 percent of black children had preauricular sinuses. In the same study the frequency of branchial cleft malformations was 0.5 percent among white schoolchildren, while no cases of this malformation were found in black children. The relative high frequency of preauricular sinuses in peoples of African origin is fur-

Table 11.2
Distribution of Nonsyndromic Ear Malformations by Race and Phenotype

Malformation	Whites		Blacks	
	No.	Rate/10,000 Births	No.	Rate/10,000 Births
Preauricular sinus	43	17.80	374	148.85
Preauricular tag	33	13.66	48	19.10
Microtia	43	1.66	5	1.99
Other malformed pinna	20	8.28	18	7.16
Branchial cleft sinus	5	2.07	6	2.39

Sources: Melnick and Myrianthopoulos (1979).
Note: Total NCPP white population was 24,153; NCPP black population was 25,126.

ther supported in a study by Simpkiss and Lowe (1961) from a survey of 2,068 consecutive infants in Mulago Hospital, Kampala, Uganda. They found an incidence of 222.43 per 10,000 births (2.2 percent), which is nearly ten times greater than that in the New York study by McIntosh (24.39 per 10,000). The incidences of preauricular skin tags (24.17 per 10,000) and other pinna malformations (9.67 per 10,000) are similar to those found in the McIntosh study (1954). Of further interest is the fact that preauricular sinuses were unilateral in 77 percent of cases according to Selkirk (1935) and in 83 percent according to Stannus (1914). The data in the literature generally indicate that for all external ear malformations the tendency is for uni-laterality.

The most recent comprehensive study of the racial distribution of nonsyndromic malformations of the external ear and branchial cleft sinuses confirmed the findings of earlier investigators. Table 11.2 summarizes these results which were derived from analysis of data collected in the NCPP (Melnick and Myrianthopoulos, 1979). The frequency of external ear malformations was 1.72 percent in blacks and 0.42 percent in whites. The primary reason for this difference is that the frequency of preauricular sinus of the ear is almost nine times greater in blacks than in whites (148.85 per 10,000 versus 17.8 per 10,000 births). The frequency of the other external ear anomalies is approximately the same in both groups.

Extensive analysis of these data fails to confirm the expected incidences in first-degree relatives which would be predicted by a multi-factorial/threshold model of inheritance. The heritability estimates for these anomalies is low in both racial groups. Pedigree analysis in approximately half of the familial probands found vertical transmission through two and three generations. It seems reasonable to consider the familial cases as possible examples of single gene determination. There might also be environmental induction of these anomalies in individuals where there is susceptibility determined by single genes. In general, however, the level of genetic determination is low, especially in blacks, which suggests that there is a strong environmental component in the etiology of external ear malformations. Unfortunately, no significant environmental or teratogenic factors could be established.

HYPOPLASIA OF THE LUNG

Hypoplasia of the lung is a morphologically heterogeneous malformation distinct from the form of lung hypoplasia associated with a short gestation and prematurity. In the analysis of data from the NCPP, it appeared that this malformation was significantly more frequent in black infants (20.6 per 10,000 births versus 6.8 per 10,000 births) (Myrianthopoulos and Chung, 1975). On closer examination, the frequency of this anomaly seemed to be inversely related to the number of prenatal visits and the weight gain during the pregnancy. Evidence from studies in rats suggests that hypoplastic lungs can be produced experimentally in rats deficient in vitamin A. It is possible, therefore, that the difference in the frequency of this malformation is primarily a function of environmental and not genetic factors.

CARDIOVASCULAR MALFORMATIONS

The frequency of congenital heart defects in black populations has been variously reported as 3.8 per 1,000 population (McLaren et al., 1979) or 6 per 1,000 population (Akman et al., 1982). In the latter study, conducted in Bogalusa, Louisiana, a disproportionate number of black children, especially boys, had congenital heart disease. In Nigeria, congenital heart disease represents approximately 5 percent of all cardiovascular disease (Adebono et al., 1978). The expected range of congenital cardiovascular abnormalities has been found in African patients who have been studied carefully. The distribution of congenital cardiac lesions in two series of patients from Ibadan, Nigeria (Jaiyesimi and Antia, 1981), and Abidjan, Ivory Coast (M'etras et al., 1979), were collected from patient populations and reviewed retrospectively. Therefore, they cannot be considered to represent the "true" population frequencies of these conditions. Nevertheless, they give us an idea of the kinds of lesions that occur in

Table 11.3
Distribution of Cardiovascular Lesions in Two African Populations

	Ibadan, Nigeria	Abidjan, Ivory Coast
Malformation	Frequency (%)	Frequency (%)
Ventricular septal defect	35.0	38.6
Atrial septal defect	7.5	13.8
Patent ductus arteriosus	22.0	7.7
Tetralogy of Fallot	10.0	8.8
Pulmonary stenosis	9.0	8.1
Coarctation of the aorta	2.0	2.3
Aortic stenosis	0.6	0.0
Transposition of great vessels	0.0	3.8
Atrioventricular canal	0.0	7.7

Source: Jaiyesimi and Antia (1981); M'etras et al. (1979).

African populations. The lesions and their proportions in each population are reported in Table 11.3. Ventricular septal defect (VSD) is the most frequent in both groups but the relative frequencies of other lesions differ in each. Investigators in South Africa conducted a survey of 12,050 children (aged two to eighteen years) in a search for congenital heart disease. The frequency of congenital heart disease in this black African population was at least equal to that in the white African population (3.9 per 1,000) They also found that VSD was the most frequent lesion, comprising 52 percent of cases. In contrast to the surveys from Nigeria and the Ivory Coast, they found no cases of persistent patent ductus arteriosus (PDA). However, they unexpectedly found five cases of situs inversus (1 per 2,410).

Only two specific cardiovascular defects were analyzed in the NCPP: PDA and VSD (Myrianthopoulos and Chung, 1975). These two defects were the most frequent lesions identified in the study. No association with race was discovered, but there was an association between birth order and PDA. More recently, increases in the reported frequency of PDA and VSD have been noted (Anderson et al., 1978). Apparently, these increases are correlated with the presence of low-birth-weight babies of shortened gestational age, more cases of which were being diagnosed in the first week of life of the infants. The increases reported may very well be merely a consequence of the rising awareness of physicians who care for premature infants and/or due to changes in the demographic characteristics of the population being studied. The increase in the frequency of PDA was much greater than that of VSD. It is not clear whether the higher rate of VSD was related to the same factor or factors as the higher rate of PDA.

SUPERNUMERARY NIPPLES, SMALL BOWEL ATRESIA, AND DOWN SYNDROME

Supernumerary nipples are extra nipples on the chest wall and/or abdomen, usually aligned with the two normal nipples. This minor anomaly is usually of only cosmetic interest. It occurs predominantly in males, and considerable evidence suggests that this condition is inherited as an autosomal dominant trait (Gates, 1947). The results of the NCPP agree with earlier studies that showed that this trait predominated in populations of African origin. The incidence of supernumerary nipples in blacks was almost thirteen times that in whites (116.6 per 10,000 versus 9.0 per 10,000 births) (Myrianthopoulos and Chung, 1975).

A population-based birth defect surveillance program was reviewed over a six-year period by Safra et al. (1976) for the presence of isolated small bowel atresia. The malformation occurred in this population at a rate of 2.7 per 10,000 live births. No familial association or drug exposure was noted. When thirty-two cases of this condition were analyzed in detail, the frequency of the isolated defect for blacks was twice that for whites, primarily because of a higher rate in black females.

Until recently the literature contained conflicting reports about the frequency of Down syndrome for U.S. blacks. Some studies offered evidence that the frequency in blacks is not significantly different from that in whites (Parker, 1950; Kashgarian and Rendtorff, 1969). The frequencies of Down syndrome in the NCPP showed no significant racial or ethnic differences. The frequency in whites was 12.4 per 10,000 births and in blacks it was 9.3 per 10,000 births. Adeyokunnu (1982) studied the incidence of Down syndrome in live births over a period of nine years in a Nigerian hospital and noted an incidence of 1 in 865 livebirths, or 11.56 per 10,000. Cytogenetic studies were performed on 386 patients. Regular trisomy 21 was found in 369 (95.5 percent) cases and translocation was found in 9 patients (2.5 percent). Six of these later cases (1.5 percent) were mosaic. A relatively high number of the cases were found in children of young mothers, but no contributory environmental factors could be identified.

OTHER SYNDROMES

Literally hundreds of syndromes have been described in the medical literature and cataloged in published collections such as the Birth

Table 11.4
Rates of Major Congenital Malformations (per 10,000 Births) by Race/Ethnicity
in the United States

Malformation	Blacks	Hispanics	Amerindian	Asians	Whites
Anencephaly	2.1	4.4	3.6	4.4	3.0
Spina bifida[a]	3.3	5.9	4.1	1.8	5.1
Hydrocephalus[b]	8.1	4.6	10.8	4.8	5.4
Microcephalus	4.8	2.8	2.6	1.9	2.1
Ventricular septal defect	14.4	13.8	19.1	21.0	17.4
Atrial septal defect	2.1	1.2	4.1	2.5	2.1
Valve stenosis and atresia	5.9	1.9	8.2	2.8	3.2
Patent ductus arteriosus	49.9	20.7	33.5	25.1	26.5
Pulmonary artery stenosis	5.4	1.4	0	1.8	1.5
Cleft palate without cleft lip	3.7	3.7	9.8	4.8	5.9
Cleft lip with or without cleft palate	4.4	8.6	17.5	12.9	9.7
Clubfoot without CNS defects	19.9	19.1	15.5	14.4	27.5
Hip dislocation without CNS defects	13.8	24.0	31.4	25.0	32.3
Hypospadius	24.6	14.9	17.5	16.5	32.7
Rectal atresia and stenosis	2.8	3.0	4.6	3.8	3.7
Fetal alcohol syndrome	6.0	0.8	29.9	0.3	0.9
Down syndrome	6.5	11.6	6.7	11.3	8.5
Autosomal, excluding Down syndrome	2.1	2.1	3.1	2.9	2.2
Total	179.9	144.4	222.0	157.6	189.8

Source: Modified from Chavez, Cordero, and Bacerra (1988).
[a]Without anencephaly.
[b]Without spina bifida.

Defects Compendium (1979) and the classical text by Warkany (1971).
Most of the syndromes are individually rare and those that are more
frequent have not been classified according to their frequency in racial or ethnic group. Even when such classification has been attempted, the reliability of the figures is likely to be suspect. Butler et
al. (1982) reported a case of Prader-Willi syndrome, said to be the
second case report of a black individual with this condition. The author noted that what appears to be a paucity of cases may be a result
of underreporting rather than a true population difference. Golden
et al. (1984) reported another two cases of Prader-Willi syndrome.
Another author (Dijkstra, 1977) claims to have reported the first case
of Goldenhar's syndrome (oculoauricular dysplasia) in a Bantu. On
the other hand, Beckwith-Wiedemann syndrome (Hamel and Yohani,
1981), Waardenburg syndrome (Sellars and Beighton, 1983; Hage-
man, 1975), and prune-belly syndrome (Aseyokunnu et al., 1982) are
all well known in African populations.

Table 11.4 is a compilation of major congenital malformations which includes frequencies per 10,000 births for Hispanics, Asians, and Amerindians in addition to blacks and whites. These data were collected by the Centers for Disease Control (CDC) from 1981 to 1986. The differences between frequencies here and in Table 11.1 may reflect changes in incidence over the past fifteen years or differences in the way the data were collected. Clearly, the collection of prospective case reports describing these and other syndromes in African and other populations is important, so that we can derive reliable and accurate incidence information. These data may be useful in genetic counseling and in helping to identify the etiology of some of these poorly understood malformations.

12

Hypertension and Diabetes

It is well established that the incidence of hypertension is variable in human populations. For example, the incidence of hypertension in blacks in the United States is higher than that of whites (Falkner, 1987). This population variability is one of the principal reasons why some investigators have proposed a genetic basis for hypertension. But ethnic differences in the prevalence of disease may be genetic, environmental, or both. And, the manifestation of genetic disorders frequently is determined by environmental factors.

As we shall see, many studies demonstrate that hypertension is more prevalent in urban populations than in rural groups. But there are exceptions. This too is understandable, for urban populations are not necessarily comparable, and rural groups differ from one other. What is classified as a rural population in New York might be classified as a congested urban area, say, in Wyoming or Senegal.

It is indisputable, however, that (1) the family histories of hypertensive patients are more likely to show evidence of hypertension than do the family histories of normotensive subjects; and (2) the blood pressures of identical twins are more highly correlated than the blood pressures of fraternal twins and of nontwin siblings (Kass et al., 1977). At this point in our knowledge, a specific genetic etiology for hypertension is unproven, even though genetic factors appear to play a role. On the other hand, as we shall see, life style, stress, social class, exercise, diet (including mineral content of water and food), age, smoking, alcohol, and altitude, to name only a few factors, have all been implicated in the etiology of hypertension.

A comparative analysis of these and other factors will be made in various populations to determine whether any or several may contribute to our understanding of population differences. Admittedly, probably no one investigation can serve as a model for population studies. Nevertheless, population comparisons may be the most fruitful means by which the possible factors causing hypertension may be elucidated.

DEFINITION OF HYPERTENSION

Probably one of the most important variables in epidemiologic studies is the delineation of hypertensive persons from nonhypertensive individuals. There are many problems in making a diagnosis of hypertension. Is hypertension defined by the level of the systolic pressure or diastolic pressure or both? Is the diastolic pressure determined by the fourth or fifth Korotokoff sound? Frohlich et al. (1988) recommended that both sounds be recorded, particularly if sounds are heard to zero mm Hg. These authors also believed that phase IV of the Korotokoff sounds (during which the pitch of the sounds changes to a distinct, abrupt muffling of sound) is a more reliable index of diastolic blood pressure. Once the sound is selected, at what diastolic level is hypertension determined? At 90 and above? At 95 and above? At 100 and above? Has the blood pressure been established by multiple examinations at one sitting or at several sittings? Kaplan (1983) suggested an operational definition, in which hypertension is

> that level of blood pressure at which the benefits (minus the risks and costs) of action exceed the risks and costs (minus the benefits) of inaction. The benefits of action include prevention of progression of the hypertension; reduction of strokes, congestive heart failure and renal damage; decreased mortality due to myocardial infarction; and recognition of other family members at risk. The risks and costs of action include assumption of the role of a patient; interference with lifestyle; biochemical aberrations with increased cardiovascular risk such as hypokalemia from diuretics and hyperlipidemia from diuretics and beta-blockers; and the costs of medication and medical care. The benefits of inaction include the continuation of a normal life and less interference with life-style. The risks and costs of inaction include an increased risk of cardiovascular disease and death, and the failure to identify other family members at risk. (Pp. 705–6)

This very complex definition of hypertension is not useful for epidemiological studies.

Various definitions of hypertension were used by Alderman and Yano (1976) in an effort to determine the effect of diagnostic criteria on the prevalence of hypertension. At an initial examination, 23.3 percent of their subjects had blood pressures greater than or equal to 160/95 mm Hg; however, less than half of these sustained that level on two subsequent occasions over the next three weeks. When an initial diastolic pressure of greater than or equal to 105 mm Hg was the criterion of hypertension, the frequency of hypertension decreased by two-thirds (from 23 to 7 percent).

Oviasu (1978) conducted a blood pressure survey in a rural Nigerian community to determine the prevalence of hypertension and the effect of reexamination of individuals who were found to be hypertensive at the initial screening. Of the 2,082 subjects (aged fifteen to fifty-nine years) who were examined at the primary study, 123 (6 percent) were classified as hypertensive. When these 123 subjects were reexamined eight to twelve weeks later, only 44 (2.1 percent) were classified as hypertensive.

It is evident that many models are used to define hypertension, and that some of the population differences in the prevalence of hypertension may merely reflect different measures for diagnosis. Further, the mercurial nature of blood pressure and its interpretation seriously compromises most epidemiological studies (Page, 1979).

MINERAL INTAKE

The mineral content of water and food can affect blood pressure.

According to Porter (1983) the association of salt deprivation with a decrease in blood pressure was first reported by Ambard and Beaujard (1904). Porter (1983) reviewed epidemiologic studies that examined the link between salt intake and hypertension. Investigations of Greenland Eskimos, Australian aborigines, and Highland populations in China confirmed an association between salt intake and hypertension during the period 1927–37. Later, investigations of the Bushmen (San) of the Kalahari, Brazilian Indians, Malaysians, and Solomon Island groups confirmed a relationship between salt and hypertension.

Hunt (1983) reviewed populations with low blood pressure and found that these groups were more active and more lean than populations in which blood pressure rose with age. They also consumed a diet low in sodium and high in potassium. But when members of these groups were exposed to Western diets, blood pressure increased with age and hypertension occurred. Hunt listed more than twenty populations with low blood pressure. Among African populations in which blood pressure does not rise with age were some rural Nigerians, Senegalese (Serer), non-Baganda in Uganda, Kenyans (Samburu, Rendille, and Turkana), Tanzanians (Masai), Ethiopians, rural Zulu, and Bushmen (San). Hunt also listed the following populations: Carajas Indians, Malaysians (Orang Asli), Micronesians (Ponape and Abaiang Islanders), Papuans and peoples of New Guinea, and Polynesians (Pakapuka Islanders). These groups were members of diverse populations who lived under a wide variety of climates from the desert to the arctic to the jungle. They resided at sea level or on high mountains, although they all lived in nonindustrialized regions. Hunt also indicated that when populations have an intake of sodium that is below 30 milliequivalents per deciliter, hypertension is rarely found, and when the sodium intake is between 30 and 60 milliequivalents per deciliter, the incidence of hypertension is below 3 percent, which is far lower than the 9–20 percent incidence in the United States.

Kerr et al. (1982) surveyed ethnic patterns of salt purchase in Houston, Texas, because of speculation that the salt intake of blacks in the United States may contribute to the development of hypertension. The sales of table salt in supermarkets in black, Hispanic, and white communities in Houston were compared. The authors concluded that sales of table salt were 50–100 percent higher in black census tract supermarkets than in those of the white census tracts. The investigators were cautious in their conclusions and stated that whether the increased sales of table salt have a cause and effect relationship to the prevalence of hypertension in the communities can only be determined by future studies. Largh and Pecker (1983) commented on dietary sodium and hypertension. They reviewed studies that indicated that rigorous sodium deprivation lowers the blood pressure in about 30–50 percent of patients with hypertension. In other patients, sodium deprivation had no effect or it even increased blood pressure. These authors maintained that there is no evidence to indicate that a widely applied moderate reduction of salt intake could counter the development of hypertension.

Langford (1983) pointed out that strong social class and geographic differences in blood pressure may be related to differences in potassium intake, or in the ratio of potassium to sodium intake. It was reported that Japanese village populations with similar sodium chloride intakes but different blood pressures had different potassium intakes. Some investigations show that blacks excrete much less potassium than whites and also consume much less potassium than whites. Further, a high potassium diet is more expensive than a low potassium diet.

Wilson (1986) studied the history of salt supplies in West Africa and differences in blood pressures today. He found that the salt available to West Africans was usually in four forms: rock or mineral salt, marine or sea salt, salt soils, and vegetable salt. Rock salt was the most common source of sodium, followed by salt soils, marine salt, and vegetable salts. Vegetable salts were the only form of salt that occurred in most parts of West Africa; however vegetable salt was not a major source because the amount of sodium in plants is small and the extraction of salt from plants is tedious. Rock or mineral salt and salt soils were mined throughout the Sahara and were traded for gold and other items by the Berbers since at least the eleventh century. West Africans also obtained salt from the coastal producers of Senegambia, but Wilson postulated that it was not a sizeable amount and salt was probably in short supply throughout most of the history of West Africa except for Senegal, Gambia, and the Sahara. Wilson then compared the blood pressures of three common ethnic groups in West Africa, the Yoruba in Nigeria, the Mandinka in Gambia, and the Serer in Senegal. The blood pressures of the Serers and the Mandinkas were generally lower than the blood pressures of the Yoruba from both urban and rural areas. Wilson further asserted that the considerable improvement in salt supplies since the 1960s in Africa—especially in urban areas—has affected the health of the population. The regions of higher blood pressure were explained in the following manner: "Since sodium is necessary for normal functioning of human cells, in areas where availability of sodium has been limited, Darwinian natural selection may have operated. Thus the ability to conserve sodium may have developed to such an extent that individuals from a line that evolved in the salt-deprived areas of West Africa now have a predisposition to salt-induced hypertension, whereas those from salt-rich areas are more resistant to it" (p. 784). Wilson also noted that African-Americans reportedly have a tendency to retain more sodium than whites given the same salt load, and possibly African-Americans who are highly sensitive to salt descended from populations in the low-salt regions of West Africa. Wilson also suggested that medical researchers should question their longstanding convention of classifying humans into racial groups for research purposes; this division assumes a genetic homogeneity that is inconsistent with modern evolutionary theory. Although Wilson's theory is provocative, we suggest there is considerable heterogeneity among West Africans that may not be explained by a historical access to salt.

A very important population study may refute some of the longstanding beliefs about the association of hypertension with high salt intake (McCarron et al., 1984). The data base of the National Center for Health Statistics, Health Examination Survey I (HANES I) was analyzed to ascertain the relationship of seventeen nutrients to the blood pressure profile of U.S. adults. The study population consisted of 10,372 individuals between the ages of seventeen and seventy-four who had no history of hypertension and modification of their diet. Nutritional factors that delineated hypertensive individuals from non-hypertensive persons consisted of significant decreases in the consumption of calcium, potassium, vitamin A, and vitamin C in the hypertensive population. The most consistent factor in hypertensive individuals was lower calcium intake. Further, higher intakes of calcium, potassium, and sodium were associated with lower mean systolic blood pressure and a lower risk of hypertension. The authors believed that nutritional deficiencies and not nutritional excesses are factors that delineate overweight or hypertensive subjects from normal individuals in the United States. And most important, they suggested that caloric restriction may be disadvantageous in that this could reduce nutrients crucial for the maintenance of normal mean arterial pressures. The investigators were careful to indicate that their findings do not prove causality. Needless to say, this study has provoked considerable controversy, but avenues for future research have been opened.

Brown et al. (1984) also believed that national pronouncements about the advisability of restricting salt intake to control hypertension in Great Britain may be premature. These authors maintained that intervention studies of decreased dietary salt were confined to small numbers of hypertensive patients investigated for short periods of time with measurements of blood pressure only; to extrapolate from the limited positive studies and give advice on lifelong dietary modification is unjustified and irresponsible. It has not been proven that the present intake of salt in Western countries causes high blood pressure; the recommended reduction of salt is arbitrary and may be harmful. (Experimentally, sodium restriction is associated with hemorrhage and renal tubular necrosis.)

To further complicate matters, an Australian National Health Service Health and Medical Research Council Dietary Salt Study Management Committee (1989) found that dietary modification to reduce sodium intake lowered systolic and diastolic blood pressure, and that the effect was attributable to the reduction in salt intake, because the lowering of blood pressure was prevented by the administration of sodium chloride to a control group.

Langford (1983) suggested that a low potassium intake may be crucial in the development of hypertension because the higher blood pressure in blacks than in whites in the United States is associated with lower excretion of potassium, probably because of a lower consumption of potassium. Langford postulated that the high cost of a high potassium diet may account for the low potassium consumption in blacks. It was suggested that potassium may reduce blood pressure by increasing sodium excretion, decreasing renin excretion, decreasing sympathetic nerve activity, or directly dilating the arteries.

The role of calcium and magnesium in hypertension was explored by McCarron (1983). Initial epidemiologic evidence associating dietary calcium with hypertension and cardiovascular disease emanated from reports of an inverse relationship between the level of water hardness and the frequency of both coronary heart disease and cardiovascular mortality in certain regions of the United States and Great Britain. Later investigations also showed a negative correlation between calcium content of water and mean blood pressure. In countries where calcium intake is high (>1,000 milligrams per deciliter) gestational hypertension occurs in fewer than 1 in 200 of pregnancies, but in societies where calcium intake is under 500 milligrams per deciliter the incidence is increased 10- to 20-fold. A relationship between osteoporosis (which is associated with inadequate calcium intake) and hypertension was also found. At the time of diagnosis, 37 percent of black women with osteoporosis had hypertension compared with a prevalence of 15 percent in white women without osteoporosis in the same age group. It was also found that the calcium intake of blacks in the United States is significantly less than that of whites at all ages and sex groups. It was postulated that this may be related to the high prevalence of lactase deficiency in blacks with the resultant avoidance of dairy products.

GENETIC FACTORS

The role of genetic factors in hypertension is difficult to isolate from familial environmental parameters. Havlik and Feinleib (1982) reviewed the epidemiology and genetics of hypertension. Familial associations were drawn from the Framingham Study by reviewing the Framingham Cohort experience with the Framingham Offspring Study. The separate effects of paternal and maternal blood pressure determined some twenty years earlier—when the parents were a similar age to their offspring—showed a statistically significant effect of parental blood pressure, except for paternal blood pressure in women. Age-adjusted correlation coefficients for parents and offspring for systolic and diastolic blood pressure showed correlations of about 0.15, which were similar to those done concurrently in the usual family study—when the blood pressure of parents and children are evaluated at the same point in time. Correlations between siblings did not show the predicted 0.5, and ranged from age-adjusted 0.143 to 0.207. There was a considerable discussion about correlations between parent and offspring, siblings, and assortative mating of parents, all of which were inconclusive. The twin studies were more impressive, as one may expect. For systolic blood pressure and diastolic blood pressure, the correlations for middle-aged twin veterans were much higher for monozygotic (MZ) than dizygotic twins (DZ): 0.55 and 0.58 versus 0.25 and 0.27. The relationship was translated into a heritability estimate (the percentage of the total variance that can be attributed to genetic factors). Using the relationship of twice the difference of the MZ and DZ twin correlations for blood pressure, Havlik and Feinleib estimated that about 60 percent of the variability is attributed to heredity. And then the usual partial disclaimer was made for heritability estimates: "This statistical analysis does not, however, avoid the possibility that the environment of the MZ twin pairs is differentially more alike than for the DZ pairs" (p. 126).

An interesting biochemical approach to the heredity of hypertension was investigated by Garay et al. (1981) in a study of Na+–K+ cotransport in subjects from France and the Ivory Coast. Outward Na+–K+ cotransport (CO) in erythrocytes from French individuals who were hypertensive was found to be excessively low (CO−) compared to normotensives (CO+) selected for a negative family history of hypertension. Of the sixty-six French unselected normotensives investigated, twenty-six (39 percent) were CO−, whereas fourteen of the eighteen unselected normotensive Ivoriens (79 percent) were CO−. On the other hand, sixty-four (80 percent) of the eighty essential hypertensive subjects examined in France were CO−, but the proportion of CO− subjects among the Ivorien hypertensive subjects was even higher. Additionally, both hypertensive and normotensive Ivorien subjects often had undetectable outward Na+ fluxes, a finding that was rare in the French study. The authors suggested that the high incidence of abnormal Na+–K+ cotransport in the Ivory Coast subjects could indicate a genetic propensity to hypertension in this population.

WEIGHT AND BODY FAT

Havlik et al. (1983) reviewed the association between weight and hypertension. Apparently, an association between blood pressure and weight can be identified early in life. The correlation coefficient increases to approximately 0.4 in young adults and then decreases at older ages. Although hypertension is an independent risk factor for cardiovascular disease, obesity is believed to be associated with cardiovascular disease through its effect on cardiovascular risk factors such as high blood pressure, increased cholesterol, and diabetes. Among people who are 30 percent above the average weight, the risk of death from coronary disease compared with those of average weight was 44 percent higher for men and 34 percent higher for women. Havlik and co-workers postulated that, based on this evidence, being overweight is either deadly by itself, or through an association with other risk factors, particularly hypertension. In fact, a relationship between weight and hypertension has been known at least since the 1930s; interestingly, weight reduction was one of the few avenues for therapy of hypertension long before the advent of drug therapy. Havlik et al. (1983) referred to the work of Stamler et al. (1978) in which the frequency of hypertension in overweight persons, ages twenty to thirty-nine, was two to three times greater than that of individuals who were average or below average in the same age group. This relationship is found in all race and sex groups; however, in each weight group, the mean blood pressure of blacks is higher than that of whites. But more important, a weight reduction program causes significant reductions in blood pressure to such an extent that drugs are often no longer needed in patients with hypertension. Tobian (1978), however, did not recommend withholding drugs in patients with moderate hypertension while the physician determines whether weight reduction reduces blood pressure. Rather the blood pressure can be reduced promptly with drugs to remove the strain on the arteries, and then a weight reduction program can be instituted by diet and exercise.

It has been previously mentioned that the prevalence of hypertension is generally low in nonindustrialized societies. Since peoples in these regions are generally more lean than those of industrialized populations, it is tempting to speculate that weight is an important factor in the incidence of hypertension.

Blair et al. (1984) examined the relationship between blood pressure and the distribution of body fat in a survey of 5,506 survey individuals between ages thirty and fifty-nine. An approximation of peripheral and centrally located body fat on both systolic and diastolic blood was ascertained by measurement of triceps and subscapular skinfolds, and the effects of race, sex, and age on the relationship between obesity and blood pressure were analyzed. Subscapular skinfold was the better predictor of both systolic and diastolic blood pressure. Age and subscapular skinfolds contributed independently to the variability of blood pressure in each race and sex group.

ALCOHOL AND TOBACCO

Individuals who drink large amounts of alcohol tend to have higher blood pressures (Friedman, Klatsky, and Siegelaub, 1982). In a Kaiser-Permanente study of about 87,000 individuals the blood pressure association was not related to demographic characteristics, adiposity, reported salt use, smoking, or coffee consumption, or the underreporting of alcohol use. The study suggested that about 5 percent of hypertension in the general population may be due to the consumption of three or more alcoholic drinks per day.

It is interesting to note that cigarette smokers have lower blood pressures than nonsmokers. It is not known whether or not the lower blood pressure of cigarette smokers is because of the thinner body build of smokers.

COLOR BLINDNESS

Morton (1975) examined the association of color blindness and hypertension in Selective Service registrants. Among 29,119 registrants with medical information (41 percent of registrants), 1,073 (3.6 percent) had hypertension, and 1,226 (4.2 percent) had some type of color blindness. There was a highly significant association between the frequency of hypertension and the prevalence of color blindness. Hypertension was present in 6 percent of color blind individuals but in only 3.6 percent of those with unimpaired color vision. Color blindness occurred in 6.8 percent of persons with hypertension, in 5.8 percent with borderline hypertension, and in only 4 percent with normal blood pressure. Types of color blindness were not differentiated.

ALTITUDE

Ruiz and Penaloza (1977) studied the relationship between altitude and hypertension. Cross-sectional surveys were performed in five small Peruvian communities; two villages were located at sea level and three above 13,000 feet. A total of 4,359 individuals were studied at sea level (1,970 males and 2,389 females) and 3,055 at the high altitude (2,189 males and 866 females). At the high altitude, the age-adjusted prevalence of hypertension—particularly systolic—was low. Diastolic hypertension was more frequent in men than in women and was commoner than systolic hypertension. The reverse was observed in populations at sea level. It was postulated that chronic hypoxia was an important causal factor in the rarity of hypertension at high altitudes, and more important than genetic mechanisms.

SICKLE CELL DISEASE

Johnson and Giorgio (1981) recorded blood pressures in 187 adult patients with sickle cell disease and compared them with an age- and sex-matched population of U.S. blacks. Blood pressures in those with sickle cell disease were significantly lower than those of the control population and did not show the expected rise with increasing age. Four patients with sickle cell disease had diastolic hypertension and two had systolic hypertension, which was significantly less than that of the black population. The authors postulated that the renal tubular defect responsible for increased water and sodium excretion may blunt the plasma volume expansion necessary for sustained hypertension and accordingly promote lower arterial pressures. Since people with sickle cell trait have hyposthenuria, blood pressure studies should also be done in this population, and if a similar difference exists persons with sickle cell trait (Hb AS) could have a second—and perhaps a more important—selective advantage over individuals with Hb AA.

THE UNITED STATES

Hypertension is different in blacks and whites in the United States (Thomson, 1980). In blacks, it develops earlier in life, it is frequently more severe, and it results in a higher mortality at a younger age, more commonly from strokes than from coronary heart disease. In fact, blacks have a threefold greater mortality from hypertensive diseases than whites and, even more striking, this disproportionate mortality increases to more than six times greater in blacks aged thirty-five to fifty-four (Falkner, 1987).

One of the earliest studies of black-white differences in blood pressure in the United States was done by Adams (1932), who analyzed a total of 28,221 blood pressure readings in 14,000 persons (8,000 whites and 6,000 blacks) in the course of routine preemployment physical examinations. (Note: preemployment epidemiologic surveys are frequently quite biased.) Mean systolic and diastolic pressure increased with age, and in all age groups, the values for blacks were significantly higher than those for whites. Many other studies in the United States have since confirmed these population differences.

One of the best-designed early studies was undertaken in Georgia (Comstock, 1957). Considerable effort was made to obtain all blood pressure readings in the subject's home environment. This objective was achieved in 91 percent of those examined. The blood pressure was obtained three times in succession; care was taken to relieve all compression in the cuff between each determination; and only the third reading was recorded. Systolic pressure was defined as the first appearance of the Korotokoff sounds; diastolic pressure was recorded as the point of disappearance (fifth sound). Both pressures were measured to the nearest even number. The number of people examined was 1,162, of which 331 were white males; 437 were white females; 168 were black males; and 226 were black females. The results of the survey were similar to those of Adams (1932), but with additional information. In young adults, females had lower mean systolic pressures than males, but among older persons, males had lower systolic pressures. This reversal of relationship of blood pressure occurred earlier in blacks (age thirty-five) than in whites (age forty-five). Blacks also had the highest mean systolic blood pressures, with no appreciable differences between the sexes. The lowest mean diastolic pressures were found in white females. Mean diastolic pressures for whites were intermediate.

Data from the Hypertension Detection and Follow-up Program (1977) were studied to explore the relationship of education to racial differences in hypertension. Standardized blood pressure, a medical history, and socioeconomic information were obtained for about 158,906 adults. Hypertension was defined as being present in those with a diastolic blood pressure greater than or equal to 95 mm Hg and in those with a diastolic blood pressure less than 95 mm Hg who reported that they were on medication for hypertension. It was found that 18 percent of whites and 37 percent of blacks were defined as having hypertension at the first examination. A higher education level was found to be inversely associated with hypertension for each sex and racial division. It was suggested that since even at the higher education levels, the adjusted prevalence of hypertension remained nearly twice as high in blacks as in whites, educational levels cannot fully account for the black-white differences.

Blood pressure was studied in young adults, aged fifteen to twenty-nine years, in a biracial community of Evans County, Georgia (Cassel, 1971). Blood pressure readings were taken seven years apart. Readings that were recorded equal to or greater than 140 mm Hg, systolic, or equal to or greater than 90 mm Hg, diastolic, or both, showed distinct race and sex differences. The frequency of pressures exceeding these limits was as follows: white males, 19 percent; white females, 13 percent; black males, 34 percent; black females, 32 percent. Similar findings were found in the incidence during the seven-year interval. Being overweight was associated with an increase in blood pressure in these groups.

Two groups in the United States have explored the association between blood pressure on the one hand, and socioeconomic status and skin color on the other. The first investigation was a survey of black residents of Charleston County, South Carolina. Using reflectance photometry, an association of high blood pressure with darker skin color was found (Boyle, 1970). In a follow-up study sixteen years later of the Charleston cohort (Keil et al., 1977), an inverse relationship was found between socioeconomic status and high blood pressure. There was no association between socioeconomic status and high blood pressure when economic status was controlled.

Harburg et al. (1978) attempted to determine whether an association between skin color and hypertension is valid. The investigation was developed along the following lines. Color consciousness, with its attendant political and economic overtones among blacks in the diaspora, was apparently the result of the subordinate status imposed on blacks by whites and Western civilization. Blacks with darker skin have suffered rejection and discrimination from lighter-skinned blacks as well as from whites. In the United States, darker-skinned blacks tend to be poorer, less well educated, and occupy a lower social status in black communities than do blacks with fairer skin. In their Detroit study, Harburg et al. found no consistent relationship between blood pressure and socioeconomic status, but there was an association between skin color and blood pressure in both blacks and whites. For blacks, the association was positive and for whites, negative. Several theories were introduced to explain the findings, including one that was most favored: genetic differences associated with skin color indicate that biochemical mechanisms in the tyrosine to melanin pathway are important in blood pressure regulation. The association of skin color with blood pressure in both whites and in blacks was explained as a possible manifestation of gene intermixture between whites and blacks.

In an editorial, Troller and James (1978) suggested that skin color in the black population in the United States may be an indicator of psychosocial processes not measured in the Detroit studies, but nonetheless related to blood pressure levels. The reasons for this premise were listed as follows: (1) Chronic elevations of blood pressure in human populations may be related in part to struggles under conditions of great uncertainty to acquire sufficient economic and social resources to "control" one's environment. (2) These struggles are less intense; or, at least are more often rewarded, for persons of higher social status. (3) In industrialized, male-oriented and -dominated societies, the struggle for mastery has been led by men. (4) In color-conscious societies, access to economic and social resources is greatly influenced by one's skin color. (5) In American society, darker-skinned black men, and those dependent on them for survival, more than most other subgroups, have been denied full access to these resources, with the result that a satisfactory degree of environmental control is rarely achieved.

Even though hypertension is more frequent in black men and women than in comparable whites, Reed (1980) found that in an extensive blood pressure screening program for high school students between fourteen and eighteen years old, the blood pressures of white youths equaled or exceeded that of black adolescents. These ethnic differences remained when age, sex, weight, and socioeconomic status were controlled.

Mexican-American men and women and Anglo men and women who lived in three socially distinct neighborhoods in San Antonio, Texas, were surveyed for hypertension (Franco et al., 1985). The overall age-adjusted prevalence rates of hypertension were similar for Mexican-American and Anglo men (10 and 9.8 percent, respectively); however, for women, the Mexican-American rate was lower (7.8 percent) than that for Anglo women (9.7 percent). After adjustment for obesity, Mexican-Americans have a tendency for lower rates of hypertension than whites of the same socioeconomic level.

Stavig, Igra, and Legra (1984) investigated hypertension among Asians and Pacific Islanders in California. Generally, Filipinos had rates of hypertension approximately equal to that of blacks. On the other hand, Japanese had low rates of hypertension.

The overall prevalence of hypertension in the United States was revised in the Final Report of the Subcommittee on Definition and Prevalence of the 1984 Joint Committee, in 1985. The purpose of this report was to arrive at a consensus on a prevalence figure and to ascertain the distribution of hypertension by blood pressure level, age, race, and sex. The prevalence rates were based on hypertension defined as blood pressure measurements greater than or equal to 140/90 mm Hg, determined by an average of three readings taken on one occasion. The study concluded that the prevalence rate for blacks was 38 percent versus 29 percent in whites, and men had a higher prevalence than women (33 percent versus 27 percent). Hypertension increased with age in blacks, whites, men, and women.

GREAT BRITAIN

Interestingly, Cruickshank et al. (1985) found similar mean systolic and diastolic blood pressures in black, white, and Asian factory workers in England, ages sixteen to sixty-four. A 78 percent response rate was evenly distributed among whites, black West Indians, and Asians. Older black women did have higher blood pressures than whites; however, the body mass indices were greater (2–5 kilograms per square meter). The authors postulated that the lack of blood pressure differences between blacks and whites in England may be due to the similarity in social class of the participants.

SOUTH AFRICA

South Africa is a fertile field for study of hypertension because of the common belief that stress factors are important in the etiology of hypertension. And one of the most important social stress factors is discrimination. Scotch (1963) reported on sociocultural factors in the epidemiology of hypertension among the Zulu. The study was based on field work conducted in two Zulu communities in South Africa. One of their communities was located in a rural native reserve, and the other was in an urban location just outside of Durban. The prevalence of hypertension was significantly higher in the urban population than in the rural group. These findings were attributed to differences in severity and variety of stress in an urban setting as against a rural environment. A very telling comparison was also made of the mean systolic and diastolic pressures by age and sex for urban Zulus, rural Zulus, Durban blacks, and Durban whites. The rural Zulus had the lowest mean blood pressure; the white Durbanians were somewhat higher; the urban Zulus were still higher; and the blood pressures of the Durban blacks was the highest of all.

Scotch (1963) made some interesting observations about the Zulu population differences. He concluded that the principal differences between hypertensive subjects and nonhypertensive individuals is that the former are unable to adapt to the demands of urban life, whereas the latter are capable of doing so. Most of the variables associated with hypertension were related to social conditions that formed the basis for nonadaptive behavior patterns for urban living. The hypertensive individuals were likely to live in an extended family, have a lower income, resort to supernatural explanations of illness and misfortune, retain traditional religious beliefs, and have a larger number of children. The reverse was true of nonhypertensive people. Additionally, nonhypertensive individuals were likely to attend the European clinic more frequently and, if they were women, to belong to the Christian church. These practices were viewed by the author as adaptive. On the other hand, if Scotch's observations of the differential behavior of nonhypertensive and hypertensive individuals is valid, the hypertensive group could be viewed as more "normal," because accommodation to an urban apartheid environment is an abnormal "human" behavior. Adaptation ("normal") and maladaptation ("abnormal") in the context of Scotch's investigation are subjective impressions, the interpretation of which depends on the observer.

Blood pressure and possible associated etiologic factors were examined in tribal and Xhosa people of South Africa (Sever et al., 1980). The blood pressures in the urban population were high and rose with age, but in the tribal group blood pressures were low and rose very little with age. Skinfold thickness, weight, and ponderal index were significantly greater in the urban population and were strongly correlated with blood pressure.

Robinson et al. (1980) commented on Sever and colleagues' work and suggested that environmental factors such as the stress of acculturation are consequential in the rural-urban differences in blood pressure. These authors reported that significantly raised blood pressures were found in Masai who had only recently left the tribe and who were still working in the vicinity. Notably, unlike the Zulu, there were no significant differences in blood pressure levels between tribal Masai and Masai who had been living and working in large urban areas for ten years or more.

To explain the high prevalence of hypertension in the urban adult Zulu compared to the rural Zulu, the relationship between blood pressure, plasma renin activity, and patterns of urinary sodium and potassium excretion rates were studied in South African Zulus and Asian Indians (Hoosen et al., (1985). The urinary sodium and potassium levels were not significantly different, and there was no association between urinary sodium excretion and blood pressure. Urinary potassium levels correlated negatively with blood pressure in rural Zulus and Asian Indians, but not in urban Zulus. The urinary sodium to potassium ratio was significantly lower in rural Zulus than in urban Zulus, but the sodium to potassium ratio of Asian Indians was not significantly different from that of Zulus. The plasma renin activity levels of urban Zulus was significantly lower in urban than in rural Zulu. The authors could not explain the differences.

Seftel (1978) examined the enigma of the rarity of coronary heart disease in South African blacks in view of the high prevalence of a major coronary risk factor: hypertension. This author also examined other conventional risk factors for coronary heart disease, such as affluence, age, hyperlipidemia, dietary excess, smoking, physical inactivity, diabetes, obesity, hyperuricemia, and hyperinsulinism. All of the foregoing risk factors with the exception of hypertension were found to be low in blacks in Johannesburg. Seftel also reviewed the clinical and epidemiological studies in Africa, Japan, and other countries that have established that hypertension that is unassociated with a Western type of nutritional-metabolic environment is weakly, if at all, atherogenic. He did caution that these findings do not mean that the postulated coronary risk factors fully account for the incidence of coronary heart disease in poor or in affluent countries.

A number of noteworthy comparative studies of hypertension were performed by Seedat and co-workers. A survey of the prevalence of hypertension in urban whites in Durban, South Africa, by Seedat, Seedat, and Veale (1980) showed that according to World Health Organization hypertension criteria (160 mm Hg or more, systolic, and/or 95 mm Hg or more, diastolic, for all age groups) the prevalence of primary hypertension in this group was 22.8 percent (25.6 percent of males; 20 percent of females). Further, the prevalence of hypertension in males under forty years of age was twice that of females of the same age. Blood pressure rose with age, and there was a greater rise in systolic than in diastolic pressure. The elevation in blood pressure was more marked in men over the age of forty and in women over the age of sixty. It is worth noting that the prevalence of hypertension in this Durban white population was also higher than that found in population studies of whites in the United States.

There were other prominent results of this South African study. About 6 percent of the white subjects had a diastolic pressure of 105 mm Hg or more. It was postulated that the high prevalence of diastolic pressures of 105 mm Hg or more, together with the high prevalence of hypertension in the white men under forty years of age, could be an essential etiologic factor for ischemic heart disease. The South African rate for ischemic heart disease in white men in South Africa twenty-five to thirty-four years old is among the highest in the world: 23.1 per 100,000, compared with 9.3 per 100,000 in the United States. These authors also pointed out that the prevalence of hypertension in the urban whites of Durban is higher than the 17 percent frequency (18.5 percent, males; 15.7 percent, females) which the United States Public Health Service Health and Nutrition Examination Survey found in a sample of 17,796 whites.

Seedat, Seedat, and Hackland (1982) summarized their various studies of hypertension in urban and rural Zulus and investigated possible biosocial factors that may be associated with hypertension. A house-to-house study was performed on 1,000 urban Zulus and on 1,000 rural Zulus. In the urban Zulus, the prevalence of hypertension was 25 percent (23 percent in men and 27 percent in women). In the rural Zulus, the prevalence of hypertension was 10.5 percent (males,

10 percent; females, 10.8 percent). In both the urban and rural Zulus, age, sex, obesity, marital status, urbanization, and number of dependents were found to be significantly associated with hypertension. In the urban Zulus, many factors were associated with hypertension which were absent in the rural Zulus: insomnia because of anxiety, cigarette smoking, alcohol intake, conditions of work, educational status, income, number of children not working, lack of recreation or sports activity, and overcrowding. In the rural Zulus, several parameters absent in the urban Zulus were associated with hypertension: a family member with hypertension and the educational status of children. Overall, there was a relationship between hypertension and social variables which were stressful.

In a random house-to-house survey of 1,000 Asian Indians in Durban, the prevalence of hypertension was found to be 19 percent (22 percent of females and 15 percent of males (Seedat, 1982). Hypertension among Asian Indians in South Africa was found to be associated with diabetes mellitus. Obesity was found to be more common in widows or separated individuals, persons from lower socioeconomic groups, and in individuals with poor education. Hindu males were more commonly affected than Muslim males. It is interesting to note that the prevalence of hypertension among Asian Indians in South Africa is higher than that of published data from India.

Seedat and Reddy (1976) compared black, Asian Indian, and white hypertensive patients. Hypertension was found to be very common in the urban blacks, and it occurred at a younger age compared to the Asian Indian and white population. Hypertension in blacks was manifested in an explosive manner, with death occurring frequently from uremia, cerebral hemorrhage, or congestive heart failure. The Asian Indian hypertensive patient resembled the white hypertensive patient in age distribution. Complications in the Asian Indian hypertensive patient were mainly cerebral thrombosis, congestive heart failure, ischemic heart disease, and uremia.

That racially identical peoples may have a higher blood pressure when moved from a rural to an urban environment is incontrovertible and is a striking example that a genetic predisposition may not be expressed in the absence of certain environmental factors, such as excess sodium intake or stress (Materson, 1985).

THE CARIBBEAN

One of the best studies of the incidence of hypertension in the Caribbean is that of Miall et al. (1962) in Jamaica. Two populations were categorized geographically. The first was a rural agricultural population who cultivated produce for home consumption and for sale in Kingston and other markets. Housing conditions were poor; the diet was low in calories and deficient in protein and animal fats; and the people were physically active. The other group was urban, living in Kingston. This group was selected with the intent of studying a population with socioeconomic standards analogous to the rural group and with sufficient stability to allow the investigation to be done. The age and sex structure of the Kingston population, however, was different from that of the rural group. There was an excess of young women in Kingston as compared with the rural population. In almost all age groups the women from the rural region had higher mean systolic and diastolic pressures than those from Kingston. Severe hypertension was also much more common in rural women. On the other hand, the age distribution and the prevalence of hypertension was not different between the two male populations.

The relationship between hypertension and renal disease was studied in these two populations. The screening technique was urinalysis for protein excretion. There were no significant differences between the urban and the rural population. With the exception of the youngest age group, however, bacteriuria was significantly more frequent in the rural females. Bacteriuria was not significant in the male populations. The mean arterial pressures of women with and without bacteriuria were investigated in both populations, studying all women from both populations who were under the age of sixty-five. In those aged thirty to sixty, the blood pressures of women with bacteriuria were higher than in those without bacteriuria. Nevertheless, although nearly 15 percent of individuals with diastolic pressures above 110 mm Hg had bacteriuria, when these individuals were removed from both study groups, the difference in the distribution of blood pressure remained unchanged. It was pointed out that the vast majority of women with bacteriuria were free from urinary symptoms and had no protein in their urine.

As has been mentioned earlier in this chapter, considerable evidence suggests that the renin profile of blacks who have hypertension is different from that of whites. Grell (1980) indicated that the renin profile of Jamaicans with essential hypertension is similar to that of whites with hypertension, in that 31 percent of Jamaicans had low renin hypertension, 45 percent had normal renin levels, and 24 percent had high renin hypertension.

In a companion study to the Jamaican study, Miall et al. (1962) surveyed whites in South Wales, Great Britain, using the same observer, the same techniques of examination, and representative populations in both countries. There was little difference in mean blood pressure between the Jamaican population just described and the Welsh population, but the proportion of hypertensive individuals was higher in Jamaica. On the other hand, Schneckloth et al. (1962) in a survey in St. Kitts showed that there were higher mean blood pressures in a Caribbean population of 1,575 black villagers compared to white populations of the same age in the Bahamas and the United States. The survey conditions were dissimilar, however. Thus these comparisons are difficult to evaluate.

A population survey for hypertension was done in St. Lucia by Khaw and Rose (1982) to compare blood pressure distribution with other West Indian investigations, and to assess the implications for health care planning. A total of 359 individuals were investigated. Mean systolic values rose with age, and diastolic pressures increased with age until the 55–64 age group in males and the 45–54 age group in females. In the age group 45–54, 9 percent of males and 19 percent of females had systolic pressures greater than 170 mm Hg or diastolic pressures greater than 100 mm Hg; in the age group 55–64, 44 percent of males and 19 percent of females had hypertension; and in the age group 65 and over, 30 percent of males and 28 percent of females had hypertension. These sex comparisons were difficult to evaluate because of the small sample sizes within the age groups. Even so, the numbers within the age groups were comparable, thus the differences may be real.

Khaw and Rose (1982) then compared the mean pressures in St. Lucia with those found in Jamaica by Miall et al. (1962), and Schneckloth et al. (1962). The mean pressures at all ages were lower in St. Lucia. The most striking differences were between the sexes. In the Jamaican investigation, the sex differences were small at the lower ages and increased significantly with age, but in St. Lucia the mean pressures at older ages were not significant. In St. Lucia, diastolic hypertension was more frequent in males in all age groups, but in Jamaica the reverse was apparent. It is important to mention that the conditions of measurement of blood pressure were similar in the three studies. Halberstein and Davies (1984) reviewed hypertension in the Caribbean with specific reference to the Bahamas and reported on an in-depth study in Bimini. Previous investigators in this region reported that about 27 to 42 percent of adults had hypertension by various measuring techniques and that hypertension is the eighth leading cause of death in the Bahamas, accounting for more than 3 percent of fatalities. In a study of 167 Bimini adults, ages 21–87, hypertension was found in 32 percent. This sample represented 11.5 percent of the total community, and 22.6 percent of the 21 and under age group. Each subject represented a separate household; 43.5 percent of Bimini's 384 households were included. Three separate readings were made for each subject, and the lowest reading was recorded. Halberstein and Davies cited a host of factors to account for hypertension in Bimini, such as the combined effect of inbreeding and genetic drift, the subtropical climate, excessive intake of salt, and other mechanisms. Nevertheless, the study was not constructed to test these multiple variables.

BOTSWANA

Truswell et al. (1972) pointed out that many populations do not show the rise in blood pressure with increasing age that is common in peoples in industrialized countries. The San of the !Kung tribe were investigated. The 800 individuals who lived in northwestern Ngamiland, Botswana, were isolated by a waterless zone sixty to one hundred miles wide. Blood pressures were studied in 152 San aged 15–83 years; there were 73 women and 79 men. In males the systolic and diastolic pressures showed a slight decrease with age. In women there was a slight increase in systolic pressure from a mean of 114 mm Hg in ages 15–19 to an average of 130 mm Hg in ages 60–69. In both groups the diastolic pressures remained essentially the same until late in life at which time there was a slight drop.

NIGERIA

Akinkugbe and Ojo (1968) examined blood pressures in 3,602 rural Nigerians (1,696 males and 1,906 females). In the male sample, 153 (9.1 percent) had a blood pressure of 140/90 mm Hg or greater. In the female sample, 214 (11.2 percent) had a blood pressure that exceeded 140/90 mm Hg. With age, an increasingly higher percentage of the population in each age group had a pressure of greater than 140/90 mm Hg. From early childhood up to 45 years a higher percentage of males had hypertension than females. After 45 years up to age 60, women had a higher prevalence of hypertension. This extensive study confirmed the observations of Abrahams and Alele et al. (1960) that hypertension was common in rural populations in Nigeria. On the other hand, Akinkugbe and Ojo (1969) found that the blood pressures were even higher in an urban population. Up to the age of 44 years, both systolic and diastolic pressures were comparable in the rural and the urban (industrial) groups. After this age the mean blood pressures were higher in the urban group (Akinkugbe, 1972).

To investigate the possible influence of occupation on blood pressure, Oviasu and Okupa (1980) compared the blood pressures of groups of African rural and urban populations in Bendel State. In the rural survey, 387 male office clerks and 1,095 male field laborers were examined. In the urban study, 916 men and 347 women were investigated. For both males and females, the mean arterial pressures of the urban sample were higher than those of the rural population. The rural clerks also had slightly higher blood pressures than the rural laborers, and the urban clerks had slightly higher blood pressures than the rural clerks.

Abengowe, Jain, and Siddique (1980) investigated the pattern of hypertension in the northern savannah of Nigeria by studying patients admitted to the Ahmadu Bello University Teaching Hospital in 1976. A total of 2,663 medical admissions were made during the period, and of these, 1,272, or 48 percent, were males and 1,391, or 52 percent were females. Hypertension accounted for 248, or 9 percent, of the hospital admissions. Males made up 101, or 41 percent, and females 147, or 59 percent, of the patients with hypertension. The difference in the sexes was significant ($P < 0.05$). The highest frequency of hypertension was in the age group, 35–44 (35 percent), and more than half (61 percent) of the hypertensive patients in the series were in the 35–44 and 45–54 age group. A high frequency of hypertension (31 percent) was observed in the lower socioeconomic group, but a lower incidence (8 percent) was found in patients of high social status. Even so, this is a biased study because the survey consisted of hospital patients.

GHANA AND SENEGAL

A blood pressure survey was performed in twenty rural Ghanaian villages to compare the incidence with studies done in Accra and in the United States (Pobee et al., 1977). Rural Ghanaians had mean systolic and diastolic pressures that were lower at all ages than the urban groups. About 2–5 percent of the subjects age 16–54 years had diastolic pressures of 95 mm Hg or higher. It was concluded that hypertension is not a significant health problem in rural Ghanaians and that large-scale hypertension case-finding and intervention programs should be confined to urban populations.

Beiser et al. (1976) studied a group of Serer in Senegal who live in pastoral settings and a group of Serer who have been rapidly exposed to urbanization. A randomized sample was obtained from 234 people fifteen years and older from the rural group and 235 individuals from Dakar. The urban sample was generally younger than the rural group. In fact, it was difficult to locate any Serer in Dakar older than age 50: only the young migrate to cities, and migration is recent. In both the urban and rural samples the blood pressure, as related to age, was relatively flat, with the possible exception of the urban females. The urban women were divided into gradations of four groups, according to their aspirations and their skills; for example, group 1 women favored Western entertainment, but lacked the tools for effective participation—they did not speak French. Group 4 women aspired to Western life styles and they spoke French. It was hypothesized that the group 1 women had the highest blood pressures because of the discrepancy between their desires and skills. Even though pressures were higher in Group 4, mean pressures in all groups were well within the normal range.

THE IVORY COAST

A survey of hypertension in the Ivory Coast was performed by Bertrand et al. (1976), studying 9,779 individuals (5,843 males and 3,936 females). Those whose systolic pressure exceeded 160 mm Hg and whose diastolic pressures were greater than 95 mm Hg were considered to have hypertension. The mean arterial pressure was 127 mm Hg systolic and 76 mm Hg diastolic in men and 124 mm Hg systolic and 74 mm Hg diastolic in women. The proportion of those having hypertension in both sexes (elevated systolic, diastolic, or both) was 14 percent. There was a progressive increase of mean blood pressure with age, but a decrease after age 55. The highest values were found in males, except for those between 10 and 15 years and after 55 years of age. Young people between 10 and 20 years occasionally had hypertension. The prevalence of youngsters severely affected between the ages of 10 and 14 years was 1.5 percent. It was postulated that the decrease in the number of Ivoriens with hypertension after age 55 could have been the result of death before this age.

THE GAMBIA, KENYA, AND UGANDA

Ree (1976) studied blood pressures in a rural Gambian population that consisted of 123 men and 174 women. Since the numbers were small, frequency distribution of blood pressures for all age groups was not determined. In this population, the prevalence of hypertension was low. The mean systolic and diastolic blood pressures for all age groups were within normal limits. They ranged in the males from 117 mm Hg, systolic; 69 mm diastolic for ages 25–29, to 132 mm Hg systolic; 78 mm Hg diastolic in the 60 and over age group. In females the mean pressures were 108 mm Hg systolic and 67 mm Hg diastolic in age group 25–29, to 133 mm Hg systolic and 76 mm Hg diastolic in the 60 and over age group.

Williams (1969) surveyed two populations in Kenya, an agricultural Kikuyu group north of Nairobi and a nomadic Samburu group in northern Kenya. At most all age levels in males and females the blood pressures in the Kikuyu were higher than in the Samburu group, and blood pressures rose significantly with age only in the Kikuyu. Shaper et al. (1969) also studied the Samburu, but for different reasons. They surveyed Samburu men after they had changed their nomadic life style by joining the army. The systolic blood pressures were higher in soldiers who were in the army as compared to nomads still living in the home district.

Shaper and Saxton (1969) examined blood pressures in a rural population in the Kasangata region of Uganda. The study population of 900 was divided into Baganda (about two-thirds) and non-Baganda. In both males and females in the Baganda, the blood pressures rose with age. In the Baganda, a systolic blood pressure of 160 mm Hg or higher and a diastolic reading below 96 mm Hg was found in 9 percent of men who were 45 years or older, and in 16.5 percent of women over age 55. In the non-Baganda, blood pressure rose with age. The blood pressure in the Baganda was similar to that of U.S. whites.

THE EASTER ISLANDERS

Cruz-Coke et al. (1964) studied the epidemiology of hypertension in Easter Islanders by analyzing two variables; namely, the level of blood pressure, and the rate of the rise of blood pressure with age in an isolated Polynesian population living in its own ecological niche and in a group of men who had moved to the mainland. Although the sample sizes were not large (Islanders, 129; Mainlanders, 50), regression analysis of blood pressure and age showed that the regression coefficients were different. The Mainlander group's blood pressure variable was significantly dependent on the age variable, but in the Islander group, the blood pressure variable was independent of the age variable. Accordingly, the rate of the rise of the diastolic blood pressure with age increased significantly in the inhabitants of Easter Island after migration to the South American continent; hypertension was absent in those who remained on the island.

These authors concluded that, regardless of age, the increase in the variance of blood pressure in a population is influenced by the intensity of the environment and the length of time spent there, and that a genotype uninfluenced by a stressful environment shows no rise of blood pressure with age.

RACIAL DIFFERENCES IN EFFECTS OF ANTIHYPERTENSIVE DRUGS

Materson (1985) reviewed black-white differences in response to antihypertensive therapy, using data from the Veterans Administration Cooperative Study Group on Hypertensive Agents. The discussion was limited to the diuretics hydrochlorothiazide and bendroflumethiazide; the β-adrenergic blocking agents propanolol and nadolol; and the angiotensin-converting enzyme inhibitor captopril. Black patients who were treated with hydrochlorothiazide had a better response than did whites and required less of the drug to achieve control of high blood pressure. Black patients who were treated with propanolol responded far less than did whites and some of the blacks had an increase in blood pressure. In a nadolol-bendroflumethiazide study, the racial differences were eliminated. In a study of nearly 400 patients treated first with captopril alone and then observed after the addition of hydrochlorothiazide, white patients appeared to respond better to the nondiuretic than did blacks. Once again, racial differences were abolished with a combination of the nondiuretic drug captopril with the diuretic.

Materson (1985) then described a multicenter cooperative study that evaluated patients who were given, on a random basis, labetalol (a drug that blocks β-1, β-2, and α-1 adrenergic receptors, but simultaneously stimulates β-2 adrenergic receptors. Propanolol lowered the blood pressure of white patients but not that of black patients; however, labetolol lowered the blood pressure of both black and white patients.

Veiga and Taylor (1986) queried whether propanolol is really ineffective in black patients. These authors maintained that, for the most part, the studies that concluded that propanolol was ineffective were of small sample size, lacked controls, and did not represent the demographics of the U.S. black population. They also pointed out that the largest U.S. study (the Veterans Administration Cooperative Study) showed propanolol to be effective in at least one-half of the black subjects, with white patients responding better, but diuretics were more effective in both races.

The differential effect, if any, of calcium antagonists is yet to be determined. These agents reduce arterial pressure by inhibiting calcium influx into the vascular smooth muscle cell, which in turn decreases smooth muscle tone and peripheral resistance (Frohlich, 1987). This investigator does list this drug as an option for blacks with hypertension. It may be too soon to infer that some of the drug effects may indicate biochemical or pharmacogenetic differences between blacks and whites with hypertension.

RENAL INSUFFICIENCY IN TREATED HYPERTENSION

Rostand et al. (1989) analyzed the clinical courses of 94 patients who had been treated for essential hypertension and who initially had normal serum creatinine concentrations, to determine the frequency with which renal function deteriorates and the causes. Despite control of blood pressure, black patients were twice as likely as white patients to have elevations in serum creatinine. The authors suggested that their observations may explain, in part, why hypertension (particularly among black patients) is a major cause of renal disease in the United States.

In an editorial comment on Rostand and colleagues' work, Klahr (1989) indicated that although good control of blood pressure has been defined as a level of less than 140/90 mm Hg, such a level may be inadequate for black patients. Differences between hypertension in blacks and whites also were summarized in the editorial. Hypertensive blacks have decreased potassium intake, reduced excretion of sodium after a sodium load, low plasma renin levels, decreased kallikrein excretion, and low activity of dopamine β-hydroxylase. Further, blacks respond more to diuretics and less to β-blockers than do hypertensive whites. Klahr suggested that the combination of low renin levels, decreased sodium excretion after a salt load, and greater response to diuretics in blacks with hypertension suggests a genetic defect in renal sodium excretion.

A COMMON DENOMINATOR FOR HYPERTENSION

A common denominator becomes apparent from the studies discussed in this chapter. The mental and social stresses of urban life have probably contributed significantly to the differential elevation of blood pressure in human populations, no matter what the presumed genetic basis for hypertension. It is, perhaps, no mere coincidence that the populations that today show some of the highest blood pressures, U.S. blacks, urban Zulus, and white South Africans are all under considerable stress for reasons that are self-evident. Racism affects not only the oppressed but also the oppressor—and is probably one of the major public health problems in the United States and South Africa.

Light et al. (1987) postulated that even though genetic factors may contribute to racial group differences in hypertension, environmental stressors may play a major role. These authors pointed out that life stress indicators, such as low socioeconomic status, social instability, crowded living quarters, and high neighborhood crime rates, were associated with greater average blood pressure or hypertension-related mortality in both blacks and whites. Nevertheless, it is no surprise that a larger proportion of blacks was found to belong to the groups exposed to these stress factors. Further, even routine life stresses may produce greater physiological responses in blacks who are already subject to socioeconomic strains.

Light et al. (1987) tested their hypothesis by monitoring heart rate and systolic and diastolic blood pressure during several stressor conditions and then during a posttask rest period in blacks and whites in the United States. The stressor conditions consisted of the cold pressor test, and three reaction time tasks: noncompetitive, competitive, and competitive plus money incentive. Blacks and whites did not differ in their blood pressures at baseline or during the preliminary casual determination. Blacks showed greater systolic blood pressure elevations over baseline levels than whites during the stressors, mainly because blacks with marginally elevated systolic blood pressures showed significantly greater stress-induced increases than did whites with marginally elevated systolic blood pressures. Light and colleagues suggested that the enhanced stressor response in blacks with marginally elevated blood pressures may be the result of higher vascular resistance during enhanced sympathetic activity, which could contribute to the higher incidence of hypertension among blacks.

Let us see what Donnison stated in 1929 about the blood pressures of South African blacks before the period of large-scale urbanization. Donnison encountered no cases of hypertension among South African blacks over more than two years, during which approximately 1,800 patients were examined. He then compared blood pressures in Europeans and Africans. Under the age of 40 years the figures for blacks and whites were similar. Above age 40, the blood pressure for

Europeans rose steadily up to age 80, after which it became lower. At 60 years of age the average blood pressure in Europeans was 140 mm Hg systolic and 90 mm Hg diastolic, but in Africans it was 105 and 67 mm Hg, respectively. Donnison developed the following hypothesis to account for these differences:

> The native African, or at least the type who lives in the native reserves . . . has probably for a large number of generations lived in a manner that has undergone very slight change. The European and American populations have, on the other hand, seen revolutionary changes in their mode of living in a very few generations as a result of the industrial revolution of the nineteenth century, the increased facilities for travel and communication, the growth of large cities, and the increasing complexity of warfare. That such developments have had their effect on the diseases of the race is undoubted. I suggest that such differences in the evolution of the two races could be held responsible for the differences in the normal standards of blood pressure; in other words that the greater mental stress required by the ordinary citizen in his everyday life, as a result of the tendencies of modern civilization, has had its effect upon the physiology as well as the pathology of the race. (P. 7)

The relative impact of genetic and environmental factors in hypertension was reviewed by Aderounmu (1981). Some very important observations were made:

> The similarities within the black racial groups are very likely to be genetic, although the pattern of inheritance remains controversial. Differences, however, occur in the prevalence and severity of the disease among the black racial groups. That these differences are due partly to environmental causes cannot be disputed because even within the same ethnic groups changing patterns are observed in the prevalence of hypertension, in the course of urbanization of rural communities, or in moving from a previously rural to an urban environment. Consequently, genetic factors may not be the only important considerations in the severity of hypertension in black subjects. (P. 597)

DIABETES MELLITUS

Diabetes mellitus is a common, heterogeneous group of disorders that appear to be the consequence of an abnormality in glucose metabolism. In its mildest form it produces only an elevation of blood glucose greater than 140 milligrams per 100 milliliters of blood under fasting conditions or a persistent elevation of blood glucose following an oral

Table 12.1
**Criteria for Insulin Dependent Diabetes (IDDM) and Noninsulin Dependent
Diabetes (NIDDM)**

	Criteria	
Characteristic	IDDM (Type I)	NIDDM (Type II)
Clinical	Thin	Obese
	Ketosis	Ketosis-resistant
	Insulin deficiency	Often treatable by diet/ drugs
	Onset, mainly childhood and early adulthood	Onset mainly after age 40
Complications	Vascular, nerve, kidney	Unusual and late
Prevalence	0.2–3 percent	2–4 percent
Family	Increased prevalence of type I	Increased prevalence of type II
MZ twin	40–50 percent concordance	100 percent concordance
Frequency in first-degree relatives	5–10 percent	10–15 percent
Insulin response to glucose load	Flat	Variable
Associated with other autoimmune endocrine diseases and antibodies	Yes	No
Islet cell antibodies and pancreatic cell-mediated immunity	Yes	No
Insulin resistance	Occasional	Usual
HLA association	Yes (D3/D4–B8/B15)	No

Source: NIH Diabetes Data Group (1979); Olefsky, (1985).

glucose load. In its more severe form it produces symptoms of thirst, polyuria, weight loss, weakness, coma, and, if untreated, death. In some instances, long-term complications occur, including retinopathy, nephropathy, neuropathy, and atherosclerosis. Extensive studies have demonstrated a relative or absolute deficiency of insulin in individuals who have diabetes mellitus.

In one of its varied forms, diabetes mellitus occurs in 5–10 percent of adults of Western European origin. The largest category of this group of diseases is called idiopathic. This class is subdivided into two forms of diabetes, type I, also called insulin-dependent diabetes mellitus (IDDM), and type II, also known as noninsulin-dependent diabetes mellitus (NIDDM). Both forms show glucose intolerance, but several criteria differentiate them. Type I was formerly called juvenile diabetes, and type II, maturity-onset diabetes. This classification is based on several criteria, including clinical, twin, family, immunologic, metabolic, and HLA association studies. Table 12.1 summarizes the criteria that distinguish type I from type II diabetes. Rotter and Rimoin (1987) cautioned that we should not assume that this classification is absolute. Some reports indicate that families of either type have more cases of the other type than that of the general population (Cahill, 1979; Gottlieb, 1980). There may be even more heterogeneity in these types than is now apparent.

Type I diabetes is uncommon in tropical regions of the world and is less frequent in African blacks than in U.S. blacks. The frequency of IDDM in European-Americans is significantly higher than in U.S. blacks (Keen, 1983). The higher frequency of IDDM in U.S. blacks than in Africans is evidently the result of gene admixture, particularly in relation to HLA-DR3 and HLA-DR4 antigens (Reitnauer et al., 1982).

Type II, or NIDDM, is more frequent in U.S. blacks than in European-Americans (58 per 1,000 versus 35.3 per 1,000) and at all ages (Health and Nutrition, 1981). It is more frequent at all ages, but the difference is most striking in people under sixty-five years of age. This distribution is illustrated from data from Health and Nutrition U.S. Center for Health Statistics (1981), which shows the prevalence of NIDDM in the United States by age and race (rate per 1,000 population):

		20–44	45–64	≥65
Black	58.0	23.4	105.2	117.9
White	35.3	10.8	47.2	92.8

In addition to these two major types of diabetes, more than sixty distinct genetic disorders associated with glucose intolerance as well as frank clinical diabetes have been reported (Rotter and Rimoin, 1987). Although each of these conditions is individually rare, from these observations it is evident that a variety of mutations at different loci can cause glucose intolerance by a number of pathological mechanisms.

Maturity-Onset Diabetes of Youth

In recent years researchers have noted another variant of NIDDM, which is inherited as an autosomal dominant trait and becomes manifest in youth (National Diabetes Data Group, 1979). This form of diabetes has been called maturity-onset diabetes of youth. Winter et al. (1987) stated that this atypical form of diabetes may account for at least 10 percent of all patients with youth-onset diabetes in U.S. blacks who live in the southeastern United States.

The condition has an unusual course. The initial signs indicate insulin deficiency, such as hyperglycemia, polyuria, polydipsia, weight loss, and ketoacidosis. Insulin is required for varying periods of time; however, the insulin requirement frequently disappears or declines a few months or years after the condition has been diagnosed. The patients do not develop clinical symptoms when insulin is withdrawn, but they do have persistent hyperglycemia. These characteristics were identified in a careful study of 129 U.S. black patients with youth-onset diabetes from pediatric diabetic clinics at the University of Florida. In this group, 12 had the clinical picture of maturity-onset diabetes of youth. This special black diabetic population had a higher frequency of HLA-DR4 antigen than did the other children with diabetes (47 percent versus 7.7 percent). Most of the patients with classic IDDM (75 percent) had either HLA-DR3 or HLA-DR4, whereas only 25 percent of those with the atypical form had one or the other of these HLA antigens. A much larger proportion of this population of blacks with youth-onset diabetes (30 percent) lacked an HLA-DR3 or DR4 antigen compared with a group of 775 whites with IDDM (5.6 percent) (Winter et al., 1987). In contrast to IDDM in youth, in blacks with maturity-onset diabetes of youth there is no increase in the frequency of the IDDM-associated HLA antigens DR3 and DR4, islet-cell autoantibodies are not found, and the level of insulin secretion is intermediate between those who do not have diabetes mellitus and patients with IDDM (Winter et al., 1987). Conditions that are similar to this atypical form of diabetes have been previously described in young Jamaicans ("J-type diabetes" and tropical pancreatic diabetes). Tropical pancreatic diabetes, however, differs from the J-type in that the affected children are usually malnourished (Keen, 1983). These conditions have also been called "third diabetes syndromes," and in some regions they may be as frequent as NIDDM (Morrison, 1982).

Cowie et al. (1989) attempted to explain why the incidence of end-stage renal disease in patients with diabetes mellitus is higher in blacks than in whites. An obvious explanation is that there is a greater prevalence of diabetes among blacks. Cowie and co-workers addressed the issue by a comparison of the type of diabetes with the risk of end-stage renal disease and found significant differences in the incidence of diabetic end-stage renal disease according to race and type of diabetes. The incidence of diabetes end-stage renal disease was 2.6-fold higher ($P \leq 0.0001$) among blacks when the higher prevalence of diabetes among blacks was adjusted with the excess risk occurring predominantly among blacks with noninsulin-dependent diabetes mellitus (NIDDM). Most black patients with end-stage renal disease and diabetes had NIDDM (77 percent), but most white patients with diabetic end-stage renal disease had insulin-dependent diabetes mellitus (IDDM) (58 percent) ($P \leq 0.0005$ for the difference between blacks and whites).

Cowie and co-workers indicated that since a preliminary study by other investigators found no difference in the prevalence of early renal disease among blacks compared with whites when both groups had comparable degrees of blood glucose and blood pressure control, some of the excess incidence of diabetic end-stage renal disease among blacks may be preventable.

13

The Health of Africans
and African-Americans

This chapter's discussion of the health of Africans and African-Americans is designed to put in perspective the place of genetic variation in relation to nongenetic disorders. Admittedly, a full analysis of the health of such diverse peoples who live on many continents would take many volumes. Hence, here we can only highlight some of the many health problems of African peoples. Health is, of course, inextricably dependent upon a number of variables, such as economics, politics, food, jobs, housing, level of poverty (or affluence), a multiplicity of environmental conditions, population size, and natural resources—to name a few.

This chapter contains a number of tables and figures on morbidity and mortality rates of various diseases: regard these figures with caution. Health and other vital statistics are only as good as the raw data, the manner in which the data are collected, the presence of bias; in short, a sophisticated infrastructure is needed for accurate statistics. Obviously, health statistics are not a major priority in poor countries that are overwhelmed with primary health care problems. Even in affluent countries, such as the United States, there are considerable inaccuracies in our census data—particularly in poor communities—and death certificates. The paucity of health statistics for blacks in South Africa is quite evident, with the exception of anthropological data, where the work of South African anthropologists is without peer.

The poorest regions of the world have the most young people. An overview of estimates of regional world populations and the percentage of distribution of those below age fifteen in 1980 are given in Table 13.1. As expected, the highest distributions of the populations younger than age fifteen are found in the most impoverished regions (Africa, followed by Central America, and South Asia), and the lowest distributions are found in the affluent regions (Europe and North

Table 13.1
Estimates of World Populations and the Percentage
below Age Fifteen in 1980 (in millions)

Region	All Ages	<15 Years of Age	%
Africa	476	215	45
North America	252	57	23
Latin America	362	143	40
Tropical South America	198	78	39
Central America	92	41	44
Temperate South America	42	13	31
Caribbean	30	11	33
Asia	2,591	993	38
East Asia			
China	1,003	370	37
Japan	117	27	23
Other	63	22	35
South Asia	1,408	574	41
Europe			
Western	154	31	20
Southern	139	33	23
Eastern	110	26	24
Northern	82	17	21
Oceania			
Australia & New Zealand	18	5	26
Other	5	2	40
Soviet Union	265	65	24

Source: Modified from Demographic Yearbook (1983), United Nations (1985).

America). (Although North America includes Mexico, we surmise that for the purpose of these statistics, Mexico was classified with Latin America.)

THE ARMS RACE AND HEALTH CARE

To set the stage for an analysis of health in African peoples, we should first review an interpretation of the effect of the arms race on health and health care throughout the world (Sidel, 1985). Sidel began his article with a poignant declaration: "Overnight, 20,000 children have died of preventable illness." He then used the analogy of a metronome:

> With every other beat—once every two seconds—a child dies of a preventable disease, a disease that could have been prevented by immu-

nisation, safe water supply, or basic adequate food supply. With each intervening beat—also once every two seconds—a child is permanently disabled, either physically or mentally, by a preventable illness and is destined to live his or her life with that disability. In other words, with each beat of the metronome a child is killed or maimed by a preventable disease. At the same time as this appalling, needless sacrifice of human lives, the world is spending with every second—with each beat of the metronome—the equivalent of $25,000 on arms. (P. 1287)

During 1985 the expenditure on weapons added up to $800 billion, or $2 billion every day, $100 million every hour, $1.5 million each minute—or $25,000 per second. The figure of $800 billion equals the gross national product (GNP) of all of the countries in which live the poorer half of the world's population—the entire debt that the poor nations owe to the rich nations. Sidel also calculated that if 10 percent of the annual military spending were used instead to decrease this debt, the debt would be eliminated in less than twenty years. He pointed out that the GNP per capita in 1980 was $170 in the populations of the least developed countries; in other developing countries the GNP was $520; and in the most affluent countries the GNP was $6,230, or 35 times that of the least developed countries. Most important, in 1980 the annual public expenditure on health per capita in the least developed countries was $1.70. In developing countries, the disbursement was $6.50, as contrasted to $244 in the most affluent countries—or 144 times that in the least developed countries.

The tale of horror continues. Sidel reported that more than 300 million children in developing countries are chronically hungry and that there are twenty-six countries in which people generally eat 90 percent or less of estimated caloric requirements. About 2 billion people have unsanitary water supplies, and in some regions less than 5 percent of the population have safe water. Health services are almost nonexistent in some poor countries. Even the wealthy are affected in many of these countries. Those who can afford it often fly to Europe or to the United States for health care. In many regions there are fewer than 10 physicians per 100,000 people, as compared with 100 or more in the most affluent countries. Sidel estimated that only about 3 percent of newborns (excluding China) are born with the assistance of a midwife or a physician. He then compared infant mortality rates and life expectancy at birth with those of the most affluent countries and appropriately referred to these data as "obscene."

THE UNITED STATES

In 1982 the total population of the United States was 231,786,000, of which 27,652,000 (12 percent) were classified as black. In 1950 the black population was 10 percent of the total.

In 1981 there were 3,629,238 total live births, of which 587,797, or 20 percent of the total, were black. The crude birth rate (live births per 1,000 population) in the same year was 14.8 for the white population and 21.6 for the black population (Health United States, 1984, U.S. Department of Health and Human Services, Public Health Service).

The age breakdown of mothers of the live births is most important from the standpoint of maternal, neonatal, and infant health. The number of live births per 1,000 women at age 10–14 years was 0.5 in the white population and 4.1 in the black population. At ages 15–17 years, the figures were, respectively (white, black), 25.1, 70.6; at 18–19 years, 71.9, 135.9; at 20–24 years, 106.3, 141.2; and at 25–29 years, 111.3, 108.3 (Health United States, 1984, U.S. Department of Health and Human Services, Public Health Service). The disproportionate number of births of black children to very young black mothers is evident.

McCormick (1985) found indirect evidence of association between low birth weight and mortality. Infants born to blacks are twice as likely to weigh 2,500 grams or less than those born to whites. Along with the difference in birth weight was a difference in mortality. Controlling for birth weight reduced or eliminated the difference in neonatal mortality associated with nonwhite race and adolescent motherhood.

Abortions

Legal abortions are now a significant method of birth control. In the white population the number of abortions per 100 live births was 17.5 in 1973 and 31.2 in 1981. Blacks were classified in an "all other" group, and probably constituted the vast majority. In this group the abortions per 100 live births in 1973 were 28.9 and in 1981, 54.4 (Health United States, 1984, U.S. Department of Health and Human Services, Public Health Service). This latter figure is instructive: contrary to common belief, a large proportion of significantly underprivileged groups can obtain abortions. On the other hand, these figures could also indicate that middle- and upper-income black women disproportionately elect abortion as a means of birth control. These statistics also represent significant nonuse or failure of contraception in both the white and the black populations.

The total legal abortions reported by the Centers for Disease Control in 1981 were 1,301,00; 1,577,000 were certified in the same year by the Alan Guttmacher Institute. The disparity in these figures occurred because the data were verified differently. The statistics from the Centers for Disease Control were acquired from two sources: central health agencies and hospitals and facilities. The Guttmacher data were provided from an annual survey of abortion providers. The data were collected from hospitals, nonhospital clinics, and physicians who were identified as providers of abortion services. To assess the completeness of the provider and abortion data, supplemental surveys were conducted of a sample of obstetrician-gynecologists and hospitals that were not in the original universe of 3,092 hospitals identified as providing abortion services by an American Hospital Association survey (Health United States, 1984, U.S. Department of Health and Human Services, Public Health Service).

Death Rates for All Causes

In 1983 the age-adjusted death rates per 100,000 resident population for all causes for white males was 701.8 and for white females, 391.5. The rate for black males was 1,024.7 and for black females, 571.5. Deaths that occurred before age one year (per 100,000) were 1,078.0 for white boys, 789.6 for white girls, 2,098.7 for black boys, and 1,581.0 for black girls. Interestingly, the age-adjusted death rate for black women was much lower than that for white men, and above age 55 years, black women also had lower death rates than white men, but black men exceeded all groups. The rates for white men 55–64 years were 1,636.5; 65–74 years, 3,849.6; 75–84 years, 8,482.4; and 85 years and over, 18,797.3. The rates for white women 55–64 years were 867.0; 65–74 years, 2,018.0; 75–84 years, 5,067.7; and 85 years and over, 14,390.5. The rates for black men were 55–64 years, 2813.7; 65–74 years, 5057.9; 75–84 years, 8552.9; and 85 years and over, 15,386.0. The rates for black women were 55–64 years, 1530.4; 65–74 years, 2934.4; 75–84 years, 5392.9; and 85 years and over, 12,273.5.

Life expectancy at birth and at 65 years of age according to race and sex also show race and sex differences, but black women after age 65 have a longer life expectancy than white males. In 1983, the remaining life expectancy at birth in years for white males was 71.6 and for white females, 78.8. Life expectancy at birth for blacks was 65.2 for males and 73.8 for females. On the other hand, at age 65, the remaining life expectancy for whites was 14.5 years for males and

18.9 years for females. The figures for blacks were 13.2 years for males and 17.2 years for females.

The United States ranked behind Japan, Sweden, Netherlands, Norway, Israel, Denmark, Australia, Spain, Switzerland, England and Wales, Canada, Greece, and France in overall life expectancy at birth according to the United Nations *Demographic Yearbook* (1983), but in the 1983 estimates, the United States only ranked behind Norway, Netherlands, Japan, Sweden, and France. The Soviet Union ranked twenty-fifth in this group, but the life expectancy of 74 years in the Soviet Union was higher than that for black men and women in the United States.

The infant mortality rates in the United States for blacks far exceed those for whites. During the period 1979–81 the number of infant deaths per 1,000 live births was 11.0 for whites and 21.0 for blacks. The region with the highest infant mortality was not the South but the eastern North Central Region, which includes Ohio, Indiana, Illinois, Michigan, and Wisconsin. The overall rate for this region was 23.8. The highest infant mortality rates for blacks were in Delaware, 27.4, the District of Columbia, 26.3, Illinois, 25.9, and Utah and West Virginia, 23.2. The United States ranks fifteenth in the world for infant deaths. The infant mortality rate of the twenty-fifth country, Cuba, was 18.5, lower than that of blacks in the United States.

One of the contributory factors to infant mortality is birth weight. The number of infants weighing 2,500 grams or less at birth per 100 total live births (1979–81) was 5.7 for white infants and 12.5 for black infants. The states with the highest rates were Wyoming, 14.8, Delaware, 14.5, District of Columbia, 14.3, South Dakota and Colorado 13.6, and Tennessee, 13.3.

The age-adjusted maternal mortality rate for complications of pregnancy, childbirth, and the puerperium in 1981 for white mothers was 6.5 (per 100,000 live births) and 22.1 for black mothers.

Death Rates for Selected Causes of Death

The age-adjusted death rates for selected causes of death are summarized in Table 13.2. Black females have a lower death rate than white males in the selected categories except cerebrovascular disease, diabetes mellitus, and homicide and legal intervention. Black males are at considerable risk in all categories except suicide and motor vehicle accidents, for which the rates are far lower than those of white males but higher than those of white females. The homicide rate of 69.2 in black males is particularly striking and far exceeds that of all the other groups combined.

Table 13.2
Age-Adjusted Death Rates for Selected Causes of Death (1981)

	White		Black	
Causes of Death	Male	Female	Male	Female
Heart diseases	268.8	129.8	316.7	191.2
Cerebrovascular disease	38.9	33.1	72.7	58.1
Malignant neoplasms	158.3	107.2	232.0	127.1
Respiratory system	57.8	18.8	84.1	20.1
Digestive system	39.3	24.7	62.1	34.5
Breast	—	22.8	—	23.7
Pneumonia & influenza	15.6	9.0	26.4	11.3
Chronic liver disease, cirrhosis	14.8	6.7	27.3	12.7
Diabetes mellitus	9.3	8.4	16.8	21.3
Accidents and adverse effects	59.1	20.2	74.7	21.6
Motor vehicle accidents	33.4	11.7	30.7	7.7
Suicide	18.9	6.0	11.0	2.5
Homicide & legal intervention	10.3	3.1	69.2	12.9

Source: Health United States, 1984.
Note: Results are given as number per 100,000 resident population.

Cancer

Horm and Kessler (1986) examined lung cancer rates in men and women in the United States using data as reported to the National Cancer Institute's Surveillance, Epidemiology, and End Results (SEER) Program. There has been a significant fall in the previously increasing incidence of lung cancer in white men: the age-adjusted incidence fell from 92.7 cases per 100,000 in 1982 to 79.3 cases per 100,000 in 1983. The incidence rates for black men did not show a similar fall. The annual incidence rates for black men varied because they are based on fewer than 1,000 cases per year. Age-adjusted male lung cancer incidence and mortality rates differed markedly between whites and blacks (Table 13.3). For example, in 1983 the incidence rate was 58 percent higher for black men than for white men (125.3). The mortality rate was also appreciably higher in black men. The lung cancer rate for women of both races did not show a decrease from 1973 to 1983; rather, the incidence rates for women showed a steady increase of about 6 percent per year.

Cancer, like other diseases, does not occur with the same frequency in all countries. Page and Asire (1985), working at the U.S. National Cancer Institute, categorized cancer rates and statistics in the United

Table 13.3
Age-Adjusted Male Lung Cancer Incidence and Mortality

	Incidence		Mortality	
Year	White	Black	White	Black
1973	72.3	103.9	61.6	74.6
1978	80.7	112.6	66.8	87.4
1983	79.3	125.3	71.2	97.3

Source: Modified from Horm and Kessler (1986).
Note: Results are given as rates per 100,000 population.

States along with some comparative studies from twenty countries with nationwide reporting systems. No African country qualified for this analysis. Of these twenty countries, the United States ranked twelfth in death rates per 100,000 population. The country with the highest death rate was Scotland with a rate of 269.8; the rate in the United States was 213.6; and the lowest was Israel with a rate of 170.5. The cancers with the ten highest causes of death in males and females and the populations are listed in Tables 13.4 and 13.5, respectively.

The most common cancer among black and white men during the period 1973–77 was cancer of the lung, with an incidence (per 100,000) of 110 among blacks and 76 among whites. Hispanic men had a significantly lower risk. The incidence of cancer of the prostate was almost as high as that of lung cancer among black men, as mentioned earlier, and this cancer was the most common cancer among Hispanic men. Cancers of the colon and rectum ranked third among

Table 13.4
International Incidence of Selected Cancers (1976) for Males

Cancer	Region and Population	Rate
Lung & bronchus	New Orleans, black	107.2
Prostate	U.S., Alameda County, black	100.2
Stomach	Japan, Nagasaki	100.2
Liver	Hong Kong	34.4
Nasopharynx	Hong Kong	32.9
Colon	U.S., Connecticut	32.3
Bladder	Geneva, Switzerland	30.2
Esophagus	Shanghai	24.7
Rectum	Canada, N.W. Territory & Yukon	22.6
Pancreas	U.S., San Francisco Bay Area, black	18.3

Source: Modified from Page and Asire (1985).
Note: Results are given as rates per 100,000 population.

Table 13.5
International Incidence of Selected Cancers (1976) for Females

Cancer	Region and Population	Rate
Breast	U.S., Hawaii, Hawaiian	87.5
Cervix uteri	Columbia, Cali	52.9
Stomach	Japan, Nagasaki	51.0
Lung and bronchus	New Zealand, Maori	48.8
Corpus uteri	U.S., Alameda County, white	38.5
Colon	U.S., San Francisco Bay Area, Japanese	27.4
Gall bladder	U.S., New Mexico, Amerindian	22.2
Thyroid gland	U.S., Hawaii, Hawaiian	17.6
Melanoma	Australia, rural New South Wales	19.1
Ovary	Israel, and European and U.S. Jews	17.2

Source: Modified from Page and Asire (1985).
Note: Results are given as rates per 100,000 population.

white and black men and fourth among Hispanic males. The third most common cancer among Hispanic men was that of the stomach; this cancer was fourth among black men and seventh among white men. On the other hand, white males were at far greater risk for cancer of the bladder than either black or Hispanic males.

The most common cancer among women of all three groups was that of the breast. It was most frequent among white women, with an incidence of 85 per 100,000, compared with 75 in black women, and 48 in Hispanic women. The second most common cancers among all three groups of women were those of the colon and rectum. The third most frequent cancer among white women was that of the uterine corpus; cervical cancer was third among both black and Hispanic women; and lung cancer was fourth among all three groups of women.

Page and Asire (1985) summarized the changing patterns of cancer, with the following conclusions:

Melanoma incidence has almost doubled for white men and women.

Breast cancer incidence has increased sharply among women of both races but is more frequent among white women.

The incidence of cancer of the uterine cervix has decreased markedly among black and white women but its incidence remains more than two times higher in black women.

Cancer of the uterine corpus, or endometrial cancer, has increased in incidence among women of both races. Its incidence is markedly higher among white women.

Ovarian cancer incidence has dropped slightly among black and white women.

The incidence of cancer of the prostate has increased among men of both races but most markedly among black men.

Bladder cancer incidence has increased for all four groups. It is higher among whites than blacks.

Leukemia incidence has decreased for all four groups.

The incidence of kidney cancer has increased slightly among all four groups.

Stomach cancer incidence has declined among black and white men. It has increased among white women.

Colon cancer incidence has increased sharply among black men since 1969. It has increased gradually among white men. Colon cancer incidence has increased slightly among white and black women.

The incidence of cancer of the rectum has increased among white men, white women, and black women; it has increased slightly for black men.

The pancreatic cancer incidence has decreased slightly among white men and remained constant among white women. It has increased significantly among black men and black women.

Lung cancer has more than doubled in the eight years covered by these data, and has almost doubled among white women. Marked increases are also seen for black men and white men. (Pp. 25–26)

Smoking, Hypertension, Obesity

The percentage of U.S. smokers has decreased in all groups since 1950, but less so among black females. The 1950 and 1980 figures included, respectively: white male, 51.3, 37.1; white female, 34.2, 29.8; black male, 59.6, 44.9; black female, 32.7, 30.6.

The frequency of hypertension in the United States has decreased since the 1960s in white females and in black males and females but it has climbed in white males. Even so, the significantly higher prevalence of hypertension in blacks as compared to whites is evident. Figures for ages twenty-five to seventy-four in 1960–62 and 1976–80 are shown in Table 13.6. (A more detailed description of hypertension is found in Chapter 12.)

Obesity is a major health problem in the United States. The frequency of obesity in the United States according to race between the ages of twenty-five and seventy-four is shown in Table 13.7. White

Table 13.6
Frequency of Hypertension in the United States, 1960–62
and 1976–80, Ages Twenty-five to Seventy-four

	1960–62		1976–80	
	Male	Female	Male	Female
White	14.8	17.1	16.3	11.3
Black	32.2	33.6	23.6	25.5

Source: Modified from National Center for Health Statistics: Health
United States, 1984.

males and white females become slightly more overweight with age.
Black males had a considerable increase in being overweight, but almost one-half of black females were classified as overweight—about
twice the frequency of obesity in the other groups. Even though
obesity appears to be a major health problem in black females, it may
not be a major factor in their mortality. The life expectancy of black
women after the age of fifty-five years is higher than that of white and
black men, and diseases of the heart are significantly lower in black
women than in white or black men. On the other hand, diabetes
mellitus is a significant cause of death in black females (see Chapter
12). Obesity may be a precipitating factor.

Poverty

There are marked differences in median income among white,
black, and Hispanic families in the United States. In 1983 the median
incomes of these groups were as follows: white, $24,603; black,
$13,599; Hispanic, $16,228. A regional breakdown of median income
among black families is even more informative: Northeast, $14,735;
Midwest, $12,734; South, $13,044; West, $16,508. Apparently, the

Table 13.7
Frequency of Obesity in the United States, 1960–62
and 1976–80, Ages Twenty-five to Seventy-four (%)

	1960–62		1976–80	
	Male	Female	Male	Female
White	24.8	24.9	25.2	26.3
Black	22.5	47.1	30.0	48.1

Source: Modified from National Center for Health Statistics: Health
United States, 1984.

Table 13.8
Weighted Average Poverty Levels Based on Money Income in 1983

No. of Persons	Income
1 Unrelated individual	$ 5,061
<65 y	5,180
>65 y	4,775
2	6,483
Householder <65 y	6,697
Householder >65 y	6,023
3	7,938
4	10,178
5	12,049
6	13,630
7	15,500
8	17,170
≥9	20,310

Source: Statistical Abstract of the United States 1984.

Midwestern region has supplanted the South as a relatively depressed region for blacks.

In the United States, families and unrelated persons are categorized as being above or below the poverty level by a poverty index as determined by the Social Security Administration in 1964 and revised by a Federal Interagency Committee in 1969 and 1980. The index is related to income and does not include noncash benefits such as public housing, Medicaid, and food stamps. The index is based on the 1961 economy food plan of the Department of Agriculture and indicates the varied consumption requirements of families based on size and composition. Poverty thresholds are updated each year to reflect changes in the consumer price index. The weighted average poverty levels based on income for families and unrelated individuals in 1983 are shown in Table 13.8.

In 1983 the following figures and percentages were given for white, black, and Hispanic families living below the poverty level: whites, 5,223,000 (9.7 percent); blacks, 2,162,000 (32.4 percent); Hispanics, 933,000 (26.1 percent). A breakdown by region confirms the plight of Midwestern blacks, even though larger numbers of blacks are below the poverty level in the South. But this is because the black population of the South exceeds that of other regions. The regional breakdown of families below the poverty level is shown in Table 13.9.

Table 13.9
Regional Breakdown of Families below Poverty Level in the United States

Region	No. below Poverty Level (× 1,000)			% below Poverty Level		
	White	Black	Hispanic	White	Black	Hispanic
Northeast	1,018	375	256	8.9	29.3	38.4
Midwest	1,399	479	62	9.8	38.0	24.8
South	1,749	1,156	274	9.8	33.1	23.3
West	1,058	153	340	10.2	23.8	23.1

Source: Modified from U.S. Bureau of the Census, Statistical Abstract of the United States; 1985.

AFRICA

Malnutrition and infectious and parasitic diseases (particularly malaria) have long been recognized as major causes of the extremely high mortality of children in the poor countries in the tropics of Africa. The overall mortality of infants is four to ten times and the mortality for children under age five is twenty to fifty times as high as that of Western Europe (Hendrickse, 1976). Hendrickse reviewed the preventive measures necessary to reduce mortality of children in these regions and asserted that the most important single safeguard against malnutrition and gastrointestinal infection is breastfeeding. Hendrickse viewed the trend to abandon breastfeeding because of marketing practices of milk product firms in Africa as one of the greatest disasters to affect children in the tropics. Be that as it may, the plight of children in Africa has long been critical, and there appears to be no end in sight. The conquest of smallpox is the only victory. Drought, famine, wars, and politics add to the perennial disasters. These disasters are reflected in a listing of countries in Africa with mortality rates of 200 and above per 1,000 live births, as shown in Table 13.10.

Liver Carcinoma and Hepatitis B

An environmental cause or association had been long suspected for hepatocellular cancer because of the longstanding observation that although this disorder is common in Africans it is uncommon among African-Americans. In fact, in the United States the prevalence of cancer of the liver is the same among blacks and whites. The availability of laboratory tests for hepatitis type B virus infection eventually led to a correlation between the hepatitis B virus and

Table 13.10
Countries in Africa with Infant Mortality ≥200

Country	1975–80
Zambia	259
Gabon	229
Guinea	216
Sierra Leone	215
Gambia	204
Nigeria	200

Source: Modified from Demographic Yearbook, 1983. United Nations, New York, 1985.
Note: Results are given as number per thousand live births.

hepatocellular carcinoma in Africa, Asia, the Pacific, and the Mediterranean region (Zuckerman, 1977). Hepatocellular cancer is relatively uncommon in North and South America and in most of Europe. Mainly because of its prevalence in Asia and in Africa, it ranks among the leading causes of cancer worldwide (Higginson, 1983). In some regions where hepatitis B and hepatocellular carcinoma are common, up to 20 percent of the healthy population are carriers of hepatitis B. About 40–70 percent of patients with hepatocellular cancer are serologically positive for hepatitis B virus (HBV) core antigen or surface antigen or both, as compared to about 1.5 percent of the controls (Higginson, 1983). There have been no reports of the progression of hepatitis A virus infection to chronic liver disease. On the other hand, the association of hepatitis B virus with hepatocellular carcinoma does not necessarily substantiate a cause-and-effect relation. Hepatitis B virus could be common in regions where cirrhosis and hepatocellular carcinoma are prevalent or where patients with hepatocellular carcinoma are quite susceptible to hepatitis B infection and the development of the carrier state. These options were explored by Zuckerman (1977), who concluded that although progress is being made, the mechanisms that are involved in the pathogenesis of liver carcinoma remain unknown.

In 1982 Zuckerman reviewed some of the criteria that he had formulated in 1977 to establish the progression of hepatitis B to primary hepatocellular carcinoma. He found that hepatitis B infection precedes the development of cancer, the tumor contains virus-specific molecules or antigens, the virus can transform cells in culture or induce cancer in experimental animals, and, finally, that immunization against hepatitis B lowers the incidence of hepatocellular cancer. The last criterion will have to await a large-scale vaccination program.

Table 13.11
Age-Standardized Incidence Rates for Cancer of the Colon
and Rectum in Men Thirty-five to Sixty-four Years of Age

Country and Population	Rate per 100,000 population
United States	
Whites	42.2
Blacks	41.6
England, Wales, Switzerland	37.3
German (Federal Republic)	30.8
Sweden	28.8
South Africa (whites)	28.7
Netherlands	22.3
Finland	15.8
Jamaica	15.7
Taiwan	14.8
Bombay	14.6
Japan and Singapore (Chinese)	13.1
Durban (Indians)	11.1
Johannesburg (Africans)	6.4
Ibadan, Nigeria	5.9
Mozambique	5.3
Kampala, Uganda	3.5

Source: Burkitt (1984).

Colon and Esophageal Cancer

The prevalence of colon cancer is comparatively high in North America, New Zealand, Great Britain, Denmark, and other European countries, and low in Africa, China, and Japan (Higginson, 1983). Iran should also be included in the low-frequency category. One of us (J.E.B.) was a pathologist in Shiraz, Iran, in 1955–61. During this period only one patient was diagnosed as having carcinoma of the colon. In this region, rectal carcinoma was also rare, but anal carcinoma was common. Age-standardized incidence rates for cancer of the colon and rectum in men thirty-five to sixty-four years old from selected countries are shown in Table 13.11. Dietary mechanisms for the variable frequency of cancer of the colon and rectum are highly suggestive from this table. Note the similar incidence of these cancers in whites and blacks in the United States, the high frequencies in Europe, and the low frequencies in black South Africans, Nigerians, Mozambiquans, and Ugandans.

Carcinoma of the upper one-third of the esophagus was common in young males in Shiraz and has been reported in high frequency in the northern part of Iran. Esophageal cancer occurs with appreciable frequency in parts of Europe, Southern Africa, the Caribbean, and northern China.

In a symposium on tumors in the tropics, Cook-Mozaffari (1982) reported that in most hospitals in Kenya, cancer of the esophagus was one of the most frequently diagnosed malignant tumors in men, but across the border in Uganda and in much of Tanzania, cancer of the esophagus is rare. Equally sharp gradients of frequency of this neoplasm were described between southern Malawi and southern Tanzania and between southeastern and northeastern Zambia. Studies from Southern Africa have shown similar gradients between the southern and northern regions of the Transkei and between Natal in South Africa and Maputo. Cancer of the esophagus in West Africa is virtually unknown, but it is found in appreciable frequency in African-Americans. Cook-Mozaffari maintained that the gradients of frequency of cancer of the esophagus in Africa developed dramatically since 1930–40. Cook-Mozaffari asserted that this increase does not appear to be a diagnostic artifact because tumors at other internal sites were being diagnosed at very much their present frequencies. Even so, dramatic increases of cancers in Africa over the years are difficult to prove. As late as the early 1960s, it was maintained that acute leukemia in children was unknown in Uganda, but in 1966 one of us (J.E.B.) was told that acute leukemia was unknown before the arrival of a pediatric hematologist at Mulago Hospital in Kampala. After 1966, and today, this disorder is common in Uganda.

Bladder Carcinoma

Cook-Mozaffari (1982) reviewed carcinoma of the bladder in Africa. She pointed out that such carcinomas are usually of squamous cell origin in Africa, in contrast to those of Europe and North America, in which the tumors are largely transitional cell in origin. The excessive frequency of squamous cell tumors has been reported in Egypt, Mozambique, South Africa, Uganda, and throughout East and Central Africa. However, in a Sudanese study, even though squamous cell carcinomas were common, transitional cell carcinoma predominated. The occurrence of cancer of the bladder is patchy in Africa (Cook-Mozaffari, 1982). It is almost unknown in central Kenya, Ruanda, and Burundi, but relatively common in coastal Kenya and southern Malawi. In these latter regions the estimated incidences are apparently the highest in the world, 40–60 per 100,000 for men. Other regions of high incidence are found in Zambia, Zimbabwe, Mozambique, and the northern Transvaal of South Africa. The rates elsewhere in South Africa are lower, and the incidence in West Africa is low or moderate. Further, Cook-Mozaffari pointed out a curious feature in the occurrence of cancer of the bladder in Africa is that elsewhere in the world the disease is five times more common in men than in women, but in Africa the rates are generally equal, and in some regions the rates in women exceed those in men.

Cook-Mozaffari (1982) maintained that the factors that have been implicated in carcinoma of the bladder in Europe and the United States (industrial carcinogens, cigarette smoking, instant coffee, artificial sweeteners) are all rare in Africa, but may explain the low incidence of transitional cell carcinoma in that region. On the other hand, the high frequency of squamous cell carcinoma of the bladder in regions where *Schistosoma haematobium* is common and the frequent occurrence of schistosoma ova in patients with bladder cancer in Egypt has led to the thesis that chronic schistosomiasis infection is a key etiologic factor. Cook-Mozaffari also asserted that even though there is a broad geographical association of cancer of the bladder and schistosomiasis haematobium, an anomalous region of moderately high frequency of cancer of the bladder exists north and west of Lake Victoria in Uganda and Tanzania in regions that are free of schistosomiasis. Even so, there are large regions that are free of schistosomiasis and bladder cancer, and a substantial number of regions of schistosomiasis where cancer of the bladder is common. There are also higher rates in women in regions where a large proportion of the men go away to work in distant mines and towns and where the women are left to work the fields, where there is a greater risk of infection.

Cancer of the Cervix

The associations of cancer of the cervix with sexual intercourse, frequency of sexual intercourse, early sexual intercourse, multiple sex partners, association with males who have multiple sex partners, a large number of children, and with herpes virus II, indicate that cancer of the cervix is a sexually transmitted disease. These associations are found irrespective of racial or ethnic groups. The epidemiological evidence in Africa and in other economically poor regions also seems to favor cancer of the cervix as a sexually transmitted disease.

Cook-Mozaffari (1982) reported a belt of relatively low incidence of carcinoma of the cervix along the line of the western Rift Valley in Uganda, Ruanda, Burundi, and Zaire, and suggested that the Lugbara who live at the northern end of this region have retained their traditional way of life, including a strict sexual code. There is apparently no correlation between the presence or absence of circumcision in the male and cancer of the cervix in the female partner in Africa.

Cancer of the Penis

Cook-Mozaffari reported a slightly higher incidence of cancer of the penis in the black population as compared to the white population. The consolidated African figures show relatively high levels of incidence, 2–7 per 100,000. In Uganda, the incidence is apparently as high as 30–40 per 100,000. Within Africa there is no association between cancer of the cervix and cancer of the penis, but this is explained by the almost complete protection against cancer of the penis that is apparently conferred by circumcision. The disease is rare in Kenya or eastern Tanzania, where circumcision is common, and the high frequencies in western Tanzania and Uganda occur among peoples who do not practice circumcision.

COMPARISON OF DISEASES IN THE UNITED STATES AND AFRICA

Burkitt (1984) and Burkitt et al. (1974) constructed an interesting comparison of certain diseases in the United States and Africa (exclusive of the communicable diseases). In addition to ischemic heart disease (see Chapter 12), appendicitis, diverticulitis, gallstones, varicose veins, deep vein thrombosis, hiatus hernia, and hemorrhoids are some of the disorders that occur at much higher frequencies in the United States than in Africa. These differences were attributable to a high-risk diet in the United States and a low-risk diet in Africa. The high-risk diet in the United States is the much-publicized excess animal fat, meat protein, high-cholesterol foods, low in fiber, high in salt, and high in sugar. Burkitt recommended, as have others, that the Western diet be altered to contain low-fat, high-fiber foods.

PRIORITIES OF THE WORLD HEALTH ORGANIZATION

The World Health Organization (WHO) identified six chronic debilitating diseases—malaria, trypanosomiasis, schistosomiasis, filariasis, leprosy, and leishmaniasis—as the principal targets for a collective international research and development program for improved and new tools for the control of tropical diseases (De Maar, 1979). The objectives of this program are as follows:

1. To determine where the need is greatest and to select diseases and technical approaches.
2. To focus on the problems and to develop priorities for research and development, for training, and for strengthening institutions.
3. An annual budget of $15–19 million, which is provided by the cooperating parties and cosponsors.

In the discussion that follows we will not necessarily adhere to the WHO formulation, but we will analyze the prevalence of malaria, trypanosomiasis, onchocerciasis, and schistosomiasis. We will also discuss acquired immune deficiency syndrome (AIDS).

Malaria

The statistics on malaria are appalling. Malaria kills about 1 million children per year. Even after many years of an extensive WHO campaign to eradicate malaria, about 480 million people are still not protected by antimalarial measures (Bruce-Chwatt, 1979). According to Bruce-Chwatt although originally about 2 billion people were exposed to malaria, eradication or control remains unchanged in Africa south of the Sahara, except in South Africa and the islands of Mauritus and Reunion, where malaria has been eradicated. Bruce-Chwatt estimated that of the 1 billion people exposed to malaria after more than twenty years of attempts at eradication, about half are on the African continent. In most regions of Africa malaria is holoendemic. Bruce-Chwatt asserted that the nature of holoendemic malaria in Africa should be understood if we are to account for the failure of malaria control in this region and to ascertain the likelihood of an effective malaria control. Since the transmission of malaria in Africa is not appreciably affected by seasonal changes or its climatic environment, this enhances malarial stability. The stability is the result of several factors: the two main vectors, *A. Gambia* and *A. funestus,* and the predominance of *P. falciparum.* That *P. falciparum* has a wide range with multiple strains, that the parasite adapts to the main vectors, and that it has a high immunogenic potential—all three of these facts contribute to the exceptionally high "basic reproduction rate," a term used by malariologists to indicate a theoretical possibility of 2,000–3,000 secondary infections from a primary case (Bruce-Chwatt, 1979).

At one time drugs were regarded as a potential means for malaria control in addition to a multiplicity of procedures to control the anopheline mosquito vector, and chloroquine was the main drug used for prophylaxis. Today, however, chloroquine-resistant falciparum malaria is present in most regions of the world where falciparum malaria is endemic: Central and South America, the Far East and Southeast Asia, East Africa, and Africa below the Equator, including isolated reports in West Africa.

Even though chloroquine-resistant disease is the most publicized of the drug-resistant malarias, all of the important drugs used to treat falciparum malaria (amodiaquine, Fansidar (sulphadoxine plus pyrimethamine, mefloxine, and quinine) are relatively ineffective against many of the strains of *P. falciparum* that are *relatively* resistant to chloroquine. We stress *relatively* because strains of *P. falciparum* that are chloroquine-resistant may still be eradicated by higher dosages of chloroquine—a most urgent point to remember in treating what could be a fatal illness (Hoffman et al., 1984). Today, the new hope for control of malaria is the development of a malaria vaccine (or vaccines). Even so, strain differences may complicate the development of effective vaccines.

Although the major malarial scourge is falciparum malaria, quartan malaria caused by *P. malariae* is more subtle and potentially dangerous. Work in West and East Africa has established that this form of malaria is a major disease in childhood (Hendrickse, 1976) because of its association with the nephrotic syndrome. The peak age incidence is later than the age incidence of the nephrotic syndrome that is seen in Europe; the patients are steroid-insensitive, and do not respond to immunosuppressive agents; the prognosis is poor; and many patients die from hypertension and renal failure. Kibukomosoke (1966) was one of the early investigators to describe the association of quartan malaria with the nephrotic syndrome in Uganda. In 1966, Kibukomosoke showed one of us (J.E.B.) at least twenty patients with the nephrotic syndrome in one pediatric ward at the Mulago Hospital in Kampala, Uganda.

Trypanosomiasis

Since Chagas disease is unknown in Africa and virtually confined to South America, this section describes only the problems of Gambian (West African) and Rhodesian (East African) sleeping sickness. About 35 million people are at risk for this disorder in Africa. Greenwood and Whittle (1980) pointed out that although the clinical and pathological features of sleeping sickness have been categorized since the beginning of the twentieth century, we still have little idea how these changes are brought about. In the Gambian form, a short while after the bite of an infected tsetse fly, a small nodule, a chancre, may appear at the site of the bite from which trympanosomes may be found.

Weeks and even months later, fever and lymphadenopathy of the posteriorcervical lymph nodes develop. There may be splenomegaly, and eventually meningitis and encephalitis are prominent. Intense pruritus develops late in the course of the disease, and eventually the patient lapses into coma. Once the central nervous system is involved, death is probably inevitable.

The clinical features of Rhodesian (East African) sleeping sickness are similar to those of West African sleeping sickness, but the progress of the disease is far more rapid. Death usually occurs within a few months, but West African sleeping sickness may last for many years. Cardiac involvement is more prominent in West African than in East African sleeping sickness, and death may follow cardiac failure or arrythmia (Greenwood and Whittle, 1980).

One of the keys to the success of the eradication of trypanosomiasis in Africa is control of flies of the genus *Glossina*. Haskel (1977) described a WHO/Food and Agriculture Organization program with an annual budget of $56 million which was designed to eliminate *Glossina* flies from 7 million square kilometers of African savannah over a period of forty years. The area is larger than that of the continental United States. Possible difficulties were outlined, such as finding enough trained personnel and the unstable political conditions. But these impediments are common for most if not all attempts at the control of tropical disorders.

Ukoli (1984) maintained that animal trypanosomiasis is the greatest barrier to the growth of the livestock industry in Africa. Africa has vast areas where livestock are decimated by the disease, which impedes agricultural development and access to meat and diary products. Protein deficiency results from the loss of meat in the diet, although such a diet does have advantages: we see a low frequency of cancers of the colon and rectum, and hypertension and coronary heart disease in populations who do not eat beef.

Onchocerciasis

Onchocerca volvulus, a filarial worm, affects about 40 million people in the world; about 30 million live in tropical Africa. It is also found in Yemen, Guatemala, Mexico, Columbia, and Venezuela (Ukoli, 1984). The disease is one of the major causes of blindness. The disease causes constant pruritus, which leads sufferers to scratch themselves continually. Indeed, an observer can easily make the diagnosis by walking down the street in a highly infected community. In Gambella, Ethiopia, where almost 100 percent of the community is infected, the dermatitis and rough skin, along with scratching, are quite in evidence. When one of us (J.E.B.) visited this region, parasitologist col-leagues stated that to their knowledge anyone who had lived at least five years in this region developed onchocerciasis. The economic im-portance of this disease is considerable even though it is not associated with high mortality; whole communities frequently have to abandon fertile land near rivers. Recent reports on the eradication of oncho-cerciasis appear hopeful (Walsh, 1986), but it is worth recalling that the World Health Organization had maintained that the eradication of malaria was within sight in the 1960s.

Schistosomiasis

The most important human schistosomes in Africa are *Schistosoma haematobium, Schistosoma mansoni,* and *Schistosoma intercalatum.* The lat-ter is not widespread in Africa and is in limited foci in Nigeria, Cameroon, Zaire, Central African Republic, Gabon, Chad, and Bur-kina Faso (formerly, Upper Volta), according to Ukoli (1984). The disease, caused by trematode flukes, is found in more than 271 mil-lion people in the world, of whom 168 million are in Africa. *Schistosoma haematobium* is found in the veins of the vesical plexus of the urinary bladder. Its distribution also includes the Middle East and India. *Schistosoma mansoni* is found in the branches of the portal veins and occurs in Equatorial and Southern Africa, focally in the Middle East, eastern South America, and the Caribbean; it is diminishing in Puerto Rico. *Schistosoma japonicum* is not found in Africa.

Schistosomiasis is not as dramatic a disease as is malaria. Ukoli (1984) believed that this is the reason why there is more emphasis on malaria control than on schistosomiasis control in Africa. Even though millions are affected, the disease is so insidious that it may be almost ignored. This is particularly true of schistosomiasis haemato-bia: one may pass through villages in Egypt where it is known that about 90–100 percent of the adults are infected and yet observe no signs of illness.

Unfortunately, the control of schistosomiasis is deceptively simple: avoid contact with all bodies of natural fresh water in Africa. This is easy for the traveler only visiting in Africa, but difficult to impossible for those who live in rural Africa. In some regions the only sources of water for bathing, drinking, washing, or farm irrigation are contami-nated streams and rivers. And in areas where dams have been built to increase farm productivity, schistosomiasis is invariably spread to re-gions that were previously uninfected. Even in the affluent suburb (Ma'adi) of Cairo, Egypt, where most Europeans and North Ameri-cans live, the immaculate lawns are watered by pipes that come directly from the Nile. And, of course, children play on these wet lawns; people wander barefooted as they lounge around filtered and chlorinated swimming pools, unmindful of the contaminated grass.

AIDS

Mann and Chin (1988) reviewed AIDS on a global perspective. Even though the origins of human immunodeficiency virus type 1 (HIV-1) are unknown, current evidence suggests that the pandemic started during the mid- to late 1970s.

Mann and Chin (1988) delineated three broad but distinct patterns of HIV-1 infection and AIDS in the world. Pattern 1 is found in North America, Western Europe, Australia, New Zealand, and many urban regions of Latin America. The sexual transmission of AIDS occurs mainly among homosexual and bisexual men; heterosexual transmission also occurs and apparently is slowly increasing. Transmission through blood is principally the result of sharing needles in intravenous drug use. These authors asserted that pediatric infection is less common, but Landsman et al. (1987) found a 1 in 50 incidence of AIDS antibody in pregnant women in a black and Hispanic community in New York. Accordingly, in this community, pediatric AIDS will be a major problem.

Pattern 2 is found in Sub-Saharan Africa and also in Latin America, particularly in the Caribbean. The sexual transmission of HIV-1 in these regions is predominantly heterosexual. Mann and Chin stated that up to 25 percent of sexually active adults and the majority of female prostitutes in some regions are infected. Transmission through blood transfusion is a major problem in areas where blood is not tested for HIV-1, and even where intravenous drug abuse is rare, transmission through unsterile needles adds to the spread of AIDS. Perinatal transmission of AIDS would also be a problem in regions of high incidence.

Pattern 3 regions include North Africa, the Middle East, Eastern Europe, Asia, and the Pacific. Apparently, HIV-1 was not present in these regions until the mid-1980s, and these regions account for only 1 percent of AIDS reported in the world. Patients were infected in these areas through contact with people in Pattern 1 and 2 areas or from imported blood. Accordingly, some countries in the Middle East and Asia have banned the importation of blood and high-risk blood products.

The Centers for Disease Control recognized the first case in the United States of a fatal new disease that came to be known as acquired immune deficiency syndrome (AIDS) (Laurence, 1985). Francoise Barre-Sinoussi, Jean-Claude Chermann, and Luc Montagnier, at the Pasteur Institute in Paris, and a group led by Robert C. Gallo, at the National Cancer Institute in the United States, allegedly independently discovered that the etiologic agent of AIDS was a retrovirus. The virus was called lymphadenopathy-associated virus (LAV) by the French group; the American group termed it human T-lymphotropic virus type III (HTLV-III). The virus has also been called AIDS-associated retrovirus (ARV). In an attempt to resolve the confusion over nomenclature, a subcommittee was empowered by the International Committee on the Taxonomy of Viruses to propose an appropriate name for the retrovirus isolates that are implicated as the causative agents of AIDS (Coffin et al., 1986). It was proposed that the AIDS retroviruses be officially designated as the human immunodeficiency viruses, with the abbreviation HIV.

In an editorial in *Science*, Marx (1986) reported that the Centers for Disease Control estimated the number of people in the United States who have been infected by HTLV-III at between 500,000 and 1,000,000. These estimates are a perennial bone of contention, and some 1988 figures ranged from 500,000 to more than 1,500,000. The groups that are at highest risk include homosexual and bisexual men, abusers of injected drugs, hemophiliacs, sexual partners of persons in the AIDS risk groups, and children born of mothers at risk. In countries where blood donors are not screened for the AIDS antibody, recipients are also at high risk.

Opportunistic infections are associated with AIDS because of viral infection of T4 lymphocytes, which leads to immunosuppression. The types of opportunistic infection differ in Africa, Europe, and the United States because of the different availability of environmental opportunistic pathogens (Biggar, 1986). Most of the information about AIDS in Africa is taken from Biggar's work. In the United States and Europe about one-half of patients with AIDS have *Pneumocystis carinii* pneumonia, which is uncommon in Africans. In Rwanda and Zaire, AIDS is associated with oroesophageal candidiasis, cryptococcal meningitis or sepsis, and mucocutaneous herpes simplex. In Europe, however, the infections in African AIDS patients have been cryptococcosis, toxoplasmosis, candidiasis, tuberculosis, and cryptosporidiosis. Biggar reported that salmonella and pseudomonas infections, disseminated strongyloidiasis, and amebic liver abscesses have also been found in African patients. Kaposi's sarcoma has long been known to be common in Africa, but investigators have been reluctant to associate this cancer with AIDS in Africa as it has been in the United States. Typically, endemic Kaposi's sarcoma is nodular, on peripheral limbs, and slowly progressive. Apparently this variety of Kaposi's sarcoma in Africa is not related to AIDS, although in Zambia and Uganda, which are both areas of high endemicity for Kaposi's sarcoma, patients have a rapidly progressive form of Kaposi's sarcoma which resembles that in U.S. patients with HTLV-III infection. This form of AIDS also has a lower than average T lymphocyte helper to suppressor ratio, but the helper to suppressor ratio in the endemic African Kaposi's sarcoma is normal (Biggar, 1986). Although it is not known whether the AIDS that is seen in Africa is old or new, anecdotal and other independent evidence supports the concept that AIDS is recent in this region, that is, since 1980. There is also no conclusive evidence that AIDS originated in Africa, because epidemics of this disease began approximately the same time in the United States and Europe.

Biggar (1986) pointed out that the question of the continent of origin of AIDS is important. It is of more than historical interest because the identification of the progenitor agent from which the AIDS agent mutated or recombined has significant implications. An analysis of the differences between the ancestor agent and the current virus or viruses could provide data about which segment of the genome confers pathogenicity. Further, if there is a nonpathogenic progenitor, this agent could be a safe source for the preparation of a vaccine.

Although the transmission of AIDS in Europe and the United States is mainly confined to male homosexuals and persons exposed to infected blood and blood products, the transmission in Africa is less well understood, because the sex distribution of AIDS in Africa is approximately equal. In the United States and Europe, however, the male to female ratio is 19 : 1. Accordingly, it is generally believed that the transmission of AIDS in Africa is principally heterosexual. On the other hand, inadequately sterilized needles, insect vectors, and other nonsexual forms of transmission are viewed as improbable because AIDS in Africa is found mainly in people who are in the sexually active age range. The high prevalences of AIDS antibody in prostitutes in Africa and in their frequent partners also favors heterosexual transmission. Information on the prevalence of homosexuality and anogenital intercourse in Africa is difficult to document. Kreiss et al. (1986) surveyed the prevalence of HTLV-III antibody in female prostitutes in Nairobi. To ascertain the prevalence of this antibody in East Africa, these authors studied the frequency of AIDS virus infection in ninety female prostitutes, forty men treated at a clinic for sexually transmitted diseases, and forty-two medical personnel. An antibody to HTLV-III was found in the serum of 66 percent of prostitutes of low socioeconomic status, 31 percent of those of a higher economic status, 8 percent of the clinic patients, and 2 percent of the medical personnel. The mean T cell helper to suppressor ratio was 0.92 in prostitutes who were seropositive and 1.82 in prostitutes who were seronegative ($P < 0.0001$). The presence of the antibody was associated with clinical and immunologic abnormalities. Notably, generalized lymphadenopathy was found in 54 percent of the seropositive prostitutes and in 10 percent of seronegative prostitutes. Although the number of HTLV-III seropositive men from the clinic was too small to analyze risk factors, all three seropositive men had one or more contacts with prostitutes. It was also pointed out that homosexuality is uncommon in Kenya, but one bisexual man was in the study population and was seropositive for HTLV-III antibody. This patient asserted that he had only one homosexual encounter. The promiscuity of his partner, however, was not recorded. None of the subjects used intravenous drugs, but a history of intramuscular medication was found in almost all of the prostitutes. Nevertheless, a similar injection history was obtained from nearly all male patients from the clinic for sexually transmitted diseases, but this group had a much lower frequency of HTLV-III antibody.

Kreiss et al. (1986) recorded that the high prevalence of HTLV-III antibody was unexpected because few cases of overt AIDS have been diagnosed in Kenya. It was pointed out that the population of Greater Nairobi is about 1 million, and even though the number of prostitutes is not known, it is estimated that there are several thousand. Although it was possible that some individuals with AIDS did not come to the attention of physicians, it was considered unlikely that a substantial number of patients would have gone unrecognized. The prostitutes also did not describe any deaths among their colleagues with symptoms that were suggestive of AIDS. The likelihood that the virulence of the retrovirus in Nairobi is diminished was rejected because the HTLV-III seropositive subjects in the study had generalized lymphadenopathy, often with serious immunologic abnormalities. Thus the pattern of pathologic response was similar to that of strains of HTLV-III in Central Africa and North America. The source of HTLV-III was not determined, but the data supported a transcontinental spread from Central Africa rather than from Europe or North America: none of the prostitutes of low socioeconomic status had had sexual contact with a non-African; and among the prostitutes, as a whole, sexual contact with men from Rwanda, Uganda, and Burundi was associated with HTLV-III antibodies. Finally, it was pointed out that the heterosexual transmission of HTLV-III with a female to male ratio of 1 : 1 raises the possibility that perinatal transmission may result in high rates of infection among infants and children.

Kanki et al. (1986) reported on a new human T-lymphotropic retrovirus related to simian T lymphotropic virus type III (STLV-III AGM) of African green monkeys, which infected apparently healthy people in Senegal, West Africa. The results of Kanki and associates emanated from their previous discovery along with investigators from the New England Primate Center of simian T lymphotropic virus type III (STLV-III) (Marx, 1986). This virus was originally found in captive rhesus macaques that have a disease similar to AIDS, and a similar virus occurs in about 50 percent of healthy African green monkeys. Some workers suggested that the human AIDS virus might have developed from the African green monkey virus. Interestingly, the group at Harvard led by Essex group found that healthy prostitutes from Dakar, Senegal, where AIDS is either rare or absent, have been infected with a virus that is related to STLV-III. The viral proteins are similar to the proteins of AIDS but are closer to those of STLV-III. The group led by Essex call their virus HTLV-IV: HTLV-IV does not kill the T helper cells, but HTLV-III and STLV-III do kill infected cells (Marx, 1986).

Clavel et al. (1987) reported in detail on the isolation and characteristics of an immunodeficiency virus from West Africa which produces signs and symptoms similar to HTLV-III (HIV-1). These investigators believed that this form of virus is identical to the before-mentioned HTLV-IV.

Although AIDS has been reported to be rampant in Africa, Konotey-Ahulu (1987) challenged some of the accepted concepts, and asserted that the high incidence of AIDS in Africa has been exaggerated. Konotey-Ahulu made a six-week tour of twenty-six cities and towns in sixteen Sub-Saharan countries, including those in which AIDS has been reported to be in high frequency. Konotey-Ahulu noted that Africa has fifty countries, but the total population of the African countries with an AIDS problem is less than 10 percent of the population of the entire continent. Yet, it is not uncommon to read that "Africa faces devastation from AIDS." (Konotey-Ahulu, however, was unable to obtain a visa for Zaire.)

An editorial in *Lancet* (1987) related that some of the confusion about high rates of infection in Uganda, Zaire, and Kenya was the result of nonspecific serological tests. Selwyn (1986) reported that the meaning of confirmatory Western blot testing among African patients has been questioned. Parasitic infections, other antigens, and cross-reactivity with a related virus have been suggested as factors that might affect HTLTV-III reactivity in sera. Nevertheless, in parts of East and Central Africa the incidence of infection ranges from 10 to

15 percent in groups that are more representative of the population (healthy adults, blood donors, and pregnant women), but lower rates of infection are found in rural areas. The *Lancet* editorial emphasized that the total number of AIDS cases reported to WHO from Africa up to June 1987 was 4,583, thus, perhaps, lending credence to Konotey-Ahulu's claim. And these figures do not in any way compare to the deaths from many other communicable and other diseases in Africa, particularly malaria. On the other hand, in areas in Africa where the adult population has incidence rates of 10–15 percent, there is no doubt that AIDS will be a major public health problem for many years to come.

Eales et al. (1987) suggested that a genetic component contributes to the susceptibility to AIDS, and Diamond (1987) reviewed this and other reports of this hypothesis. The genetic factor postulated is group-specific component (Gc) (see Chapter 5), the three common alleles of which are *1F, 1S,* and *2.* The allele *Gc1F* is supposedly positively associated and *Gc1S* and *Gc2* are negatively correlated with the risk of infection after exposure to the AIDS virus, the severity of clinical symptoms, and progression to AIDS. None of the patients with AIDS in the study of Eales et al. (1987) was homozygous *Gc2.* On the other hand, no subjects who were seronegative for HIV were *Gc1F* homozygous. Accordingly, *Gc2* probably protects against AIDS, but *Gc1F* predisposes to the disease. Sub-Saharan Africa has one of the highest frequencies of *Gc1F* and one of the lowest prevalences of *Gc2* (Diamond, 1987, Eales et al., 1987).

Nzilambi et al. (1988) studied the prevalence of HIV over a ten-year period in rural Zaire to ascertain changes in incidence. In 1985, 659 serum samples collected in a remote rural province of Zaire in 1976 were studied for HIV. Five (0.8 percent) were positive. A follow-up study in 1985 showed that three of the five seropositive patients had died of illnesses suggestive of AIDS and two remained healthy but seropositive. In 1986 a serosurvey was done using a cluster-sampling technique in the same region. The seroprevalence was 0.8 percent in random residents. The authors indicated that the stability of HIV infection in rural Zaire over a long period contrasts with epidemic spread of HIV in major African cities, and that the traditional village life in the rural province carries a low risk of infection. On the other hand, the disruption of traditional life styles and the social and behavioral changes that accompany urbanization are probably major factors in the spread of AIDS. Thus AIDS resembles hypertension (with its low prevalence in rural areas and high prevalence with urbanization), but the mechanisms for the increase are different.

As has been pointed out by many authors and as reviewed in an editorial in *Lancet* (1988), in Africa, the likelihood of sexual transmission of HIV apparently is governed by the likelihood of exposure to an infected partner, as well as the specific sexual acts performed with the infected partner. Even though systematic studies of sexual practices in Africa are not available, investigations of African males with AIDS have revealed that they had more sexual partners and/or contacts with female prostitutes than did controls. (Undoubtedly the proliferation of the AIDS epidemic in the United States among homosexuals was attributable to the large number of sexual partners of those with HIV infection.) Cofactors, such as an activated immune system as a result of chronic infections and other sexually transmitted diseases, may be associated with increased HIV transmission.

But enough. Research on AIDS is so rapidly changing that any chapter or article on the subject is outdated as it is written.

DILEMMAS OF HEALTH CARE

The dilemmas facing health care givers in Africa and in other poor regions of the globe are enormous. Most of the parasitic diseases are preventable by personal hygiene, safe water, the provision of latrines or toilets that do not contaminate the groundwater, adequate food, housing, education, and a litany of measures that are far beyond the economic ability of impoverished countries. Politics, war, exploitation, Western agribusiness, neocolonialism, apartheid, greed, countless coups, and other foibles of humankind are also major factors. In a presidential address to the Royal Society of Tropical Medicine and Hygiene, Browne (1978) referred to Lord Rosenheim's assertion that, if there were a moratorium on research in tropical diseases, what is already known could be applied for the next twenty years and result in widespread improvement in health. But if research were to cease, this would drive the teachers and clinicians in tropical medicine from the field. On the other hand, if research continues at the present pace, sponsors of tropical research will be unable to supply sufficient funds.

Browne (1978) maintained that further research necessitates a higher degree of specialization; for example, articles on the molecular biology of parasites already occupy more and more space in tropical journals. But, he noted, investigators in highly specialized fields will lose sight of the practical objectives of their work in the light of human needs. This was not so in the 1950s and 1960s. Then specialists in tropical diseases had to go to the field and live there for varying periods of time; they could not avoid acknowledging the plight of their research subjects. But today, specimens may be flown to laboratories and molecular biologists may unravel DNA homology or discordance within and among species without even leaving their affluent countries. And then their African students return home and become frustrated when their facilities are less modern than those in the West.

Browne (1978) also remarked that research in tropical medicine cannot be isolated from agriculture, rural development, and water supply. Those who do research in increasingly narrow fields will be blind to the pressing needs in other fields or even the ramifications of their own discoveries. Many, if not most, tropical disorders can be eliminated by economic development. Malaria was reduced and eliminated from most parts of Europe long before scientists discovered its etiology and transmission: the secret was development and drainage of the land that was necessary for the agricultural revolution in the West.

14

Genetic Counseling and Its Adaptation to Varying Needs

The currently accepted definition and goals of genetic counseling are comprehensive. Persons who need counseling have many and varied needs, and no single approach will suffice. This chapter presents some approaches to genetic counseling that should facilitate communication, especially when the clients have different ethnic, religious, or socioeconomic backgrounds. Admittedly, there are troublesome situations in which no technique is entirely satisfactory. In these areas research is needed. Ways must be found to assist individuals or families who have special ethnic, socioeconomic, or language needs. With each discovery of procedures for individual, newborn, or prenatal diagnosis and for therapy, the number of families in need of access to these techniques increases.

THE GOALS OF GENETIC COUNSELING

An ad hoc committee of the American Society of Human Genetics developed a comprehensive definition of genetic counseling (1975). This definition emphasizes communication—in particular, the communication of facts about the disorder in question to couples at risk—concerning the genetics of the condition and the options for avoiding the birth of or for treating affected children. Families should also be helped to identify their personal goals and value systems so that they can make decisions in their best interest. Inherent in this complex process is the requirement to provide ongoing support for the family or individual at risk while they work through their psychosocial problems, emotionally adjust to the disorder, and eventually make their decisions.

The most difficult aspect of this complicated and often prolonged process is that of conducting the counseling procedure to arrive at an outcome that suits the desires and goals of the family, not those of the counselor. Information, ideas, concepts, and options must be presented in a balanced fashion. Counselors must avoid directing a family or an individual toward a particular decision. This approach is justified primarily on ethical grounds. It is obvious at the outset that counselors are not, and really cannot be, completely neutral about the information they present: in other words, counselors cannot be "nondirective" in the strict meaning of the word. Nevertheless, each counselor must strive for this ideal.

There are common goals in all genetic counseling, but the method of achieving these objectives differs depending on the following factors:

1. The type of disorder, with special attention to its psychological and emotional burden
2. The natural history and prognosis of the disorder
3. The recurrence risk for the condition
4. The value system and goals of the clients
5. The psychosocial and socioeconomic status of the client

An indispensable characteristic of effective genetic counseling is flexibility: the ability to tailor the counseling strategy for an individual or a family in a way that will take these five factors into account.

Genetic counseling properly performed is a multifaceted, multifactorial process. Its success as a communicative, educational, and psychotherapeutic process hinges on a variety of factors, which include the motivation, emotional state, and educational background of the counselee, and the background, training, and motivation of the counselor. The setting of the counseling session is also important to successful communication.

The counseling client's motivation and emotional state are crucial. An unmotivated or poorly motivated person usually will not seek counseling in the first place. An uninterested, depressed, anxious, or hostile individual will not be receptive to any information and will receive the information ineffectively, and perhaps inaccurately. In some of the large-scale screening programs for the hemoglobinopathies, counselors often personally deliver the test result to the client—which means that the result of the screening test is unavailable without counseling of some type being delivered. In other settings, counselees must return to the testing site to learn the test result, which is delivered to them just before or during the course of the counseling session. Yet these nonvoluntary methods of counseling can be problematic. For example, most individuals (more than 70 percent, according to a small survey at Howard University) believe that they are not carriers of the sickle cell gene, or any other hemoglobinopathy gene. When they receive the news that they carry an atypical or abnormal gene, they often experience significant anxiety, and sometimes express hostility toward the counselor.

One study (Murray et al., 1974) demonstrated a positive correlation between educational level and the score on a written twenty-item test of knowledge following sickle cell trait counseling. This result may not be directly extrapolated to the expected counseling result in all genetic disorders, but it does indicate that in at least this one condition, the educational level of the counselees plays a role in the amount of information acquired during counseling using a standardized format. Most counseling programs have had general uniformity in the educational level of the clients and/or their families. Genetic counseling clients generally tend to be middle-class people with at least a high school education (Carter et al., 1971). In screening and counseling programs for Tay-Sachs disease, for example, the study population was extremely well educated, with the average person having a college education or beyond (Childs et al., 1976). However, at least one study suggested that the information these individuals acquire after one counseling session tended to be poor.

Some of the variability in the educational and psychological outcome of genetic counseling is almost certainly the result of differences in the background and training of the counselor. The combined effect of these factors is difficult to measure, especially when one tries to determine their long-term effects.

THE COUNSELOR'S ROLE

The content and style of genetic advice depend on the background and training of the genetic counselor. For example, the traditional role and training of the physician may be responsible for the tendency of the M.D. counselor to be more directive and authoritative than the Ph.D. counselor (Sorenson, 1973; Headings, 1976). On the other hand, a counselor with a strong background in behavioral sciences (such as a psychologist or social worker) is much more likely to focus on psychological and social aspects of the counseling process, and to have a less comprehensive knowledge of the disease process in question. Although there does not appear to be general agreement on what the characteristics of the ideal genetic counselor ought to be, one psychologist (Money, 1975) felt strongly that the counselor's primary training and background should be in the behavioral sciences (psychology or psychiatry), while a background and knowledge of medical and genetic principles might be acquired secondarily. Money based this advice on the belief that the psychosocial component of the counseling process is most often neglected or is less effectively handled. Some experienced counselors assert that the psychotherapeutic aspect of counseling is much more important than the communication of facts about inherited disease. Kallman (1969), who held this view, called counseling "short-term psychotherapy." Ideally then, good counseling means effectively communicating the genetic facts to a client, while paying close attention to the client's emotional needs that are related to his or her reaction to the impact of the genetic disease.

It is obvious that effective counselors must be highly motivated and intensely interested in all phases of the counseling process, even if they are not fully competent in all aspects of the process. There may be times when a counselor's interest in counseling may wane significantly. If this should happen, someone else should take over in the best interest of providing effective counseling. Motivated, intelligent clients quickly detect a negative attitude or a lack of interest on the part of the counselor. Effective communication is very difficult under these circumstances, and counseling, therefore, would be ineffective.

PHASES OF THE COUNSELING PROCESS

There are four readily discernible phases in the counseling process:

1. Characterization
2. Education
3. Evaluation
4. Follow-up evaluation

Each phase is comprehensive in scope and does not necessarily proceed in the order listed. For example, education may have begun before the client arrives for counseling. Evaluation of the educational process and of the emotional state of the counselee occurs during all phases of counseling.

It is assumed, however, that an accurate diagnosis of the disorder has been made and the Mendelian or empirical recurrence risk has been determined before the first counseling session. At the very least, the counselor will have determined whether the condition does or does not have a significant genetic contribution.

Characterization

If one wishes to communicate information and exchange ideas effectively with another individual, it is crucial to know some of the characteristics of that individual. The same holds true for the genetic counselor and the counselee. Here are some characteristics important in counseling:

1. Emotional state, for example anxiety, hostility, depression
2. Educational level
3. Expectations about genetic counseling
4. Socioeconomic status
5. Religious beliefs and/or ethical value systems
6. Cultural values, such as family structure and the value of children
7. Information and attitudes already held about the disorder

We listed assessment of the emotional status of the counselee first, because it so strongly influences and shapes the counseling process. Some informal evaluation of the counselee's mental state should be made and recorded (privately), at least at the beginning and end of each counseling session. This will help the counselor avoid ignoring sometimes subtle indicators of emotional distress that would interfere with effective counseling. One can ask the counselee directly how he or she feels, but responses to this question may often be misleading. Individuals may not feel comfortable with the counselor during the first session and so may consciously or unconsciously conceal true feelings of hostility, depression, or anxiety. An observant, empathetic, and experienced counselor should be alert to the clinical signs of anxiety, hostility, or depression, which might interfere with effective communication.

Evidence has already been presented suggesting that educational level is directly correlated with the amount of information retained by people advised about the pathogenesis, origin, and genetics of sickle cell trait when this is measured by a brief multiple-choice questionnaire. In general, it would appear logical to assume that the higher the educational level of the clients, the better equipped they should be to process, assimilate, and understand relatively complicated and unfamiliar biological concepts. This is generally true, except for the ability to understand and apply the statistical principles of odds or recurrence risk. Very few people—even well-educated individuals—are familiar with and comfortable with probability. The counselor must work very hard and use special techniques to communicate these ideas.

Clients of different socioeconomic backgrounds expect different outcomes from genetic counseling. To meet the counselees' needs, the counselor must have some idea of what the individuals, parents, or families expect from genetic counseling. One approach to this problem is to begin the counseling session by asking directly just what the clients expect to result from the counseling experience. Some clients are not primarily interested in knowing about the disease in question or about the genetics of the disorder. Instead, they want to know only what will happen to them or their family, or whose "fault" it was that their child "has a terrible disease."

Knowledge of the socioeconomic status of a family enables the counselor to do a better job of helping the parents of a handicapped child make plans for the child's medical program and for special educational needs. The counselor will also have a better appreciation of the economic burden imposed on the family by a handicapped child, and will be able to take specific steps to find governmental assistance where it is needed. In one sense this should not be necessary, because all families with handicapped children should have access to good-quality support for the needs of their children, regardless of their income. This is true in many states, and in many countries, but, unfortunately, not in all.

Religious, ethical, and cultural value systems can be critical elements in an individual's or a family's decision to take a particular course of action or to reject an option that might seem positive to others. This may be especially true where prenatal diagnosis and termination of pregnancy are concerned. If counselors determine at the outset that a family has strong Catholic or orthodox religious beliefs that forbid abortion, even when an abnormal fetus has been identified, they should not be surprised when this option is rejected. Religious, ethnic, and/or cultural beliefs may also strongly color a family's reaction to the birth of a deformed child. The skilled counselor may detect feelings of extreme guilt based on the idea that the mother and father are being punished for sins committed in the recent or distant past.

Cultural attitudes and practices should be considered because, like religious values, they may influence an individual's responses to reproductive options or the interpretation of medical or genetic information. This is especially vital when modification of traditional reproductive practices may be necessary. For example, studies of the attitudes of black and Hispanic males reveal an almost universal rejection of artificial insemination as an alternative reproductive option (Buckhout, 1971). In these same cultural communities children are highly valued, so abstaining from child bearing can hardly be considered a viable reproductive option for many at-risk couples. In some communities, sickness is considered a normal part of human existence, and the idea of a chronically ill child for them may not seem so threatening. The counselor may have to bring these values to the surface by direct questioning.

Most clients have preconceived ideas about the disorder in question, even if they have never seen an affected person. These ideas or concepts are often erroneous and should be recognized in advance so that they can be eliminated and/or corrected. To correct misconceptions held by the counselee, it is helpful to point them out and then state clearly what facts are known about the disorder.

Education

Providing information to the client in clearly understood terminology with appropriate illustrative materials lies at the heart of the counseling process. This is a special challenge when the clients are from different cultures or when their language is different from that of the counselor. If they do not clearly understand the facts about the genetic disease, including its manifestations, prognosis, genetic mechanisms, and recurrence risk, they can hardly be expected to make a rational decision about what they want to do. This phase of the counseling program is so important that some individuals in the field call this process "educational counseling" rather than "genetic counseling."

The educational principles involved are well known, but are not easy to apply. Factual information must be presented in easily digestible units in language appropriate to the person or persons being advised. One should have clearly defined learning objectives; in other words, the specific ideas or concepts that are to be communicated to the counselee should be clearly delineated and presented in advance. The counselor can keep a checklist handy to be certain that the essential concepts, at least, have been presented and discussed.

Naturally, the depth of inquiry and the extent of discussion about the condition vary according to the client's background, level of education, and interest. This kind of question-and-answer format is helpful, because it almost always stimulates counselees to begin asking questions, which should lead to optimal interaction between the counselor and counselee.

In some screening programs (for example, Tay-Sachs and sickle cell), clients have already been exposed to a fair amount of information about the condition in question and mainly need to have that information reinforced. They might also wish to go into more depth in discussing the disease, its genetics, and the prognosis.

Families who seek counseling because they already have an affected child possess some knowledge of the condition and almost always have specific questions about the child's condition, particularly about recent advances in treatment and prognosis of the disorder.

It is a challenge to translate complex biological, genetic, and medical information into language that is readily understood by individuals who may be poorly educated, and who have little grasp of the most elementary biological concepts, especially of the functioning of the human body. The communication of these ideas to functionally illiterate persons poses an even more difficult problem. Special studies are required to solve this challenge.

Evaluation

Both formal and informal methods of evaluating the counseling process may be necessary to ensure that counseling is achieving its major goals of communicating information about a particular disorder, and helping the client or clients cope with it. Most counselors use informal methods or cues to ascertain whether the person is absorbing and understanding the information being presented and also how they may be coping emotionally. The counselor makes a subjective assessment of how he or she thinks the counselee is receiving the material and uses clinical clues to assess the client's emotional state. It is also essential, at least periodically, to have some formal, preferably written, method of evaluating the educational and psychotherapeutic aspects of counseling.

The simplest and most easily administered type of evaluation is a written, short-answer, multiple-choice test consisting of ten to twenty questions. This will determine whether a counselee has learned and can recall the counseling information. One should be able to define specific facts about the disorder so that the individual or family may have the essential information needed to make a reproductive decision as well as to have some understanding of the disease and its potential burden.

The basic difficulty with a specific written evaluation is that the test is relatively inflexible. It is ideal for persons who have little or no background information about the disorder in question, but it may be inadequate for clients who already have a good deal of information about and an understanding of the condition when they arrive for

counseling. One might take the position that if the counselee already
has information sufficient for decision making, the counseling process should focus on those elements of psychological adjustment and
planning that are also important in the counseling experience.

A formal oral evaluation might be substituted for a written evaluation for the following reasons:

1. An oral evaluation of the information acquired during counseling by clients who might be functionally illiterate is more reliable
   than a written exam.
2. The person who administers the evaluation can see immediately
   whether or not questions are clearly understood; individuals will
   be able to let the evaluator know that they do not understand.
3. The evaluation process is more flexible, and can be adjusted to
   the level of knowledge and understanding of the client.
4. The evaluator will have a more reliable idea of the emotional
   status of the client.

Despite its obvious advantages, formal oral evaluation has some
distinct disadvantages:

1. Oral evaluation is less efficient and takes more time and effort.
2. The evaluator may consciously or unconsciously help or interfere with the client's efforts to respond to the questions asked.
   The same person would have to administer the instrument
   every time to control for this factor, which might prove inconve-
   nient.
3. Even if the same evaluator does the questioning, his or her
   attitudes and method of presenting the questions may vary from
   day to day according to his or her state of mind, level of fatigue,
   or general motivation.
4. Clients may be reluctant to reveal their lack of information and
   so might refuse to be evaluated.

There are definite advantages and disadvantages to each approach to
evaluation, but one should determine what type or what combination
of approaches best suits the needs of the clients, the counseling program, and the disease being considered. In the final analysis, one
might have to develop oral evaluations for relatively uncommon disorders and use a written format for relatively common clearly defined
disorders, such as cystic fibrosis, sickle cell anemia, or Down syndrome.

Most, if not all, experienced counselors recognize how important it is to have some insight into a counselee's psychological state. Yet this goal is discussed relatively infrequently in the genetic counseling literature. Except for the Health Orientation Scale (Wooldridge and Murray, 1988) which was devised to evaluate the psychological status of individuals with sickle cell trait, no instrument for evaluating the emotional status of affected individuals or parents of children with genetically determined disorders is available. Standardized tests for measuring mental disturbances such as depression exist, but most (such as the Minnesota Multiphasic Index) are too cumbersome to be used routinely.

Another area has been neglected: the psychological status of people who are being counseled not because they are affected or have a child affected with a genetically determined condition, but because they have been found to be carriers of a mutant gene that in double dose would be responsible for an abnormality. It usually has been assumed that gene carriers who are not at risk to be affected, or who are not part of an at-risk mating, suffer little or no psychological trauma. Nevertheless, stigmatization is common in persons who are carriers for sickle hemoglobin (sickle cell trait).

Experienced counselors routinely attempt to evaluate subjectively the way that counseling affects the counselee's emotional status. As in any psychotherapeutic transaction, the counselor must have a clear sense of "self" so that personal "hang-ups" do not interfere with a clear appreciation of the needs and conflicts of the individual and/or the family being counseled.

At the beginning and after each counseling session the counselor as well as other health professionals should record their perceptions of the client's emotional status. They should note whether signs of hostility, anxiety, depression, or denial are present, and whether these signs are increasing or decreasing. One measure of counseling success is whether or not these signs of emotional stress have been successfully revealed, confronted, and resolved.

FOLLOW-UP COUNSELING

The counselor is vitally interested in what the counselees do after they have been counseled. Most reports that follow up on genetic counseling have focused almost exclusively on the reproductive outcome. These studies generally have shown that couples with a moderate to high risk of recurrence (25–100 percent) of a genetic disorder have fewer offspring than couples with a low genetic risk (10 percent and less) (Carter, 1967; Stevenson, 1961). These investigators also showed that couples at risk to have offspring with genetically determined disorders with a perceived high burden had fewer offspring than did couples who are at risk to have children with a perceived low burden. Burden (Murphy, 1973) is defined in terms of the expected cost of the disease, not only in money but also physical and emotional pain, labor, death, and moral conflict. For example, the burden of color blindness or postaxial polydactyly is trivial, but that related to trisomy D (Patau syndrome), anencephaly, or Tay-Sachs disease is very heavy. The duration of the burden is also vital. Of course, the burden of disease may be perceived differently by individuals, families, and by people whose cultural and religious backgrounds are unique.

Even though the evidence from the follow-up studies cited suggested an association between clients' reproductive behavior and genetic counseling, the documentation that reproductive behavior is a direct result of the information that the counselees have been given or related to their contact with the counselor was largely circumstantial. Follow-up evaluation, as described here, is concerned as much with the couple's interactions and personal adjustment to the presence of the disease in the family as it is with determining the factual information they still recall from the counseling session.

The basis of the decision that couples make about reproduction is quite complex and includes emotional, cultural, religious, and economic factors (Griffin et al., 1976). Counseling information appears to vary in the role it plays in the reproductive decision that is finally made by a couple. What is most important is that all couples in the Griffin study who decided to have children after being counseled, recalled and understood the recurrence risks involved with each pregnancy, in contrast to those couples who decided not to have additional children. This is consistent with findings of other authors (Emery, 1975).

At the very least, the counselor who follows the comprehensive definition of counseling should schedule return visits at intervals appropriate for the psychosocial needs of couples at risk. Couples clearly need support and guidance as they continue the decision-making processes involved with caring for their affected child so that child will achieve its maximum potential.

GENETIC MECHANISMS AND RECURRENCE RISKS

When it is evident that a client's undue tensions about genetic issues have been resolved or reduced, the counselor can describe the genetic mechanisms involved and explain the recurrence risk for the condition. It is helpful to use various audiovisual methods to illustrate genetic mechanisms, especially the meaning of recurrence risk. Flipping coins, using playing cards, or throwing specially color-coded dice have all proved useful in explaining and illustrating simple Mendelian modes of inheritance. More elaborate devices can be adapted to illustrate empirical recurrence risk. For example, toy roulette wheels can be adapted to this purpose by placing specially designed covers over them consistent with recurrence risks in the disorder. Colored balls in containers might help families visually understand what 1 in 10, or 1 in 50 recurrence risk means. Of course, one need not dwell on this kind of exercise if parents have already made the decision not to have any more children. Nevertheless, parents do sometimes change their minds, and at the very least they should be introduced to this idea. A few well-framed questions about the genetics and recurrence risk that have just been reviewed should help the counselor determine whether or not these concepts have been understood. This strategy is especially valuable for those couples who do wish to have additional children and who must decide what reproductive options they will choose. This is also an especially sensitive and vital area of the counseling process, because some options might upset clients who may not have realized that if they have additional children there is no way to avoid a high risk of having another affected child.

Difficulties may arise in discussions of prenatal diagnosis and artificial insemination, because some people have religious or cultural prohibitions against abortion, and a majority of males of all ethnic groups do not accept artificial insemination (Buckhout, 1971). As noted earlier, more than 90 percent of black and Hispanic men reject artificial insemination as an option. Furthermore, even though many women are willing to consider artificial insemination, 80 percent of them reject it if the sperm will be donated by a male other than their husband. On the other hand, a good many women find artificial insemination an acceptable option while their husbands reject it. This can lead to a rift in the relationship if the wife perceives her husband's attitude as being primarily directed toward blocking her desire for self-fulfillment as a mother. Unless the man is able to be fulfilled as a father, he will usually feel left out if artificial insemination is chosen as an option. Only a man with unusual ego strength will be able to accept this option.

Counselors must be especially aware of situations where the couple is split on the course of action to be taken. The couple should be assured that the counselor will continue to work with them until the conflict has been resolved. Where the rift is deep, psychiatric consultation may be sought, or the couple may be referred to a specialist in marital relations. The birth, or the possible birth, of a defective child into the family may exacerbate a relationship that had already begun to degenerate.

At the conclusion of counseling sessions, the couple may voluntarily state that they have made a decision to have children or not to have children, but it is very important to encourage them not to fix this decision in their minds, but rather to discuss it together, with close friends or relatives, or with their priest, or minister, or rabbi, or other religious leader if they have a formal religious affiliation. Additional counseling sessions allow the couple to discuss their decision with the counselor at that time as well as to consider any other questions that have come up in the interim.

PLANNING THE CARE OF THE AFFECTED CHILD

The question of plans for treatment and long-term care of an affected child may arise during the first interview, during the diagnostic evaluation, or during the initial counseling session. Frequently, other specialists who have seen the child have not provided the parents with the opportunity to discuss the long-term aspects of their child's problem. When children are retarded the parents usually want to know what can be done about special schooling or whether they should consider institutionalization. Even when the physician has outlined a program of care and projected the developmental program of the child, the parents still need to be directed to specific professional agencies or institutions where they will be able to get assistance in providing the best opportunity for their child's development. In many parts of the world where medical resources are limited, there are no agencies or institutions to provide for the special needs of retarded or handicapped children. And such agencies or institutions are often either limited or lacking in personnel and facilities in the United States.

PSYCHOSOCIAL CONSIDERATIONS
IN THE UNITED STATES

It is apparent that at each stage of the counseling process psychological and sociological considerations must be kept prominent. Counseling is expected not only to inform and instruct but also to help families cope with a child with a birth defect. Parents who learn that they are carriers of mutant genes have to contend (Falek, 1972) with this often unexpected knowledge. They also confront negative societal attitudes toward children with genetically determined handicaps.

When recurrence risks are presented to parents, the counselor must make a special effort to determine their psychological state because it might significantly influence their interpretation of the risk figures (Pearn, 1973). Clients who are emotionally positive may interpret a 1 in 4 or 25 percent risk as "good news," but to a depressed person, a 1 in 20 or 5 percent risk (a low genetic risk) may seem unacceptable. Evidence suggests that many people who seek counseling for Down syndrome and sickle cell anemia have feelings of anxiety, hostility, and depression. The counselor must be able to identify these feelings in parents who seek counseling; parents must be helped to recognize them and cope with them. Parents of genetically handicapped children may experience other negative feelings, such as denial, guilt, grief, mourning, and the psychology of defectiveness (Shore, 1975).

DEALING WITH IRRATIONAL GUILT

Parents who have given birth to children with genetically determined defects often experience deep-seated feelings of guilt. Almost always they will harbor conscious or subconscious feelings that something they did directly caused the problem. Many are certain they are being punished for something they did, or did not do, in the past. Deeply religious people are especially prone to these feelings. They frequently believe that all events in the world are under the control of the "Almighty"; they are loath to accept the idea that the birth of their abnormal child was a chance happening. In some cultures family members may be convinced that they have been "cursed" by someone they have wronged or angered. Parents who experience a second abnormal birth in the family may experience feelings of "cosmic guilt" in which they perceive themselves as having been singled out for special punishment (Fletcher, 1972) These feelings are accentuated if the parents were at all ambivalent about having the child.

Guilt feelings can become irrational and overwhelming in some families where the mother, father, or both are already suffering from neurotic tendencies, or if they are having an especially difficult time coping with the child. No matter how the counselor tries to work with the parents' efforts to relieve the guilt, it may be fruitless. Psychiatric or pastoral consultation may be necessary if genetic counseling is to be successful.

Another reason why families must deal with guilt, especially irrational guilt, is because it can lead parents to devote inordinate energy and family resources to the care of the defective child. In extreme cases, healthy siblings may be ignored to the extent that they may be physically and emotionally damaged. This kind of behavior motivated by overwhelming guilt must be thwarted early if the unaffected child or children are to avoid being damaged. Where this kind of behavior is suspected, a home visit by the counselor or a specialist in family dynamics is necessary. Some parents involved in this emotionally destructive guilt do not recognize it, and special effort must be expended to make them aware of it.

THE PROBLEM OF STIGMATIZATION

Inherited or genetically determined disorders often elicit social stigmatization. Sometimes they are viewed with considerable disdain in a sociocultural sense, because parents transmit the disorders to their children. The person affected by a genetically determined disease may develop ideas of unworthiness or inadequacy, because the effect of the genetic defect is extrapolated to the whole person. These feelings exist even where the gene does not produce any obvious external abnormalities (for example, hemophilia).

The feelings of defectiveness may also be incorporated by the parents of a child affected by an autosomal recessive disorder, or by the mother of a child with the Down syndrome. If latent feelings of unworthiness already exist, the knowledge that they carry a mutant gene, or that they might have transmitted a chromosomal anomaly, can result in feelings of stigmatization in the parents of an affected child. The potential for these feelings is so strong in many individuals that they refuse to be tested to confirm the fact that they are mutant gene carriers.

This feeling can be observed in the black fathers of children with sickle cell disease, and when the husband of a mother with HbAA has a child with HbAS. He may refuse to be tested because it would prove that he had passed the mutant gene on to this child. Individuals who behave in this fashion often justify their behavior by stating "I'm fine. Nothing's wrong with me!" even when it has been emphasized in counseling that sickle cell gene carriers are generally healthy. These individuals would feel stigmatized by the idea that some part of them is defective or abnormal. If the counselor is aware that such feelings can exist—even in carrier parents—then such emotions are much more likely to be discovered, confronted, and modified, or placed in the proper perspective, so they will not interfere with the counseling process.

HEALTH PROFESSIONALS IN GENETIC COUNSELING

The methods of genetic counseling and the techniques that have been discussed here require the skills and knowledge of more than one health professional. It is impossible for a single specialist to possess all the skills and knowledge to diagnose the extremely wide range of disorders that have a significant genetic contribution, and the medical, psychological, social and economic considerations that must be understood.

Experienced genetic counselors (Epstein, 1973; Hsia, 1974) advocate the team or group approach to counseling because it makes the optimal performance of all functions and activities related to genetic counseling more likely. A major problem with this system, however, is to coordinate the services of these professionals.

Currently, the actual counseling often is conducted by a non-physician—a nurse, a specially trained genetic assistant, or a psychologist. The reason advanced for having a non-M.D. health professional do the counseling is that physicians without special psychiatric training tend to be overly directive, and they lack the sensitivity of other specially trained health professionals. It also appears that clients often are more open and communicative with health professionals other than physicians.

THE ECONOMICS OF GENETIC SERVICES

Pyeritz, Tumpson, and Bernhardt (1987), Bernhardt et al. (1987), and Bernhardt and Pyeritz (1989) examined the economics of genetic services. In the first of the series of articles, the authors cited a most important quotation from L. R. Dice in his presidential address to the American Society of Human Genetics (Dice, 1952). "It will rarely be practical for an heredity clinic to charge fees for its services. . . . an heredity clinic cannot usually be self-supporting." Pyeritz, Tumpson, and Bernhardt (1987) pointed out that improvement in clinical genetics services "requires a sensitivity to the special problems of all the major participants in genetic services: the providers, the patients and their families, and the payers [society]" (p. 556). In a time analysis of a clinical genetics service, Bernhardt et al. (1987) found that income from clinical practice covered 37 percent of the clinical portion of personnel costs, and that cognitive clinical genetics services are labor-intensive, yield low payments per service hour, and are not financially self-supporting. To improve the economic status of genetics clinics, it was suggested that administrators should increase charges for services; bill for all services provided to family members; charge for all genetics professionals' time, including that of counselors and social workers; and even request payment at the time of service. Bernhardt and Pyeritz (1989) expanded on the cost deficiencies and added the necessity of seeking state, federal, and foundation support for services. A perennial problem is that there is no Current Procedural Terminology (CPT) insurance code specifically for genetic counseling, and third party payers may not reimburse for the service or their policies are inconsistent. Medical assistance and crippled children's programs often do not cover genetic counseling, and insurance companies reimburse at a lower rate or less often for counseling than for genetic services.

The job of providing comprehensive genetic counseling is a challenging one. The complexities that arise in advising families are even greater when counseling is done for individuals or families from Third World or poor countries. Special efforts and knowledge are required to ensure that the counseling techniques are adapted to special circumstances, and to unique cultural factors.

15

Human Genetics, Ethical Issues, and Public Policy: An International Perspective

Technological developments in medicine have introduced complex problems in ethics, law, and public policy. Some of these advances include techniques for prenatal and postnatal diagnosis of genetic disorders; potential therapy through genetic engineering; the salvage of newborns, many of whom have limited prospects for life without sophisticated medical, psychological, educational, and social support; transplants of human and other primate organs; artificial organs; the prolongation of dying through life-support systems; in vitro fertilization; and research on embryos and fetuses.

Rapid communication and travel imply that technological advances in one country could be used everywhere. Nevertheless, technological transfer is not assured, and scientific research has been restricted and censured by authorities throughout history. More recently, prohibitions against fetal research in the United States instituted as a result of political pressure from theologians, scientists, and others have delayed the application of in vitro fertilization techniques that were developed in Great Britain; public-funded research on fetal tissues is still banned as of 1990.

In many Third World countries, complex problems of the prevention and treatment of communicable and nutritional diseases, and the economics of health care, may assign very low priority to many technological advances. Even so, these scientific discoveries will not disappear; eventually some or all of them will enter into public policy debates in even the poorest of countries. Genetics programs should be coordinated with other health care priorities, and should take into consideration specific characteristics of the disorder and of the populations in which it is found.

The legacy of National Socialism in Nazi Germany and of the eugenics movement in the United States necessitates caution in the enactment of genetics programs among minority groups, who may view efforts at prenatal diagnosis with selective abortion of affected fetuses as subtle genocide. This is particularly true if the minority groups are poor, undernourished, and undereducated and have high infant and mortality rates, and if there is little prospect for change.

Undoubtedly, the acceptance of prenatal diagnosis depends on the severity of the genetic disorder that is to be prevented. As we have seen, the morbidity and mortality of sickle cell disease are variable and quite unpredictable: some patients die early, while others have long, productive lives. But β-thalassemia is different: most patients who are homozygous for β-thalassemia do not live past the second decade; frequent transfusions and iron chelation therapy are common. Bone marrow transplantation, however, does offer hope. Children with Tay-Sachs disease, on the other hand, usually die before the age of five, and treatment is prohibitively expensive.

The presence or absence of severe mental deficiency is an important consideration in genetics programs. Abortion of fetuses with genetic disorders associated with severe mental deficiency is far more acceptable than is abortion of fetuses who would not be so affected. If severe mental deficiency is associated with multiple congenital abnormalities, the indications for abortion are even more evident.

An examination of the implications of the new medical technologies in affluent and in poor countries would take many volumes. This chapter will investigate only the effect of public policy on hemoglobin genetics programs in selected countries characterized by political, economic, ethnic, and religious problems—with the hope that some principles may emerge that could form the basis for genetics programs. Many of the opinions are speculative: definitive solutions may never be possible; however, some lessons learned in the United States may have application elsewhere. Even so, considerable research and planning are indicated in diverse societies. On the other hand, technological developments are so rapid that poor countries may soon find themselves overwhelmed, as we are in the United States.

THE UNITED STATES

In the United States, genetic and other medical technology has had a significant effect on legislation, the courts, and public policy. First, the social movement for mandatory sterilization of individuals with mental deficiency and other disorders (or, of those who merely happened to be poor) followed the development of relatively safe techniques for tubal ligation and vasectomy. Second, abortion on demand became legal when simple procedures were developed for the initiation of abortion in the first and second trimesters of pregnancy, and also because women were already obtaining abortions in large numbers— often under unsafe conditions. Third, the development of accurate screening techniques for a variety of genetic disorders led to state, federal, and private education, counseling, and screening programs. Fourth, the courts have generally supported suits against physicians and other health workers who fail to inform their clients about recent advances in the prevention of the birth of children with genetic and other disorders, or who failed to counsel properly or perform laboratory tests accurately. On the other hand, the courts have supported the right of the state to refuse public support for abortion even though public monies provide funds for research on prenatal diagnosis—with the principle objective of aborting affected fetuses.

The development of new technology is almost always followed by delays in societal acceptance. But today, many of the applications of the new technology are anathema to many groups in the United States, and perhaps the United States is not exceptional. Even so, genetic technology is here to stay. It is naive to believe that it can be ignored.

To understand current issues in genetics programs, it is useful to review certain historical, ethical, legal, and technical issues. We will begin with eugenics and end with prenatal diagnosis for genetic disorders.

The Eugenicists

With the rediscovery of Mendelian genetics, many scholars, philanthropists, theologians, legislators, jurists, bigots, racists, and others believed that a wide variety of disorders thought to be genetic could be eradicated by mandatory sterilization. The eugenicists believed that the world was divided between the fit and the unfit, that the fit should be encouraged to reproduce and the unfit discouraged—by coercion if necessary (Ludmerer, 1972).

The background for legislation proposed by the eugenicists for mandatory genetics laws today is supported by several landmark Supreme Court cases. *Munn* v. *Illinois* (1876) established that:

1. Every statute is presumed to be constitutional. The courts ought not to declare a law to be unconstitutional unless it is clearly so. If there is doubt, the will of the legislature should be sustained.
2. Members of society part with some rights or privileges, which as

individuals not affected by relations to others they may retain.
3. This power may, of course, be abused. But the people must resort to the polls for protection against abuses by the legislature, not the courts.
4. Under the police power of the state, private interests must be subservient to the public interest.

The Ethical Conundrum

Munn v. Illinois is based on utilitarian ethics. Utilitarians believe that laws and social arrangements should be so constructed as to place the interests of everyone, as much as possible, in keeping with the interests of society. Education should be so constructed as to instill in all an association between one's own happiness and that of the whole. There must be implanted in everyone an impulse to promote the public good. But one of the major disadvantages of utilitarian ethics is its potentially adverse effect on a minority that is discriminated against. Williams (1972) emphasized that not all societies or segments of society can adopt a utilitarian ethic and that when the future is dim a person must be mad who would sacrifice the present. Unfortunately, some ethicists attempt to translate utilitarianism into a system of social or political decision making. Williams dissented. He believed that utilitarianism should be treated as only a system of personal morality.

Mandatory Sterilization

In October 1926, the Supreme Court decision in *Buck v. Bell* (1927) affirmed the right of society to mandate sterilization. Carrie Buck, an allegedly feeble-minded white woman, was an inmate of the State Colony for Epileptics and the Feeble-Minded. She was the daughter of a feeble-minded mother in the same institution and the mother of an illegitimate, allegedly feeble-minded child. Justice Holmes stated, in part:

> We have seen more than once that the public welfare may call upon the best citizens for their lives. It would be strange if it could not call upon the best citizens for their lives. It would be strange if it could not call upon those who already sapped the strength of the State for these lesser sacrifices, often not felt to be such by those concerned, in order to prevent our being swamped with incompetence. It is better for all the world, if instead of waiting to execute degenerate offspring for crime, or to let them starve for their imbecility, society can prevent those who are manifestly unfit from continuing their kind. The principle that

sustains compulsory vaccination is broad enough to cover cutting the Fallopian tubes. . . . Three generations of imbeciles are enough.

It was later shown that Carrie Buck was only mildly retarded, that her education stopped at the fourth grade, and that her illegitimate daughter was not feeble-minded but a very bright child who died early in life (Coogan, 1953).

Eventually, more than thirty states passed mandatory sterilization laws that were usually directed against persons who were mentally retarded, were epileptic, or had criminal tendencies. Many of these laws were eventually repealed, and, today, thirteen states have such legislation on their books. Nevertheless, even though a state may have repealed mandatory sterilization legislation, sterilization can still be ordered by the court by citing precedence in other states and *Buck* v. *Bell*. The next section discusses more recent mandatory genetics legislation.

Genetics Legislation

Guthrie (1961) described a microbiological assay for blood phenylalanine that could be adapted to mass screening. This test was subjected to field trials in many states and reported to be efficacious. Within a short time, most states passed mandatory phenylketonuria screening laws under the premise that if phenylketonuria were treated early with a low phenylalanine diet, mental deficiency could be prevented. This book is not the place to go into the many difficulties with this program and with problems of mass testing before appropriate techniques had been developed. It should be mentioned, however, that in many states that have mandatory legislation, no provision is made for education of the public or for counseling or for treatment and care of those infants who are found to be affected. Further, at least six states have very successful voluntary screening programs (Childs, 1975).

Let us now review the sickle hemoglobin story. In the early 1970s a solubility test for sickle hemoglobin was commercialized and actively promoted by a major pharmaceutical company with the encouragement of some physicians, scientists, and laypeople. This test did not distinguish sickle cell trait from sickle cell disease. Many community health workers, physicians, and community activists began screening for sickle hemoglobin in the black community. Educational booklets were issued by private organizations and even by the National Institutes of Health which confused sickle cell trait with sickle cell disease. Congress also made this error: the first sentence of the National Sickle Cell Anemia Control Act stated that approximately 2 million blacks in the United States have sickle cell anemia. These practices led to the following events (Bowman, 1984):

1. The creation of serious psychological distress in carriers for sickle hemoglobin
2. Erroneous interpretation of a cause-and-effect relationship between sudden death and sickle cell trait by medical examiners
3. Discriminatory mandatory sickle hemoglobin screening laws
4. Needless increases in health and life insurance rates for carriers of sickle hemoglobin
5. The exclusion of people with sickle cell trait from participation in athletics, even though some of the best athletes in professional football, basketball, and track have sickle cell trait
6. Selective screening of blacks in industry for sickle hemoglobin, even though sickle hemoglobin is found in high frequency in many populations other than Africans and their descendants
7. The exclusion of people with sickle cell trait from flight duty in the armed forces (this restriction has recently been modified)
8. Selective screening of black flight attendants for sickle hemoglobin, with the subsequent firing of carriers of the trait (many airlines now no longer practice such discrimination)

Abortion

Several important court decisions involve abortion, a subject that generates considerable controversy. The abortion issue pits race against race, religion against religion, spouse against spouse, children and parents against one another, and is the occasion for conflict among members of Congress, presidents, and the courts, to name a few. It would take several volumes to analyze the complex issues of abortion in the United States. We will not discuss here the morality of abortion, whether abortion on demand should remain the law of the land, when life begins, what "a person" is, and some of the other disputed issues. Here, we are concerned with developing a background for a discussion of how the abortion polemics may affect genetics programs and public access to prenatal diagnosis. *Roe* v. *Wade* (1973) established that a woman may have abortion on demand in the first two trimesters of pregnancy. Second trimester abortions must, however, be done in the hospital. But *Roe* v. *Wade* does not concern us here. The 1973 case, *Maher* v. *Roe* (Humber and Almeder, 1979), has considerable implications for genetics programs, particularly those in which prenatal diagnosis is indicated.

Even though women now had a right to have an abortion, many indigent women discovered that although they had a right to an abortion, and Medicaid paid for the care of the pregnancy and the child, most states would not pay for an abortion. For example, in *Maher* v. *Roe* (1973) the Connecticut Welfare Department authorized payment for abortion under the Medical Assistance Program of the Social Security Act under the following conditions:

1. In the opinion of the attending physician the abortion is medically necessary. The term "medically necessary" includes psychiatric necessity.
2. The abortion is to be performed in an accredited hospital or licensed clinic when the woman is in the first trimester of pregnancy.
3. The written request for the abortion is submitted by the woman and, in the case of a minor, by the parent or guardian.
4. Prior authorization for the abortion is secured from the chief of medical services, Division of Health Services, Department of Social Services.

Justice L. Powell spoke for the majority in denying state aid for abortion in *Maher* v. *Roe*. Powell maintained that *Roe* v. *Wade* did not declare an unqualified constitutional right to an abortion. Justice H. Blackmun, Justice W. Brennan, and Justice T. Marshall dissented, from which we quote, in part:

A distressing insensitivity to the plight of the impoverished pregnant woman is inherent in the Court's analysis. The stark reality for too many, not just "some" indigent women is that indigency makes access to competent licensed physicians not merely "difficult" but "impossible." As a practical matter, many indigent women will feel that they have no choice but to carry their pregnancies to term because the state will pay for the associated medical services, even though they would have to have abortions if the state had also provided funds for that procedure. This disparity in funding by the state clearly operates to coerce indigent women to bear children they would not otherwise choose to have, and just as clearly this coercion can only operate upon the poor, who are uniquely the victim of this form of financial pressure. Mr. Justice Frankfurther's words are apt: To sanction such a ruthless consequence, inevitably resulting from a money hurdle erected by the State, would justify a latter day Anatole France to add one more item to his ironic comments on the "majestic equality" of the law. The law in its majestic equality, forbids the rich as well as the poor to sleep under bridges, to beg in the streets, and to steal bread.

Another important stage in a court test of federal funding for abortion came in September of 1976. Representative H. Hyde (R-Illinois) strongly opposed the ruling in *Roe* v. *Wade*. He also had been the principal instigator for a constitutional amendment to nullify *Roe* v. *Wade*. On September 30, 1976, when the first version of the Hyde Amendment was enacted by Congress, consolidated cases, which challenged the amendment, were filed in the District Court for the Eastern District of New York. The plaintiffs were Cora McRae (a Medicaid recipient in the first trimester of pregnancy who wished to have an abortion), the New York City Health and Hospitals Corporation (which operated sixteen hospitals, twelve of which provided abortion services), and others who acted to enjoin the enforcement of restricted funding of abortion (*Harris* v. *McRae*, 1980). The Supreme Court ruled that:

1. . . . Title XIX does not require a participating State to pay for those medically necessary abortions for which federal reimbursement is unavailable under the Hyde Amendment.
2. The Hyde Amendment is Constitutionally valid. . . . The Hyde Amendment, like the Connecticut Welfare Regulation in *Maher*, places no governmental obstacle in the path of a woman who wishes to terminate her pregnancy, but rather, by means of unequal subsidization of abortion and other medical services, encourages alternative activity deemed in the public interest. The present case does not differ factually from *Maher* in so far as that case involved a failure to fund nontherapeutic abortions, whereas the Hyde Amendment withholds funding of certain medically necessary abortions.

The abortion saga is far from over in Congress and in the courts. Many actors are in the wings and on the stage. So far human geneticists and other scientists have managed to avoid the fray, but not for long.

Wrongful Birth—Wrongful Life

Since many genetic disorders may now be diagnosed before birth, and because of the ruling in *Roe* v. *Wade*, women now have the option of abortion of affected fetuses. Whenever a pregnant woman has an increased risk of having a child with a genetic disorder, she should be so informed by her physician. With this knowledge, parents and children have instituted "wrongful birth" and "wrongful life" lawsuits. In a wrongful birth action, the physician is sued by the parent for a perceived failure of contract involving prevention of conception or birth. In a wrongful life action, the child (now an adult) or the parent sues with the claim that the affected child would have been better off by not having been born. (The unborn state is presumed preferable to a miserable life.)

Wrongful birth cases have involved both the birth of normal children and the birth of children with genetic and nongenetic disorders (Shaw, 1984). Suits that have been instituted because of the wrongful birth of normal children generally have failed, because the courts reject the view that the birth of a healthy child, whether planned or not, is a disaster. But there have been exceptions. In the case of *Cockrum* v. *Baumgartner* (1981) in Illinois, a physician's negligent performance of a sterilization operation resulted in an unplanned child.

The categories in which wrongful birth or wrongful life action has been instituted include the following: failure to prevent conception; denial of the right of a woman to have an abortion, or failure to abort; failure to inform client of the risk of birth of a child with a genetic disorder, or failure to counsel properly; failure to perform proper tests, improper performance of tests, misdiagnosis of test results, failure to inform of test results; failure to recommend or perform amniocentesis, with the resulting birth of a child with a genetic disorder.

Dilemmas and Conflict

A vocal minority in the United States lobbies vigorously to repeal abortion on demand and for most other reasons. The courts have supported the right of the state to refuse public support for abortion even though public support provides research on techniques of prenatal diagnosis that may lead to abortion. Thus, any movement to stop such research may lead to some cessation of abortion for the poor, but it could also truncate research that would eventually benefit all. But a majority in the United States supports abortion in specific instances, including the prevention of the birth of children with genetic and other disorders. On the other hand, those of us who are more fortunate must protect the rights of the most disadvantaged members of our society to be born, if the parents so desire. Too many of us are obsessed with perfection—an impossible and an undesirable goal. If the United States is to remain a pluralist democracy, there must be accommodation. Unfortunately, today, the opposing groups will not countenance converse views. The only reasonable solution that we envision is individual choice.

One of the premises of this chapter is that the precepts of eugenicists and right-to-life advocates are such effective contravening forces that neither extreme will predominate. On the side of the modern eugenicists, we find Blumenthal (1977), who believed that there should be "world-wide improvement of physical health and fitness by selective breeding of the more fit human physical types." We query who is to select "the more fit human physical types" on a worldwide, or even a local basis. Blumenthal continued, on the subject of planned parenthood,

> In view of the importance of both biological and social parenthood for mankind the following rules might apply: (a) All persons must be licensed to beget or rear a child; (b) Some might be licensed to beget but not to rear and others to rear but not to beget; (c) To raise the quality of human stock (primarily the mental) most children would result from artificial insemination with superior donors by cloning superior people; (d) All persons capable of being parents but not licensed to beget children would be temporarily or permanently sterilized by law, whether married or not. This would eliminate most illegitimacies. (P. 63)

And as we have seen, right-to-life groups would not allow abortion under any circumstances.

But, single-interest pressure groups are characteristic and essential for the proper functioning of a pluralist democracy (Dahl, 1982). On the other hand, Dahl indicated that although these groups are essential in a democracy, their independence also allows them to do harm. Dahl, in his perceptive analysis of a pluralist democracy, went into considerable detail about the characteristics and behavior of organizations that are in conflict in a society such as the United States. Organizations that are successful in imposing their will in the domain of public policy cannot represent the broad concerns that are found in large groups, for to do so would seriously handicap their purpose. Dissent within the groups is counterproductive; a middle ground cannot be tolerated. Those who attempt to mediate or to reach a compromise often find themselves besieged by both extremes. But unless consensus is achieved, conflict frequently becomes brutal. We reject the premises of eugenicists and the right-to-life groups, even though, interestingly, both groups can present quite persuasive arguments in support of their respective positions. Here we will approach a middle ground with full knowledge that such a public policy leaves us exposed to both extremes.

Someone once said that public policy is what those who have power decide. Marx maintained that in societies other than a classless society (which is, as yet, nonexistent), morality is decided by the ruling class and is used to control those who are without power. In the United States, abortion on demand was once a criminal act; today it is legal. Slavery was once constitutional. Public policy is flexible; otherwise it is ignored. Mandatory sterilization, mandatory genetic education, mandatory genetic testing, mandatory genetic counseling, and mandatory abortion for genetic or for social reasons are either public policy or they could be legislated. On the other hand, abortion may be once again declared illegal by constitutional amendment or by the Supreme Court. Prenatal diagnosis for genetic disorders could be negated. Research in genetics could be abrogated. These Draconian measures may one day be the "public will," or the will of a legislature that has been bludgeoned by a minority.

As Dahl (1982) pointed out: "The principle that any large group can rightfully dominate any smaller group for no reason except that it is more numerous is not deducible, as far as I can tell, from democratic ideas or, for that matter, from any reasonable moral principles" (p. 87).

Although abortion for genetic defects in the United States should not necessarily be decided by a popularity contest, it might be useful to refer to a National Opinion Research Center Poll (quoted by Shery, 1978), which showed that in 1975, 85 percent of respondents approved of legal abortion if there was a defect in the fetus. When the questionnaire was broken down by religion, 84 percent of Protestants and 77 percent of Catholics approved abortion if there was a defect in the fetus; and Jews were generally supportive of abortion.

Today, women have the choice of not having children with a long list of genetic disorders. But this can lead to problems. Although many genetic programs are voluntary, the day may come when couples who have children with genetic disorders may be maligned to such an extent that they will be coerced to abort fetuses with a variety of genetic and other defects or to be sterilized. If we come to that point—and we hope we do not—where does one draw the line? General rules cannot be made about all genetic disorders. Children with Tay-Sachs disease die at an early age; children with sickle cell disease have a variable life expectancy, and abortion for sickle cell anemia eliminates many who could have lived long, productive lives. But should "productive" be a criterion for life? If so, a large number of humans without genetic disorders could have been eligible for abortion. General acceptance of abortion for genetic disorders could lead to abortion for mere ugliness; or for minor defects, such as cleft lip;

or for sex determination of a child even when X-linked disorders are not an issue. On the other hand, if abortion on demand is legal, then abortion for ugliness cannot be prohibited.

Abortion for genetic disorders in the fetus may affect siblings who have the same genetic disorder. Many children, particularly adolescents, go through periods when they believe that they are unwanted and unloved. Suicide may be the end result. How do adolescents with sickle cell disease feel when they learn that their mother has aborted a fetus with sickle cell disease?

Consider the plight of the pregnant woman who is bearing a fetus with a genetic disorder. She should not be made to feel guilty if she does not want to have a child with a genetic disorder. Yet, some right-to-life advocates would force this woman to bear the child, and then oppose provisions for public support of the child. The plight of the poor in the United States is already critical without adding the burden of prohibitive hospital and home care costs to those who can least afford it. We quote from Reich (1982):

> Since October 1981, the poor and near-poor in America have lost more than $10 billion in federal support. Some 661,000 children have lost Medicaid coverage; 900,000 poor youngsters no longer receive free or reduced-price lunches; 280,000 no longer receive free or reduced-price breakfasts; 150,000 poor working families have lost eligibility for government-supported day care; 200,000 fewer pregnant women, new mothers, infants and children are getting special federal coupons for milk, juice and other diet supplements. One million people have been dropped from the food stamp rolls. In addition, 890 school districts have cut back on special education programs. (P. 32)

But if the poor are neglected, scientists are not. Public monies are spent on research in genetics and for genetic education screening, and counseling programs for untreatable genetic disorders, many of which can be prevented only by selective abortion after prenatal diagnosis. Public monies are also spent on research to improve techniques of prenatal diagnosis, with the potential end result of selective abortion. Poor people are encouraged by state and by federal genetics programs to participate. They are led to the brink of the final decision and then must be told that here support ends. A demented solution would be to deny public funds for research in genetics, but to abrogate public monies for research in genetics would also abolish research in many areas of biology, biophysics, biochemistry, and medicine. Research in some fields of education, social work, psychology, and other social sciences would also be compromised. Thus the prohibition of public funds for research in genetics would lead to a ludicrous chain reaction. Since this is unlikely, we believe that scientists have a moral obligation to insist that the fruits of their research be made available to all.

Genetics programs should be viewed as one segment of a multiplicity of health problems, even in the United States. An undue emphasis on genetics programs at the sacrifice of other health needs is not useful. In the United States, for example, about 35 million people live in poverty, and infant and maternal mortality rates in blacks are about twice those in the white population.

Access to prenatal care is generally believed to affect maternal mortality, at least in poor populations. If prenatal diagnosis is to be effective, pregnant women should seek prenatal care early. In the white population, 75 percent of women seek prenatal care within the first trimester, but in the black population, only 56 percent seek care in the first trimester. The second-trimester figures are, of course, better: 92 percent of white women seek care by this period, and 85 percent of black women. For the purpose of effective prenatal diagnosis, however, the first-trimester figures are important, since the second-trimester figures give no indication at what time in the second trimester that prenatal care is sought. If the woman seeks care late in the second trimester, there may be no time for completion of the necessary prenatal diagnostic procedures.

It is thus evident that if the new genetic technology is to be made available to all in the United States, alleviation of poverty and significant improvement in education and access to prenatal and other health care must become general public policy.

On the other hand, there are those who question whether equitable health care is possible with present constraints on costs in the health care system. Systems of health care rationing have been proposed, and some have speculated that many of the recent advances in medicine may one day be available to only those few who can afford it. Health care resources are stated to be limited in the United States. This is nonsense. Health care resources are not scarce. Our society—meaning our executive, legislative, and judicial branches of government—allocates priorities. In a report by the Children's Defense Fund, Washington, D.C. (1982), entitled "A Children's Defense Budget: An Analysis of the President's Budget and Children," many billions of dollars were found in the budget that could have been allocated to alleviate the dreadful plight of poor children in our affluent society. Some representative samples follow:

1. There was an average increase per year of $800,000 in direct residential expenses for the White House. If this one item were

eliminated, 40 percent of the child abuse preventive services that President Reagan had cut in the budget could be restored.

2. Elimination of the depletion and expensing tax breaks for oil companies—savings, $4.3 billion.
3. Cancellation of agricultural tax benefits that work to the advantage of corporate or foreign owners—savings, $1.3 billion.
4. Elimination of military unit cost overruns—savings, $8–10 billion.

These few examples of many billions of misspent dollars will suffice to establish our premise that we have not yet reached the point in the United States where we cannot afford medical services for all in need. After all, a society is best judged by how its most disadvantaged members are treated.

But what about other countries? In all nations, including the United States, the development of a rational public policy for human genetics is dependent on a host of factors, such as whether the country is poor, intermediate, or affluent; the incidence of other major health problems; the presence or absence of minority religious, ethnic, or racial groups that have been subject to discrimination; the availability and complexity of techniques for prenatal diagnosis; the accessibility of prophylactic, medical, or surgical therapy for genetic disorders, and the cost; the psychosocial and economic effect of the disorder on the family; and the economic effect on society.

Predictably, the development of genetics programs is most advanced in Western countries, but as we have indicated, once technology has been developed its use frequently becomes widespread, even if its usage may be contrary to rational health policy. Here, we will only briefly describe some programs for thalassemia in Canada, Great Britain, Italy, Greece, and we will share some of the considerations for genetic policy for sickle hemoglobin that one of us (J.E.B.) gleaned from recent travels to Saudi Arabia and Nigeria.

Many Western countries have human genetics programs, and some of these are similar to those of the United States. These countries include Great Britain, France, Italy, Greece, the Scandinavian countries, and West Germany. In Canada, prenatal diagnosis for genetic disorders is available as part of governmental health programs. In Great Britain, genetics programs are supported by the National Health Service. Some of these services screen for thalassemia in Greek Cypriot, Asiatic Indian, and Pakistani populations; for sickle cell disease in various groups; and for biochemical, chromosomal, and other disorders in the general population. In Italy, prenatal diagnosis for thalassemia is organized to such an extent that there have been significant reductions in the birth of children with thalassemia within recent years; this has been accomplished despite the opposition of the Catholic Church to abortion. Denmark, a country with a relatively homogeneous population, has a long history of the maintenance of extensive family genetic records and has well-organized genetics clinics. Isolated genetics programs are developing or are in existence in affluent and poor countries outside of North America and Europe.

CANADA

Scriver et al. (1984) described a prevention program in communities in Quebec at highest risk for β-thalassemia. The populations were of Greek, Italian, Asian, and Oriental, and French Canadian descent in which the thalassemia deme is present. A total of 6,748 persons were screened over a period of twenty-five months from December 1979 to December 1982, using mean corpuscular volume/Hb A_2 indices. This group included 5,117 senior high school students. The rate of participation in the high school sample was 80 percent. (In Canada, parental permission is not needed for screening of students.) The prevalence of β-thalassemia carriers was 4.7 percent, with a tenfold variation among the various ethnic groups. After the screening of high school students was completed, 60 carriers and 120 unaffected individuals were surveyed. Most of the carriers (95 percent) told their parents and 67 percent informed their friends of the results of the tests; and 38 percent of the carrier's parents and 18 percent of the noncarrier parents were also tested. Eleven fetal diagnoses were performed during the study period, either by fetoscopy and globin chain analysis or by amniocentesis and DNA analysis. There was one spontaneous abortion following fetoscopy and there were seven live births. The economic cost of the program (judged as cost per case prevented) was $6,700, which was slightly less than the average cost of treatment of one patient in one year.

Interestingly, a preprogram survey showed that even though 88 percent of those surveyed favored a program, only 31 percent considered fetal diagnosis an acceptable option. Nevertheless, all of the couples at risk took the option of fetal diagnosis. Other important facts emerged. The couples who were at risk claimed that they would not have considered pregnancy without the availability of fetal diagnosis. Access to prenatal diagnosis reduced the incidence of thalassemia disease. Only one couple with an affected fetus refused abortion. The authors concluded that their findings indicated acceptance of the program, relative absence of stigmatization of carriers, acceptance of the efficacy of fetal diagnosis, and cost-effectiveness.

GREAT BRITAIN

The influx of Africans and their descendants, Asians, and Greek Cypriots to Great Britain and their use of the National Health Service undoubtedly have stimulated the institution of screening programs for sickle hemoglobin and the thalassemias and the making available of prenatal diagnosis for these disorders. Weatherall et al. (1985), in a review of prenatal diagnosis for common hemoglobin disorders, remarked that as the high neonatal and childhood death rates from malnutrition and infection are controlled in developing countries, disorders of hemoglobin will create an even greater drain on health resources. Along these lines, Weatherall explored the most reliable and economic approach to organize comprehensive services for the prevention of these common disorders by prenatal diagnosis.

Weatherall's group (Old et al., 1984) attempted to ascertain the feasibility of prenatal diagnosis in Cypriot and Asian Indian populations in the United Kingdom by using seven restriction fragment length polymorphisms (RFLPs) in the β-globin gene cluster. A total of forty-two Asians and twenty Cypriots were studied, along with their families. It was discovered that 76 percent of the Asian and 35 percent of the Cypriot families had DNA polymorphisms that would have allowed prenatal diagnosis of a fetus that was homozygous or compound heterozygous for β-thalassemia. Further, in the majority of the remaining families, there would have been at least a 50 percent chance of a successful diagnosis of an unaffected or a heterozygous fetus. The contrasting success rates in these populations were explained by a greater diversity of RFLP haplotypes in Asians as compared to the Cypriots. In Cyprus, the common β^+-thalassemia mutation was found more often on an RFLP haplotype, which is also the most common β^A haplotype. Accordingly, it is frequently impossible to distinguish the normal from the β-thalassemia chromosome in parents who are heterozygous. Weatherall et al. (1985) indicated, however, that this problem has been largely overcome by the discovery of a strong linkage disequilibrium between the AvaII gene polymorphism and the IVS-1 110 mutation, which is common in Cypriot and other Mediterranean groups. Weatherall et al. maintained that the inclusion of this polymorphism in the analysis would have increased the likelihood of prenatal diagnosis to between 80 and 90 percent in the Cypriot population in Great Britain. (It should be mentioned that Weatherall's group used the term *Cypriot* without qualification. The Cypriot population in Great Britain is mainly a Greek Cypriot group that had its largest migration to Great Britain during and after war on Cyprus in 1974 between the Greek and Turkish Cypriots.)

There is no doubt that the prenatal diagnosis of β-thalassemia is feasible in Great Britain. And, as we shall see, in other countries where the health infrastructure is well developed, prenatal diagnosis of thalassemia is quite practicable. In countries with major public health problems and widespread poverty, however, the attainment of significant prenatal diagnosis for genetic disorders is highly questionable; more about this point later.

ITALY

Extensive surveys for abnormal hemoglobins and thalassemias have been reported in Italy (Tentori and Marinucci, 1983). Our analysis will concentrate, however, on the outcome of a screening program for the prevention of thalassemia in the province of Latium, Italy. The average frequency of heterozygous α- or β-thalassemia was 2.4 percent. This figure was ascertained from a study of 289,763 students. In a region of 17,000 square kilometers, a single center was able to examine about 50,000 students per year, which is about 80 percent of all intermediate schoolchildren. Since 1980, in this region, at-risk couples of child-bearing age were analyzed. Some of these couples (51 out of 161) had already had an affected child and came to the center during or before a new pregnancy, but the majority (110 out of 161) had no affected children and came for counseling before conception as a consequence of the school screening program. Of the 94 prospective couples, 35 were former students identified in the school screening program or through one of their close relatives. Of significance is that from January 1980 to April 1983, 37 of the 110 prospective couples became pregnant and six of thirty-one monitored pregnancies had a homozygous fetus that was aborted. Further evidence demonstrated that the population approved of the screening program and at-risk couples accepted prenatal diagnosis and abortion.

GREECE

Loukopoulos (1985) reported on the experience of the prenatal diagnosis of thalassemia at the University of Athens since 1974. The number of couples who requested prenatal diagnosis increased steadily over the years. The percentage of couples who requested prevention was 50 (50 out of 101 subjects) in 1977–78 and increased to 78 percent (356 out of 454) in 1984. Further, there was a steady rise in the proportion of couples who sought prenatal diagnosis after a second diagnosis, or more. The yearly number of newborns with thalassemia major showed a significant decrease by almost 50 percent in Greece. The estimated expense of caring for newborns with thalassemia far exceeded the cost of carrier identification and prenatal diagnosis.

Loukopoulos also reported on 5,000 couples at risk for thalassemia who took advantage of prenatal diagnosis. The evaluation was made possible by a WHO-sponsored registry in several countries in which the populations were at high risk for thalassemia. The number of couples who requested prenatal diagnosis rose steadily over the period of the study. The total number of tests registered from June 1974 through December 1983 was 5,617. About one-half were performed in Greece, Italy, Cyprus, and in the United Kingdom for Cypriot and Asiatic Indians. The remainder were from thirty laboratories all over the world.

Although Canadian, United Kingdom, Italian, and Greek studies showed ready acceptance of prenatal diagnosis for thalassemia, the substantiation of the acceptance of prenatal diagnosis for sickle cell anemia in the Middle East and in Africa is yet to be demonstrated. Falciparum malaria, tuberculosis, schistomiasis, yellow fever, hook worm, trypanosomiasis, typhoid fever, cholera, Loa loa, malnutrition, and other disorders may so overwhelm the health care delivery system that genetic disorders may not have a high priority.

SAUDI ARABIA

The acceptability of prenatal diagnosis for sickle cell anemia in Moslem countries is not known. One of us (J.E.B.) can report on a discussion at an international meeting in Riyadh, Saudi Arabia, in 1984. It was pointed out that the sickle cell disease found in Saudi Arabia is far more mild than that found in many other parts of the world. (See Chapter 7.) Accordingly, prenatal diagnosis with abortion would not have the priority that it may have in countries in which the more severe forms of the disease are common. On the other hand, some couples may still not wish to have a child with sickle cell disease. Accordingly, some of the participants suggested that perhaps widespread population screening programs for sickle cell disease should be instituted so that couples would know their sickle hemoglobin status before marriage. (Illegitimacy is almost unheard of in Saudi Arabia.) At that point the couple could decide whether or not to avoid marriage. It was also noted that since divorce in Saudi Arabia is relatively easy, a husband could divorce his wife if both have sickle cell trait. The divorced people could then marry others who are not carriers for sickle hemoglobin or another β-chain variation. And most important, abortion is rejected by many Moslem theologians, even though the Koran is ambiguous about induced abortion in early pregnancy. The approach of genetic screening and genetic planning without prenatal diagnosis appears to be the present course for Saudi Arabia. Even so, Islam, like Christianity and Judaism, has many religious divisions. Other Moslem countries or different Moslem groups may choose divergent approaches.

AFRICA

In many countries in Africa, complex problems of the prevention and treatment of communicable and nutritional diseases are compounded by the birth of thousands of children with sickle cell disease each year. As we have seen, the frequency of this disorder at birth may be as high as 1 in 25 to 1 in 50 in some regions. Investigators in Africa, the Middle East, and the Far East have contributed to the development of technology for ascertaining abnormal hemoglobins and thalassemias, and have cooperated with Western scientists in anthropological studies using restriction-enzyme analysis of DNA—the techniques of which are also used for prenatal diagnosis. Accordingly, scientists in these countries are now confronted with ethical, economic, and legal issues similar to those facing some Western countries.

Various Western scientists have suggested that in African countries, programs for selective abortion of fetuses with sickle cell disease may be indicated. Some African investigators retort that this policy would be a misuse of scarce health funds, and that such proposals are merely subtle efforts by Western governments to control the growth of non-Western populations. Konotey-Ahulu (1982) astutely pointed out that with respect to sickle cell disease, for at-risk couples alone, 140,000 amniocenteses per 1 million conceptions would be required to detect all cases of hemoglobinopathy and that there is a population increase of 1 million every four months in this region.

In many regions in Africa, ethical prohibitions against abortion are similar to those found in other countries. Nevertheless, there is wide divergence of views.

Even newborn screening for sickle cell disease may be impractical in some sections of Africa. Several years ago, one of us (J.E.B.) visited the sickle cell clinic at the University of Ghana Medical School, which was founded in the early 1960s by Dr. Konotey-Ahulu. On the day of the visit, the pediatric hematologist had about 120 patients. In response to a naive query about newborn screening for sickle cell disease, the hematologist replied that he had already far more patients than he could handle. And this was during a period when strict gasoline rationing adversely affected access of patients to the sickle cell clinic.

At a 1985 sickle cell conference in Lagos, Nigeria, Nagel (1985) pointed out that in some parts of Nigeria and Burkina Faso (Upper Volta), falciparum malaria kills most of the children with sickle cell anemia before they are two years of age. Thus a newborn screening program for sickle cell disease should follow or at least coincide with malaria control. But this is not the situation in Lagos. Here, the economic development of Lagos and other measures has led to a decrease in the endemicity of falciparum malaria. There are many adults with sickle cell disease in this region, many of whom were unaware of their disease until adulthood. Accordingly, a newborn screening program in Lagos may be practical.

Our group at the University of Chicago also assisted Professor Kaptue-Noche in initiating a sickle cell screening program for newborns in Yaoundé, Cameroon, where the environment differs from that of the severely malarious regions of Nigeria or Burkina Faso. (Nevertheless, antimalarial prophylaxis for travelers is still indicated in most parts of Africa south of the Sahara.)

Some countries, such as Nigeria, have formed Sickle Cell Anemia Clubs on a nationwide basis (Proceedings of International Congress on Sickle Cell Anemia, Lagos, Nigeria, 1985). The genetic activities in Nigeria presently consist of voluntary screening of individuals and couples who desire to be tested, and the care of patients with sickle cell disease. Extensive educational programs are planned, with the objective of developing supportive services for patients and their families. The couples who wish to have prenatal diagnosis in Nigeria and who are affluent usually go to Great Britain for this procedure. Even though hematologists and geneticists in many African countries have long been involved in genetic screening programs for purposes of population genetics, there are presently no plans for massive sickle hemoglobin population screening in Nigeria. The logistics and expense could be prohibitive. Nigeria is a country of about 100 million people and about 25 percent of the population are carriers for at least sickle hemoglobin.

The new medical technology has stimulated intense public debate.

Public policy is evolving on issues in which the questions are easy, but the answers are either complex or apparently insoluble. The public debate and solutions offered are frequently based on the social and cultural and religious background of the participants, on the presence or absence of discrimination by ethnic or religious majorities against minorities (or minorities against majorities, as in South Africa), by economics, and by philosophical preferences such as those espoused by deontologists and divine command theologians, utilitarians, situation ethicists, and moral relativists, to name a few. Most if not all of the philosophical arguments have been based on precepts of Western-oriented theologians and philosophers. The views of non-Western scientists, theologians, and philosophers are rarely heard.

The development of hemoglobin genetics programs continues at a rapid pace in many countries, but the end is not near. It is likely that poor countries that are overwhelmed by a multitude of public health problems will give population screening for reproductive purposes and prenatal diagnosis a low priority. On the other hand, public policy is often irrational.

16
Conclusion

We have seen that African peoples are heterogeneous populations. The genes of Africans followed slave trade routes and were disseminated also by voluntary migrations to the Middle East, Europe, and the Americas. Ancestral identity with Africa is acknowledged in Europe and the United States in the form of racial classification, but African origins in the Middle East are generally unimportant and not delineated. The social, political, economic, and racialist bases for racial classifications in the Americas and in South Africa were dealt with in some detail.

African peoples in Africa are quite variable, and are frequently genetically different from population to population by an assortment of genetic markers—probably even more so than indigenous peoples of other continents. But the environment of Africa is also diverse, from the perpetual snows of Mount Kilimanjaro, to the savannas, to the deserts, to the rain forests and the dense jungles, to the temperate climate of Nairobi, Kenya, or Kampala, Uganda, and to the intense heat and humidity of Mombassa, Kenya, on the Indian Ocean. And so the peoples differ, like their environments.

Even though the slave trade figured prominently as a source of Africans in continents outside of Africa, we cannot ignore the early migrations of Africans to the Americas so elegantly described by Von Wuthenau and Sertima (and virtually ignored by historians, anthropologists, and population geneticists). The evidence appears firm. Surely additional research is needed in this important area. Were the genes of the early African and Middle Eastern travelers to the Americas lost in the indigenous populations? What about the antiquity of devastating diseases in the Americas which were indigenous in Africans and Middle Easterners, malaria, smallpox (now extinct), syphilis, tuberculosis? And add to this list yellow fever, and schistosomiasis from Africa.

Thé delineation between environmental and hereditary variation (or both) in anthropometry and skeletal variation is difficult. A major tool, radiological analysis of bones, is probably a technique of the past, because of the dangers of radiation exposure. Unfortunately, the literature on metrical variation is complicated by the lack of standardization of measurement indices. On the other hand, since these indices are utilized by a variety of disciplines, standardization will probably never come to pass, nor would it be practical. Anthropometric measurements are not only important in anthropology and population genetic studies, but body measurements are significant in such diverse areas as the clothing industry, aircraft design, and architecture (contrast the height of doorways and ceilings in Japan to those in Sweden). Birth weight is a crucial parameter in neonatal and infant mortality, and in the predictability of cognitive and physical development of the child and adult. Growth differences are valuable measurements of nutritional status.

We made no mention of the inheritance (if any) of physical skills, mainly because of the paucity of such studies. Height, for example, may determine whether one has the potential to become a great basketball player or whether one will be excluded from Air Force pilot training. The best long-distance runners were formerly from Northern Europe, but now East Africans, particularly Kenyans and Ethiopians, often dominate the field. The descendants of West Africans in the Americas have never been serious international competitors at long distances, but black women and men excel, along with East Europeans (particularly women), at short distances. Some of this talent is undoubtedly due to opportunity and training—particularly in sports such as golf, swimming, diving, and tennis—but physical ability is probably associated with body build and physiology, which may have a hereditary component. Comparative studies are needed in these areas of physical skills.

Anthropometry has been used, unfortunately, by racialists to denigrate ethnic groups, particularly blacks, and to falsely equate certain facial types with criminality. On the other hand, there are elegant studies by physical anthropologists to delineate population variation and selection, by forensic anthropologists, and by biological and medical anthropologists to identify or associate skeletal variation with medical conditions.

The never-ending polemics about the correlation between brain size and intelligence has been laid to rest, we hope, by the valuable work of Tobias of South Africa and Gould in the United States. Africans and their descendants are forever in debt to these scientists. But, unfortunately, many individuals will remain unconvinced.

Skeletal variation and facial features are a rich source for analyzing population differences between Africans and other groups, but such variation is also a model of population variation within Africa, particularly between South Africans and other African populations south of the Sahara.

The reasons for nasal shape and its relationship to the environment have long been the subject of debate. We have added little to the discussion, because nasal differences among Africans are some of the most diverse in the world. All forms are found in Africa. Most of the relatively broad noses are found in some West Africans, indigenous Australians, and some Polynesians, but this nasal shape is in the minority in most of Africa.

The possible polygenic inheritance of cleft lip with and without cleft palate was analyzed, and studies to date favor this method of inheritance. The highest frequencies were found in Japanese, followed by European-Americans, and then African-Americans.

The teeth have been a major focus of anthropological investigation, possibly because they are resistant to disintegration over time. Significant population differences were described in the time of eruption of the third molar teeth. Most African populations show earlier eruption of third molar teeth than that found in Europeans. Further, missing third molar teeth are most frequent in Chinese followed by Eskimos, Amerindians, African-Americans, and they are infrequently, if ever, absent in Africans. Interestingly, most populations (with the exception of Africans) appear to be on the way to missing third molar teeth. Presumably, the later development of third molar teeth encounters jaws that are already "set," which could account for the pain of "late" eruption and their consequent early loss. This would be less of a problem in Africans.

It was also pointed out that persistent milk molar teeth and absence of the third molar is important surgically. Dentists should never extract a second milk molar until they have evidence that there is a premolar substitute.

The distomolar is a fourth molar tooth; it is usually rudimentary in Europeans, and is produced by an unusually prolonged dental lamina. The distomolar is presumably more frequent in Africans and is often fully developed. Since third molar teeth may be absent in Europeans, but rarely absent in Africans, one would expect that if fourth molars are missing one would indeed anticipate them to be more often absent in Europeans than in Africans. And they are.

The significance of shovel-shaped maxillary central incisors in Mongolian populations and their relative rarity in whites and blacks is not known, but at least this distinctive variation (we do not call it an anomaly because it is found in almost 100 percent of Mongolian populations) may be considered a marker for this group. Although mesial rotation of the maxillary central incisors has not been described in blacks we have seen this variation in African-Americans. Rotation of central incisors has a higher frequency in Amerindians and Japanese than in whites.

Another tooth variant rare in blacks is Carabelli's cusp—a fifth cusp in deciduous and permanent upper molars. The significance of this much-studied variant is not known. An obvious advantage is that it is an extra grinding surface, but this is only speculation.

The mylohyoid bridge consists of anomalous bone formation that transforms a part of the mylohyoid groove on the medial aspect of the mandible into a canal. At one time this variant was viewed as a Mongolian marker, but the Khoisan of South Africa have one of the highest frequencies of this variation. The Khoisan also have high frequencies of the torus palatinus (a bony ridge in the midline of the hard palate), and the torus mandibularis (a hyperostotic anomaly of the mandible), variants that are also high in some populations of Asian descent. Other facial configurations of the Khoisan resemble Mongoloid populations, but there is no obvious connection between these groups.

On the other hand, brachymesophalangy V (shortening of the middle phalanx of the fifth digit) is much more frequent in Mongolian populations than in Europeans or Africans. Singer, Kimura, and Gajisin (1980) found no one with this anomaly in various South African populations. Hence, skeletal resemblance of Mongolian and some South African populations appears to be confined to the skull. Inexplicably, none of the similarities between some South African populations and Mongolian populations have been found in other African populations. But, as we have seen, the affiliation of indigenous South Africans and West Africans has an ancient hiatus.

Triquetral-lunate fusion is quite common in populations of African origin, but is rare is most other populations (Europeans, Asian Indians, Amerindians, Mexican-Americans, and Puerto Ricans). This anomaly is thus most useful in physical anthropology, and could be of considerable assistance in forensic identification of bodies. The overwhelming prevalence of perforation of the corono-olecranon septum in Africans (47 percent) as contrasted to Europeans (6 percent) could also have forensic significance, when combined with other skeletal markers.

Anterior femoral curvature is infrequent in African-Americans. The femurs of European-Americans were intermediate between U.S. blacks and Amerindians, who had the most curvature. It is believed that since femurs of blacks were also longer, denser, and of greater diameter than those of other populations, such femurs are much more resistant to the bending forces that are imposed by body weight and muscular action. We found no studies of anterior femoral curvature in other African populations; these would be important because the body features of many other African populations differ from those of West Africans and their descendants.

Fractures of the hip are more common in whites than in blacks, and fractures of the hip in elderly white women are more common than fractures of the hip in black women. Much has been made recently of the role of calcium and estrogenic hormones to account for the differences; however, anatomical variation may also be a factor in population differences. Walensky and O'Brien (1968) found that femurs from black women were longer, more dense than those of white women, and possessed a larger neck-shaft angle and smaller angle of inclination. Studies of diverse African populations could determine whether these population disparities persist.

Reportedly, tear duct size is larger in the skulls of Africans compared to those of Europeans, and the alleged rarity of dacrocystitis in African-Americans may be in part explained by differences in skull morphology. Otolaryngologists should confirm this important skeletal finding, which has medical implications.

Skin color has been the major visual marker used to differentiate Africans and their descendants from other groups. But the use of this obvious feature is fraught with error. Even though Africans have been called Negro, Negroid, and black as a "racial" designation, many indigenous Africans do not have black skin (in fact the color black is rarely seen); some have complexion ranging from light brown to that of Southern Europeans. But the skin of most Africans south of the Sahara is dark to light brown. Even so, many other populations have skin colors that rival or are even darker than that of most Africans, such as southern Asian Indians, Native Australians, and isolated Pacific Island groups (so-called Negritos).

Skin color, of course, is most readily affected by sunlight, and accurate measurement for comparison purposes requires sophisticated instruments, which are not standardized, one against the other. And even though skin color has been viewed as a model of quantitative inheritance, the mode of inheritance is unknown. On the other hand, the results of measurements of skin color indicate phenotype, not genotype. Eventually DNA analysis should add to our understanding of this important physical feature.

Curiously, we still are not certain about the advantages of one form of skin color over another. The vitamin D selective effect, though attractive, is counteracted by the disadvantage of the heat effects of dark skin in the tropics, and the lack of vitamin D intoxication of light-skinned peoples in the tropics. (But most light-skinned peoples do tan.) Light-skinned peoples are at an increased risk of developing basal cell carcinomas, which are usually not fatal, squamous cell carcinomas, which may be fatal, and melanomas, which are frequently fatal. Thus even though the selective advantages of skin color in various climates may not be well established, there are at least sufficient data to state that light skin is advantageous in northern climates and dark skin in tropical climates. But these differences have become less important with the use of protective clothing, medication (vitamin D and vitamin D-fortified products), education about the hazards of actinic rays, and sun screens for the skin.

There is no doubt, however, about the disastrous effects of albinism in tropical Africa and other similar regions in the form of severe sunburn, blisters, solar elastosis on exposed parts of the body, solar keratoses, chronic ulcers, and basal cell and squamous cell carcinoma. And most tragic are the findings of Okoro (1974) that no albino over the age of twenty years in Nigeria was free from malignant or premalignant skin lesions, and that mortality of albinos is markedly increased after age thirty. Here, widespread public health education of albinos and the parents of albino children should be instituted, because most of the complications are preventable by shielding of exposed surfaces of the skin. But even this well-intentioned recommendation is quite impractical, as anyone who has had young children will attest. And protective measures will, unfortunately, add to the stigmatization that albino children experience—except in the occasional society where albinos are revered.

Melanomas are rarely of public health significance in Africans, but the relationship between pigmented patches on the soles of the feet and melanomas should be further investigated. There may even be enough evidence now to suggest that all pigmented patches on the soles of the feet in Africans should be removed.

Although dermatoglyphics have long been the subject of investigation by physical anthropologists and workers in forensic science, Penrose and Leosch (1970) maintained that the traditional methods of palmar dermatoglyphic classification should be modified for pathological comparisons and genetic and population genetic studies. These investigators recommended that all loops should be described. (A whorl is included, because a whorl always consists of two loops.) All triradii should be enumerated, and the most important ones should be specified. Arch formations, cusps, multiplications, and vestige fans may be neglected because they are not topologically significant, as are the exits of main lines. Even so, this information may be appended to the description.

Population studies of digital dermatoglyphics in Africa were painstakingly reviewed by Vecchi (1981). Several previous studies found geographical clines of digital dermatoglyphic patterns in the form of a decrease in the frequency of whorls, an increase in the frequency of loops from north to south in West Africa, and an increasing frequency gradient of loops from west to east in North Africa. Vecchi confirmed only some of these gradients. He reported an increase in loop frequency and a decrease in whorl frequency in a north-south direction for West Africa and in an east-west direction for North Africa. Vecchi attributed his findings to different criteria for the selection of samples. He also found that if the entire Sub-Saharan region is taken into account, the distribution of digital dermatoglyphics is far more complex than a simple north-south gradient of increasing loop frequency and decreasing whorl frequency.

Even the Pygmies are heterogeneous for digital dermatoglyphics. The prevalence of whorls among the Babinga of the Central African Republic are among the highest in Africa, but the Efe (Bambuti) Pygmies of the Ituri Forest have a low frequency of whorls. Accordingly, digital dermatoglyphics confirm the genetic heterogeneity of Africans.

Studies of digital dermatoglyphics of African-Americans delineated similarities with some African groups, as one would predict. Not surprisingly, the dermatoglyphics of U.S. blacks were closer to those of Africans other than those of Pygmies or the San. Interestingly, although there were statistically significant differences in the type and frequency of digital patterns between European-Americans and African-Americans, there were also differences among groups of African-Americans studied by different workers. Unfortunately, such differences complicate the evaluation of dermatoglyphics in clinical situations. Yet, even if there were no variation in digital dermatoglyphics among African-Americans, this would be inexplicable, for we already know that African-Americans are also a genetically heterogeneous population.

Polymorphic variation involves blood groups, serum proteins, red cell enzymes, hemoglobins, HLA, chromosomes, DNA segments and haplotypes associated and unassociated with disease. Variants of these markers may or may not cause disease in homozygous, autozygous, heterozygous, or hemizygous forms. They are not only are essential to our understanding of genetic and nongenetic disorders, but they are crucial in delineating population differences, individual identification, linkage analysis, the mapping of the human genome, admixture analysis, selection, genetic drift, population migrations, and eventually, evolution. Disease susceptibility, morbidity, and mortality are inextricably connected with polymorphisms. Thus polymorphisms provide clues to our very survival or demise as *Homo sapiens*. They are not to be treated with derision, racism, sexism, or a host of the other abuses that have been characteristic of some studies through the years. They are to be treasured as keys to our past, and, perhaps, to our future.

The G-6-PD deficiencies also vary in severity from a congenital nonspherocytic hemolytic anemia to no effect unless certain drugs or fava beans are ingested. In general, however, a Mediterranean form differs from that found in Africans and their descendants. The type common in Iran has a clinical effect similar to that found in Mediterranean peoples. Favism has not been described in Africans and their descendants, and the clinical picture after the ingestion of primaquine and other hemolytic drugs is more severe in Mediterranean peoples and Iranians than that found in Africans and their descendants. And most important, primaquine prophylaxis at a dose of 45 milligrams of the base once a week while in the malarious area, or 15 milligrams of the base daily for two weeks, or 45 milligrams of the base weekly, on return from a malarious region, has little effect on Africans and their descendants, but may kill Iranians or Mediterranean peoples. The effects of hemolytic drugs and fava beans are variable in these groups.

The severity of sickle cell disease varies among populations and in individuals. For some reason sickle cell anemia in populations of the Middle East and India is less severe than the sickle cell disease in Africans and their descendants and Mediterranean peoples. Within populations of high severity, some patients live for long periods with clinical symptoms; others die at an early age. Considerable research is being done to explain the population and individual differences with the hope that the elucidation of the reasons for clinical variability may provide clues to the amelioration or the cure of sickle cell disease. Research along these lines must not be restricted to the hemoglobin molecule or β-globin DNA in the region of the *hemoglobin S* mutation. Answers to the conundrum may be found elsewhere in the β- or α-globin DNA, elsewhere in the genome, or in the environment, or all of the foregoing.

Sickle cell disease, the α- and β-thalassemias, and glucose-6-phosphate dehydrogenase deficiency are the most important abnormalities of red cell structure or synthesis that are found in high frequency in most Africans and their descendants. These variants and disorders are also found in populations that have long been exposed to holo- or hyperendemic falciparum malaria. Current dogma is that persons who are heterozygous for sickle hemoglobin, α- and β-

thalassemia, and G-6-PD deficiency are relatively resistant to the lethal effects of falciparum malaria, and thus falciparum malaria is the
selective force that maintains the frequency of these abnormalities in
human populations.

Nevertheless, important medical cautions are too frequently ignored by those who associate population selection with individual
resistance to falciparum malaria. Persons who are heterozygous for
sickle hemoglobin, G-6-PD deficiency, β-thalassemia, and who have
single and double gene deletions at the α-hemoglobin locus can and
do develop falciparum malaria. We want to emphasize this point because we have seen patients who are heterozygous for sickle hemoglobin and other red cell variants and who have been advised by
reputable hematologists that they need not take antimalarial drugs
when they go to malarious regions. These misguided patients often
return—if they are lucky—with falciparum malaria. It is also rarely
mentioned—in the enthusiasm to accept the malaria hypothesis—that
children with sickle cell anemia are highly susceptible to the lethal
effects of falciparum malaria. Accordingly, malaria prophylaxis in
holoendemic and endemic regions is *imperative* in children with sickle
cell disease. On the other hand, our African colleagues caution
against the use of antimalarial prophylaxis in young children without
sickle cell disease in endemic regions, because long-range prophylaxis
is impractical unless there is a clear medical indication, and also because malarial prophylaxis may prevent the development of natural
immunity. Of course, susceptible children and adults who are foreign
to malarious regions and who visit there for short periods should take
antimalarial prophylaxis.

Evidence against the malaria hypothesis has been presented, but
until a better thesis is introduced to explain the high frequency of
sickle hemoglobin, we had best be content with the hypothesis.

The inability to digest lactose properly is related to the level of the
enzyme lactase, but in most mammals lactase is present in infancy and
diminishes with age. High levels of lactase are found in only a minority of human populations. All populations that originate in nonmilking zones in Africa, Southeast and East Asia, New Guinea, and the
Americas have a high frequency of lactose malabsorption, and all
groups who have a long history of consuming lactose-rich milk have a
low prevalence of lactose malabsorbtion, with the exception of the
Nilotic peoples. Since milk is a very important food for young children, it is most important to emphasize that lactose intolerance does
not necessarily mean that milk is clinically harmful: *individuals with
lactose intolerance usually can drink at least one glass of milk at a sitting with
no symptoms.*

For some reason, the incidence of dizygotic twinning is highest in African peoples. The twinning incidence in western Nigeria, for example, is about four times that of Europeans, and the dizygotic rate among Japanese is about one-third that of Europeans. The high incidence in Africans is unfortunate because twinning and multiple births are significant factors in perinatal and infant mortality, and we suspect that the maternal mortality of women with multiple births is higher than that of women with single births.

High rates of twinning are associated with several factors, which include heredity, a previous history of twin birth, the mother's history of being a twin, having four or more births, maternal age, socioeconomic status, and diet (in Nigerians). Ghanaian twins have the same survival rate as Italian twins, but the rate is higher than that of English twins. The frequency of female-female twins in Ghana is also lower than in European populations for unknown reasons. Oddly, the prevalence of dizygotic twinning has been declining in all populations studied.

Congenital malformations are the most common type of genetically determined abnormality in human populations. At least a third of congenital malformations are identified after the neonatal period and within the first year of life. African-Americans have a higher frequency of congenital malformations than whites, but this is because they have a higher frequency of minor malformations. The overall incidence of major malformations is similar in both groups, but whites have a higher frequency of multiple malformations than blacks. Some malformations are more frequent in particular ethnic groups, such as polydactyly (in blacks) and congenital hip dysplasia (in whites). Blacks have a greater frequency of malformations at birth, but by age five the differences between the incidence of malformations in blacks and whites disappears. The role of the increased incidence of prematurity as a contributor to malformations is unknown.

The evaluation of congenital malformation statistics is difficult because there are no standards, and many congenital malformations are discovered long after the neonatal period.

The etiology of hypertension is an enigma. (We exclude organic causes.) There is an old dictum in pharmacology: Judge the efficacy of medical therapy by the number of drugs or modalities that are recommended—the larger the number the less we know. We feel that this applies to essential hypertension, and fall back on an old cliché—which could very well be valid—there are probably multiple causes. There is no doubt, however, that without exception rural peoples have very low incidences of hypertension, and that the blood pressure in these populations does not increase with age. When rural peoples move to urban areas, the incidence of hypertension climbs, almost without exception. This compelling observation does not exclude a genetic basis for hypertension, for the obvious, much-repeated reason: the manifestation of genetic disorders is frequently environmentally determined. But what do we mean by environmentally determined in this situation? Of course, we do not know. Stress or change in diet are two obvious general explanations, and both have been implicated with persuasive evidence. Even so, Africans and their descendants have disproportionately higher frequencies of hypertension than do most other populations. Yet Africans also have some of the lowest frequencies of hypertension—if they remain rural. It is inescapable that the resolution of reason(s) for the rural versus urban differences in the prevalence of hypertension could lead to a better understanding, and, perhaps, amelioration of this disorder.

Insulin-dependent diabetes is uncommon in tropical regions of the world and occurs less often in Africans than in African-Americans. The frequency of insulin-dependent diabetes in European-Americans is significantly higher than it is in African-Americans. Presumably, the higher prevalence of insulin-dependent diabetes in African-Americans than in Africans is attributable to European admixture, particularly in relation to HLA-DR3 and DR4. Maturity-onset diabetes of youth is special, and is found in at least 10 percent of all patients with youth-onset diabetes in African-Americans from the southeastern part of the United States.

Genetic disorders are generally dwarfed by the appalling health situation of Africans and their descendants be they in "Mother Africa" or in their adopted countries. It would take several volumes to describe the reasons for this "universal truth." Our philosopher friends tell us there are no "truths," much less "universal" ones. Wilson (1978) analyzed how the political and economic structure of the United States affected the condition of African-Americans, and downplayed the significance of "race." Many disagree. But no matter what the cause, maternal, newborn, infant, population mortality, and disease morbidity wreak havoc among African-Americans, and any genetics program that does not take these social conditions plus inadequate to second-class education, poverty, joblessness, drugs, AIDS, and other misfortunes into account deals with an unreal world, and is doomed to failure.

In Africa, except among the affluent, infectious and communicable diseases are rampant. Malaria, tuberculosis, measles, hepatitis, cholera, yellow fever, schistosomiasis, and dracunculosis are associated with undernourishment and malnutrition that are often legacies of human's inhumanity to humans. Sidel's poignant article on the effect of the arms race on children is indelible.

The march of scientific advances is ceaseless, and it should be. Scientific advances in genetics will be utilized, no matter how poor the society, and often at the sacrifice of health problems that should have a higher priority. But try to tell this to a mother who has a child with a genetic disorder, such as sickle cell disease. "Yes, children are dying from preventable measles, or malnutrition, or malaria, but my child is sick. What should I do, Doctor, what about my future children? Do I have to bear this pain again?"

Will prenatal diagnosis be widely utilized in Africa? Konotey-Ahulu has elegantly outlined the difficulties on many occasions. And AIDS now complicates the picture in some regions. But the poor in Africa are not alone. Blacks in the United States live in pockets of under-development, and face the same problems as their African ancestors with respect to the availability of genetic services. They do not have the options that their more affluent neighbors enjoy in the United States. Their choices of genetic disease prevention often do not include access to the latest genetic technology; they are still limited to the old options of abstinence, divorce, contraception, and artificial insemination.

The adaptation of genetic counseling to special circumstances, and for different racial, ethnic, and religious groups is difficult. An intimate knowledge of the individual, the family, the community are crucial to effective communication and assistance. Unfortunately, genetic counselors are in very short supply in the United States, and even more so in most other countries. Further, counseling involves a team approach and physicians are but one factor in the process. Psychologists, social workers, the clergy, educators, nurses, and physician assistants all have a role to play, depending upon the circumstances.

The discovery of glucose-6-phosphate dehydrogenase deficiency by Carson led to effective prophylaxis and treatment of many drug-induced hemolytic anemias and favism, to the basis of some congenital nonspherocytic hemolytic anemias, to the delineation of population differences and another possible mechanism of natural selection, to the discovery of other hemolytic anemias caused by deficiencies of red cell enzymes of the pentose-phosphate and Emden-Meyerhof pathway of the erythrocytes, to a better understanding of malaria prophylaxis and therapy, to an important locus of linkage analysis on the X chromosome to the pioneer studies of Lyon and Beutler on X-chromosome inactivation, and to a better understanding of biochemical and clinical genetic variability and a ranking with the blood groups and hemoglobins, to the investigation of multiple alleles in genetic disorders on the X-chromosome and on autosomes.

Technological developments in medicine, and particularly in medical and molecular genetics, have introduced complex problems in ethics, law, and public policy which have an international perspective. Decisions on the use (and abuse) of advances in genetics will undoubtedly depend on the country, region, and the cultural, social, and religious makeup of the people. Unfortunately, public policy often evolves or is thrust upon society with little consideration of the need or priorities of the target populations. Examples have been given in the United States, and fortunately many of the less industrialized countries still have time to learn from our errors of mass population screening without appropriate education, the dissemination of erroneous information, the development of screening programs without integration into the health care system, the initiation of programs without taking into consideration drawbacks such as poverty, high maternal and infant mortality, poor prenatal and postnatal care, and disparate educational and job opportunities.

Each country will establish its own policies with respect to the utilization of genetic technology and the increasing opportunities offered by discoveries in molecular genetics to pre- and postnatal diagnosis and treatment of a wide spectrum of genetic disorders.

One day, cancer, heart disease, diabetes, sickle cell disease, thalassemias, Tay-Sachs disease, cystic fibrosis, Down syndrome, AIDS, and countless other disorders will be ancient history because of advances in molecular genetics, chromosomal variation, biochemical and other polymorphisms, which contribute to an understanding of our marvelous individual and population variation.

The rationale for mapping the human genome, for example, was outlined by McKusick (1989). Indeed, as Mckusick pointed out, the final objective of the human genome initiative has already begun. Numerous genetic disorders, such as Huntington disease on chromosome 4, adenomatous polyposis of the colon on chromosome 5, cystic fibrosis on chromosome 7, retinoblastoma on chromosome 13, and many other disorders, have been located. Prenatal and premorbid diagnoses, carrier detection, and therapy—including gene therapy—logically follow. McKusick also rightly pointed out that even with all of this knowledge, we will not know what makes us uniquely human, and, most important, knowing the sequence of the human genome will not resolve the problems that affect the interactions of individuals and nations. Unfortunately, debauchers of science, politics, religion, philosophy, ethics, economics, and other disciplines are common. Our inhumanity to each other—particularly to those who are "different"—will probably endure. What a waste.

Appendix A

GENETIC VARIATION AND DISORDERS MORE COMMONLY FOUND IN BLACKS (OTHER THAN THOSE EMPHASIZED IN THE TEXT)

Abnormal separation of sutures
Arcus corneae
Branchial cleft abnormalities
Café au lait spots
Calcinosis, tumoral, with hyperphosphatemia
Clubbing of digits
Commissural lip pits
Delayed tooth eruptions
Dermatosis papulosa nigra
Earlobe absent
Ear pits
Epicanthus (frequency between Europeans and Mongoloids)
Futcher line
Gigantomastia (Nigeria)
Gustatory sweating
Hairy pigmented nevi
Hereditary hypertropic osteoarthropy
Hermaphroditism (Zimbawe; South African Zulu)
Hypokalemic alkalosis (Bartler syndrome)
Insulin resistance and acanthosis nigricans
Isovaleric acid (inability to smell)
Keloid
Low-set ears
Lunatotriquetral synostosis
Lung hypoplasia
Lymphangioma of alveolar ridge
Melanocytoma

Metatarsus adductus
Multiple myeloma
Nailbeds, pigmentation
Nevus of Ota
Olivopontocerebellar atrophy III
Osteochondrosis deformans
Porphyria cutanea tarda (black South Africans)
Postural defects
Prehalicin fistula
Protrusio acetdeuli (Pygmy) (somatomedin C deficiency)
Raindrop depigmentation
Scaphocephaly
Sclerostosis (South African Bantu)
Supernumerary nipples
Tongue, pigmented fungiform papillae
Umbilical hernia
Urethral meatal stenosis
Vitiligo
Xanthism (rufous albinism)

Sources: McKusick, V. A. 1988. *Mendelian inheritance in man*, 8th ed., Baltimore: The Johns Hopkins University Press; Bergsma, D. (ed.) 1979. *Birth defects compendium*, 2d ed., New York: Alan R. Liss, Inc.; Stanbury, J. B., J. B. Wyngaarden, and D. S. Fredickson, J. L. Goldstein, and M. S. Brown, eds. 1983. *The Metabolic basis of inherited disease*. 5th ed. New York: McGraw–Hill; Emery, A. E. H., and D. Rimoin. 1983. *Principles and practice of medical genetics*, vols. 1–2. Edinburgh: Churchill Livingstone.

Appendix B

GENETIC VARIATION AND DISORDERS INFREQUENT IN BLACKS

Abetalipoproteinemia
Abdominal aortic aneurysm
Acatalassemia
Alpha-1-antitrypsin deficiency
Analphaliproteinemia
Anencephaly
Aneurysm, intracranial, berry
Angiokeratosis diffuse (Fabry disease)
Ankylosing spondylitis
Baldness
Cleft lip and palate
Cleft uvula
Congenital dislocation of the hip
Cystic fibrosis
Dens indente
Dens evaginatus
Dentogenesis imperfecta
Dysautonomia
Familial Mediterranean fever
Gilles de la Tourette syndrome
Gingival fibromatosis and digital anomalies
Glucosyl ceramide lipoidosis: Gaucher disease
GM_2 gangliosidosis with hexosamidase A deficiency (Tay-Sachs disease)
Hair, α-protein polymorphism
Hartnup disease
Hemophilia A
Musk, inability to smell
Neural tube defects
Nevi flammeus

Nieman-Pick disease
Osteochondritis dessicans
Otosclerosis
Pentosuria
Phenylketonuria
Porphyria variegate
Pseudocholinesterase types; E (1) variants
Psoriasis vulgaris
PTC nontasting
Pyloric stenosis
Spongy degeneration of the brain
Torus palatinus
Vesicureteral reflex
Wernicke-Korsakoff syndrome
Wilson disease
Xeroderma pigmentosa
Xyy

Sources: Same as Appendix A.

Appendix C

DEVELOPMENTAL INDICES IN AFRICAN-AMERICAN CHILDREN

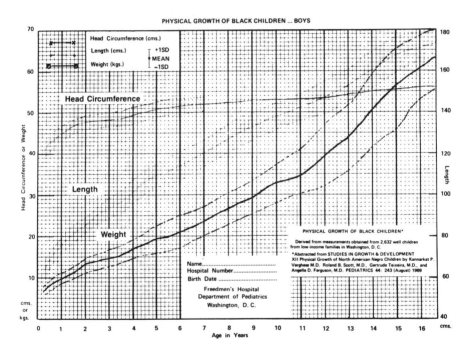

Appendix C. Figure 1. Physical growth of black children: boys.

Source: Freedmen's Hospital, Department of Pediatrics, Washington, D.C.

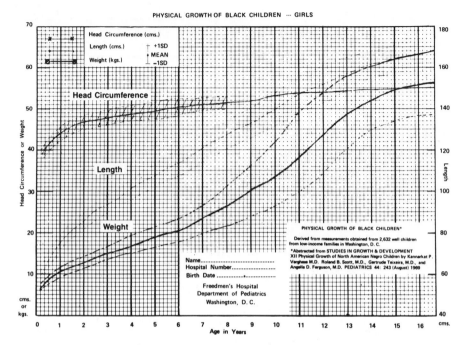

Appendix C. Figure 2. Physical growth of black children: girls.

Source: Freedmen's Hospital, Department of Pediatrics, Washington, D.C.

Appendix D

POLYMORPHIC TRAITS OF EXCEPTIONALLY HIGH OR LOW FREQUENCY IN PEOPLES OF AFRICAN ORIGIN

	Alleles or Phenotypes	Population Data
Polymorphism		
Cerumen	Sticky/dry, flaky *AR*	Dry, flaky type—high frequency in Ainu (Japan), low frequency in Africans
PTC taster	Inability to taste PTC (tt nontasters) *AR*	Nontasters—low frequency in Africans
Red-green color vision	Inability to recognize red & green colors (*XLR*)	Low frequency in blacks & Africans
Chromosome		
Y chromosome length	Long Y	Long Y chromosomes—high frequency in Japan
Pericentric inversions	Inv (9)	Increased frequency in blacks
	Inv (i)	Increased frequency in blacks
	1qh partial inv	Lower frequency in whites
	9 qh complete inv	Increased frequency in blacks
Size heteromorphisms	13, 14, 15, 21, 22	Increased frequency in blacks
Blood Group		
ABO blood groups	$A_{1,3}$	Increased frequency in Africans (Aint & Abantu)
Secretor system	Se/se; Se/Se (positive)	Increased frequency of Se gene in San, Australian aborigines, Eskimos, & Amerindians.
P system	P_1	Increased frequency in Africans

(*continued*)

	Alleles or Phenotypes	Population Data
Blood Group		
Rh system	Rh positive (*R⁰ cDe*)	High frequency in blacks, Africans (0.5–0.9%)
V antigen	*V* (accomp. *cDe, cde*)	Increased frequency in Africans especially in the Sudan
Duffy system	*Fy* (silent gene) *AR* *Fya*	Very high frequency in blacks Very low frequency in blacks
Kidd system	*Jka*	Increased frequency in blacks
MNSs system		
U/u system	U+ U− (*AR*)	Low frequency in blacks Highest frequency in Zaire Pygmies
M₁	*M₁+*	Increased frequency in blacks, Africans
Hunter	*Hu+*	7% frequency in U.S. blacks 22% frequency in West Africans
Henshaw	*He+* (linked to MNSs)	2–3% in American blacks 2.7% West Africans (Also N+ & S+)
Dombrock	*Dᵃ+*	Low frequency in Africans
Cartwright	*Yᵗᵃ+*	Not found in blacks
Isoenzyme		
Phosphoglucomutase	*PGM₂2* *PGM₁ (1 Twa)* *PGM₁ (2S)* *PGM₁ (1F)* *PGM₁² (AR)*	Only found in blacks Twa Pygmies North Twa Pygmies South Twa Pygmies and Tutsi Blacks—lower frequency
Peptidase		
PEP A	PEP A, 2–1; PEP A 2	Increased frequency in blacks
PEP C	PEP C-O (*AR*)	Frequency in Babinge Pygmies
PEP D	*PEP D³*	Increased frequency in blacks
Red cell acid phosphatase (ACP)	*ACP₁ᶜ* *ACP₁ᴿ*	Low frequency in blacks Increased frequency in U.S. blacks Common in S. African blacks (Babinge Pygmies, Khoikhoi and San people, Khoisan)
Aconitase (ACONS)	*ACONS 4, 2, & 6*	Higher frequency in Nigerians
Adenylate kinase	AK 2-1	Lower frequency in U.S. blacks (1%)

	Alleles or Phenotypes	Population Data
Isoenzyme		
		Absent in Ghana, Nigeria, & Lacondon
		Low to absent in Vietnam
Red Cell galactokinase (GALK)	GALKP (lower GALK activity)	Higher frequency in blacks
Pharmacogenetic		
Alcohol dehydrogenase	ADH_2 Ind	Increased frequency in blacks
Debrisoquin hydroxylation 4-hydroxylation	heterozygotes extensive metabolizers	Increased frequency of heterozygotes in Nigeria
Red cell glutathione peroxidase (GPX)	$GPX1*1/GPX1*2$	Increased frequency of $GPX1*2$ in Punjabi & African-Americans, Jamaicans, & Jews
α-N-acetyl-D glucosaminidase (NAG)	high and low activity	Higher mean levels of activity in blacks than whites
Acetyl transferase	Rapid & slow inactivators	Orientals high frequency rapid inactivators
		Libyan Arabs slow inactivators
Paroxonase (PON)	PON^1-low activity (AR) PON^2-high activity	Africans & orientals high frequency of high activity alleles
Serum Protein		
Haptoglobin	Hp^0 (AR)	High frequency in the Gambia (22.5%)
	$Hp\ 2^m$ (related to Hn^0)	Frequent in black populations
	Hp^1	Higher in African populations except for San & Ethiopians
	Hp^2	Higher in whites, Pygmies, & San
Transferrin	TfD_1	Increased frequency in blacks, Australian aborigines (12%), & Bi Aka Pygmies
Group specific component	Gc^{Chip}	Frequent in Chippewa Indians (8%)
Vitamin D-binding globulin	Gc^{Ab}	Frequent in blacks, Australian aborigines (10%) & found in West & Central Africa

(continued)

	Alleles or Phenotypes	Population Data
Serum Protein		
	Gc^{1F}	Increased frequency in U.S. blacks
Ceruloplasmin	Cp^{A}	10 times as frequent in blacks; rare in whites
	CpA^{Wap}	Macushi & Wapishana Amerin-
	$CpAB$	dian increased frequency— U.S. blacks (19%)
α-1-antitrypsin (Pro-	Pi^{Gam}	Found in Gambia, West Africa
tease Inhibitor-Pi)	Pi^{M3}	Increased frequency in U.S. blacks
Thyroxin-binding glo-	$TBG\ S$ (slow)(XLR)	Frequent in blacks
bulin	$TBG{-}2/TBG{-}1,2$	Frequent in blacks
Plasminogen	$PLGN*A$	Orientals highest & blacks next highest frequency
Salivary protein poly- morphisms		
Salivary proteins	$Pb^{2}/AMY\ 1^{3}$	Deficient in Kenyans/absent in Kenyans
Parotid size variant	Ps^{0}(null allele)(AR)	Frequent in blacks, twice as frequent in whites
Milk protein polymor- phism	β-casein	Frequencies in blacks reverse that of whites
	β-E casein	Present in Kikuyu
HLA polymorphism	HLA-A28, Aw30	Increased frequency in blacks
	HLA A2-B12	Highest frequency in blacks
	HLA B 15 related anti-	High frequency in Kenya and
	gens	Tanzania
	HLA SV; Bu	Frequent in blacks; low in whites
	HLA D-DRw 13	Most frequent in blacks (30%)
	HLA D-Dw2, DR2, DW19	Frequent in blacks
B locus haplotypes much more fre- quent in blacks	A30, cw⁻, Bw42, Dw⁻, Dr3, Drw52, DQw 2 Bw71, Dw3, DR3, DRw52, DQw2 Bw71, Dw⁻, DR3, DR52, DQw2	
Properidin factor B	Bf F, Bf S, BfF1, Bf S1	In S. Africa coloured have fre- quencies intermediate be- tween blacks & Asian Indian populations
	$Bf^{SO.7}$	High frequencies in Saudi Arabs

	Alleles or Phenotypes	Population Data
Serum Protein		
C4 (complement)	C4A6, C4A5, C4A4 C4B4, C4B3 & C4BQ0	Frequencies differ between U.S. blacks & whites
C8 (complement)	*C8^{A1}*	Increased frequencies in blacks
DNA (restriction frag- ment length)		
Mitochonrial DNA	*Hpa* I morph 3 *Ava* II morph 3	Patterns differentiate Sengalese from San & Bantu
D14S1 probe (EcoRI)	>40 alleles	DNA fragments 14.3–32.5 kbp
HRAS-1 probe (Taq 1)	18 alleles	DNA fragments 1.85–4.5 kbp Differences between populations is statistically significant
Coagulant of F.IX	Two alleles, type 1 & 2	Type 2 allele lower frequencies in U.S. blacks

AR—Autosomal recessive inheritance; *XLR*—X-linked recessive inheritance. Unless otherwise indicated alleles are inherited in a codominant fashion.

References

Abbas, H., and M. Ahmad. 1983. Persistence of high intestinal lactase activity in Pakistan. *Hum. Genet.* 64:277–78.

Abengowe, C. U., J. S. Jain, and A. K. M. Siddique. 1980. Pattern of hypertension in the northern savanna of Nigeria. *Trop. Doct.* 10:3–8.

Abrahams, D. G., and C. A. Alele. 1960. A clinical study of hypertensive disease in West Africa. *W. Afr. Med. J.* 9:183–93.

Adam, A. 1961. Linkage between deficiency of glucose-6-phosphate dehydrogenase and colour-blindness. *Nature* 189:686.

————. 1962. A survey of some genetical characters in Ethiopian tribes. I. Glutathione stability and glucose-6-phosphate dehydrogenase activity in red blood cells. *Am. J. Phys. Anthrop.* 20:172–73.

Adams, J. M. 1932. Some racial differences in blood pressures and morbidity in groups of white and colored workmen. *Am. J. Med. Sci.* 184:342–50.

Adelola, A., K. R. Kattan, and F. N. Silverman. 1975. Thickness of the normal skull in the American blacks and whites. *Am. J. Phys. Anthropol.* 43:23–30.

Aderounmu, A. F. 1981. The relative importance of genetic and environmental factors in hypertension in black subjects. *Clin. Exp. Hypertens.* 3:597–621.

Adeyokunnu, A. A. 1982. The incidence of Down's syndrome in Nigeria. *J. Med. Genet.* 19: 277–79.

Adeyokunnu, A. A., and J. B. Familusi. 1982. Prune belly syndrome in two siblings and a first cousin: Possible genetic implications. *Am. J. Dis. Child.* 136:23–25.

Ad Hoc Committee on Genetic Counseling. 1975. *Am. J. Hum. Genet.* 277:240–43.

Ahmad, M., and G. Flatz. 1984. Prevalence of primary adult lactose malabsorption in Pakistan. *Hum. Hered.* 34:69–75.

Akinkugbe, O. O. 1972. *High blood pressure in Africans.* Edinburgh: Churchill-Livingstone.

Akinkugbe, O. O., and O. A. Ojo. 1968. The systemic blood pressure in a rural Nigerian population. *Trop. Geogr. Med.* 20:347–56.

Akman, D., J. B. Berenson, C. V. Blonde, L. S. Webber, and A. R. Stopa. 1982. Heart disease in a total population of children: The Bogalusa Heart Study. *South. Med. J.* 75:1177–81.

Alderman, M. H., and E. Yano. 1976. How prevalence of hypertension varies as diagnostic criteria change. *Am. J. Med. Sci.* 271:343–49.

Allison, A. C. 1954. Protection afforded by sickle-cell trait against subtertian malarial infection. *Br. Med. J.* 1:290–94.

———. 1955. Aspects of polymorphisms in man. *Cold Spring Harbor Symp. Q. Biol.* 20:239–255.

Allison, A. C., and D. F. Clyde. 1961. Malaria in African children with deficient glucose-6-phosphate dehydrogenase. *Br. Med. J.* 1:1346–9.

Alving, A. S., C. F. Johnson, A. R. Tarlov, G. J. Brewer, R. W. Kellermeyer, and P. E. Carson. 1960. Mitigation of the haemolytic effect of primaquine and enhancement of its action against exoerythrocytic forms of the Chesson strain of *Plasmodium vivax* by intermittent regimens of drug administration: A preliminary report. *Bull WHO* 22:621–31.

Ambard, L., and E. Beaujard. 1904. Causes de l'hypertension arterielle. *Arch. Gen. Med.* 1:520–33.

Anderson, C. E., L. D. Edmonds, and J. D. Erickson. 1978. Patent ductus arteriosus and ventricular septal defect: Trends in reported frequency. *Am. J. Epidemiol.* 107:281–89.

Anh, N. T., T. K. Thuc, and J. D. Welsh. 1977. Lactose malabsorption in Vietnamese. *Am. J. Clin. Nutr.* 30:468–69.

Antonarakis, S. E., C. D. Boehm, G. R. Serjeant, C. E. Thersen, G. J. Dover, and H. H. Kazazian. 1984. Origin of the βS globin gene in blacks: The contribution of recurrent mutation and/or gene conversion. *Proc. Nat. Acad. Sci. (U.S.A.)* 81:853–56.

Antonarakis, S. E., H. H. Kazazian, S. H. Orkin, C. D. Boehm, and P. G. Waber. 1983. DNA polymorphism in the β-globin gene cluster: Use in clinical, molecular, and evolutionary studies. In J. E. Bowman, ed., *Distribution and evolution of hemoglobin and globin loci*, pp. 105–18. New York: Elsevier.

Antonarakis, S. E., J. A. Phillips, and H. Kazazian. 1982. Genetic diseases: Diagnosis by restriction endonuclease analysis. *J. Pediatr.* 100:845–56.

Arens, W. von. 1950. Uber die angeborene Synostose zwischen dem Os lunatum und dem Os triquetrum. *Fortschr. Geb. Rontgenstr. Nuklearmed. Erganzungsband.* 73:772–74.

Arnold, J. M., M. Diiop, M. Kodjovi, and M. Rozier. 1980. L'Intolérance au lactose chez l'adulte au Sénégal. *C R Soc. Biol. (Paris)* 174:983–92.

Aruwa, R. M. W., J. G. O. Harding, and Z. H. Verjee. 1983. Glucose-6-phosphate dehydrogenase (G6PD) deficiency in some Kenyan ethnic groups. *East Afr. Med. J.* 60:30–33.

Ashkenazi, S., F. Mimouni, P. Merlob, and S. H. Reisner. 1983. Neonatal bilirubin levels and glucose-6-phosphate dehydrogenase deficiency in preterm and low-birth-weight infants in Israel. *Isr. J. Med. Sci.* 19:1506–8.

Austin, D. M., and J. Ghesquiere. 1976. Heat intolerance of Bantu and Pygmoid groups of the Zaire River Basin. *Hum. Biol.* 48:439–53.

Australian National Health and Medical Research Council Dietary Salt Study Management Committee. 1989. Fall in blood pressure with modest reduction in dietary salt intake in mild hypertension. *Lancet* 1:399–402.

Azen, E. A. 1978. Genetic protein polymorphism in human saliva. *Biochem. Genet.* 16:79–99.

Azen, E. A., and C. Denniston. 1980. Polymorphisms of Ps (parotid size variant) and detection of a protein (Pms) related to the Pm (parotid middle band) system with genetic linkage of Ps and Pm to Gl, Db, and Pr genetic determinants. *Biochem. Genet.* 18:483–501.

Azen, E. A., and P. L. Yu. 1984. Genetic polymorphism of CON 1 and CON 2 salivary proteins detected by immunologic and concanavalin A reactions on nitrocellulose with linkage of CON 1 and CON 2 genes to the SPC (salivary protein gene complex). *Biochem. Genet.* 22:1–19.

Azevedo, E. S. , M. C. Da Silva, A. M. Lima, E. F. Fonseca, and, M. M. Conceicao, 1979. Human aconitase polymorphism in three samples from northeastern Brazil. *Ann. Hum. Genet.* 43:7–10.

Baird, M., I. Balazs, A. Guisti, L. Miyazaki, L. Nicholas, K. Wexler, E. Kanter, J. Glassberg, F. Allen, and P. Rubenstein. 1986. Allele frequency distribution of two highly polymorphic DNA sequences in three ethnic groups and its application to the determination of paternity. *Am. J. Hum. Genet.* 39:489–501.

Baine, R. M., D. L. Rucknagel, P. A. Dublin, and J. G. Adams. 1976. Trimodality in the proportion of hemoglobin G Philadelphia in heterozygotes: Evidence for heterogeneity in the number of human alpha chain loci. *Proc. Nat. Acad. Sci. (U.S.A.)* 73:3633–36.

Baker, P. T. 1958. Racial differences in heat tolerance. *Am. J. Phys. Anthropol.* 16:287–305.

Bakken, A. F. 1977. Temporary intestinal lactase deficiency in light-treated jaundiced infants. *Acta Paediatr. Scand.* 66:91–96.

Barnes, D. S. 1969. Tooth morphology and other aspects of the Teso dentition. *Am. J. Phys. Anthropol.* 30:183–93.

Barrai, I. 1968. Dermatoglyphics in Babinga Pygmies. *Atti. Assoc. Genet. Ital.* 13:92–94.

Bartholomew, C., and O.-Y. Pong. 1976. Lactose intolerance in East Indians of Trinidad. *Trop. Geogr. Med.* 28:336–38.

Bayless, T. M. 1976. Recognition of lactose intolerance. *Hosp. Pract. [off.]* 11:97–102.

Bearn, A. G. 1966. Wilson's disease. In J. B. Stanbury, J. B. Wyngaarden, and D. S. Fredrickson, eds. *The metabolic basis of inherited disease*, 2d ed., p. 761. New York: McGraw-Hill.

Bearn, A. G., B. H. Bowman, and F. D. Kitchen. Genetic and biochemical consideration of the serum group-specific component. *Cold Spring Harbor Symp. Quant. Biol.* 29:534–42.

Beaudet, A. L., G. L. Feldman, S. D. Fernbach, G. J. Buffone, and W. E. O'Brien. 1989. Linkage disequilibrium, cystic fibrosis, and genetic counseling. *Am. J. Hum. Genet.* 44:319–26.

Beiser, M., H. Collomb, J.-L. Ravel, and C. Nafziger. 1976. Systemic blood pressure studies among the Serer of Senegal. *J. Chronic Dis.* 29:371–80.

Bell, J. 1951. On brachydactly and symphalangism. In L. S. Penrose, ed., *The treasury of human inheritance*, vol. 2, pt. 1. Cambridge: Cambridge University Press.

Bell, K., H. A. McKenzie, W. H. Murphy, and D. C. Shaw. 1970. Beta-lactoglobin (Droughtmastre): A unique variant. *Biochim. Biophys. Acta* 214:427–36.

Bender, K., R. Frank, and H. W. Hizeroth. 1977. Glyoxalase I Polymorphism in South African Bantu-speaking negroids. *Hum. Genet.* 38:223–26.

Bender, K., G. Mauff, and H. W. Hitzeroth. 1977. No evidence for linkage disequilibrium between Bf and GLO in African Negroids. *Hum. Genet.* 38:227–30.

Bennett, L. 1984. *Before the Mayflower*. Middlesex, England: Penguin.

Bernhardt, B. A., and R. E. Pyritz. 1989. The economics of clinical genetic services are not self-supporting. *Am. J. Hum. Genet.* 44:288–93.

Bernstein, S. C., and J. E. Bowman. 1980. G6PD/Malaria hypothesis: A balanced or transient polymorphism. *Lancet* 1:485.

Bernstein, S. C., J. E. Bowman, and L. Kaptue-Noche. 1980a. Genetic variation in Cameroon:Thermostability variants of hemoglobin and of glucose-6-phosphate dehydrogenase. *Biochem. Genet.* 18:21–37.

———. 1980b. Interaction of sickle cell trait and glucose-6-phosphate dehydrogenase deficiency in Cameroon. *Hum. Hered.* 30:7–11.

Bertin, T., J. E. Harris, R. E. Ferrell, and W. J. Schull. 1978. The Nubians of Kom Ombo: Serum and red cell protein types. *Hum. Hered.* 28:66–71.

Bertrand, E., F. Serie, I. Kone, M. Le Bras, J.-L. Boppe, B. Beda, M. O. Assamoi, and J. Y. Thomas. 1976. Etude de la prévalence et de certains aspects épidémiologiques de l'hypertension artérielle en Côte d'Ivoire. *Bull. WHO* 54:449–54.

Betke, K., E. Beutler, G. J. Brewer, H. N. Kirkman, L. Luzzatto, A. G. Motulsky, B. Ramot, and M. Siniscalco. 1967. Standardization of procedures for the study of glucose-6-phosphate dehydrogenase. *WHO Tech. Rep. Ser.*, p. 366.

Beutler, E. 1959. The hemolytic effect of primaquine and related compounds: A review. *Blood* 14:103–39.

———. 1972. Glucose-6-phosphate dehydrogenase deficiency. In J. B. Stanbury, J. B. Wyngaarden, and D. S. Fredrickson, *The metabolic basis of inherited disease*, 3d ed., pp. 1358–88. New York: McGraw-Hill.

———. 1978. *Hemolytic anemia in disorders of metabolism*. New York: Plenum Medical Book Co.

———. 1983. Glucose-6-phosphate dehydrogenase deficiency. In J. B. Stanbury, J. B. Wyngaarden, D. S. Fredrickson, J. L. Goldstein, and M. S. Brown, eds., *The metabolic basis of inherited disease*, pp. 1629–53. New York: McGraw-Hill.

Beutler, E., and M. C. Baluda. 1964. The separation of glucose-6-phosphate dehydrogenase deficient erythrocytes from the blood of heterozygotes for glucose-6-phosphate dehydrogenase deficiency. *Lancet* 1:189–192.

Beutler, E., R. J. Dern, and A. S. Alving. 1954. The hemolytic effect of primaquine. III. A study of primaquine-sensitive erythrocytes. *J. Lab. Clin. Med.* 44:177–84.

Beutler, E., M. Robeson, and E. Buttenweiser. 1957. The mechanism of glutathione destruction and protection in drug-sensitive and non-sensitive erythrocytes. In vitro studies. *J. Clin. Invest.* 36:617–28.

Beutler, E., M. Yeh, and V. F. Fairbanks. 1962. The normal human female as a mosaic of X-chromosome activity: Studies using the gene for G-6-PD deficiency as a marker. *Proc. Nat. Acad. Sci. (U.S.A.).* 48:9–16.

Beutler, E., and C. West. 1974. Red cell glutathione peroxidase polymorphism in Afro-Americans. *Am. J. Hum. Genet.* 26:255–58.

Beutler, E., and A. Yoshida. 1973. Human glucose-6-phosphate dehydrogenase variants: A supplementary tabulation. *Ann. Hum. Genet.* 37:151–56.

Bianchi-Porro, G., F. Parenta, and O. Sangaletti. 1983. Lactose intolerance in adults with chronic unspecified abdominal complaints. *Hepatogastroenterol.* 30:254–57.

Bienzle, U., O. Ayeni, A. O. Lucas, and L. Luzzatto. 1972. Glucose-6-phosphate dehydrogenase and malaria: Greater resistance of females heterozygous for enzyme deficiency and of males with non-deficient variant. *Lancet* 1:107–11.

Biggar, R. J. 1986. The AIDS problem in Africa. *Lancet* 1:79–82.

Blair, D., J. B. Habicht, E. A. Sims, D. Sylvester, and S. Abraham. 1984. Evidence for an increased risk for hypertension with centrally located body fat and the effect of race and sex on this risk. *Am. J. Epidemiol.* 119:526–40.

Blum, H. F. 1961. Does the melanin pigment of human skin have adaptive value? *Q. Rev. Biol.* 36:50–63.

Blumenfeld, O. O., and A. M. Adamany. 1978. Structural polymorphism within the amino terminal region of MM, NN and MN glycoproteins (glycophorisms) of the human erythrocyte membrane. *Proc. Nat. Acad. Sci. (U.S.A.)* 75:2727–37.

Blumenthal, A. 1977. *Moral responsibility: Mankind's greatest need.* Santa Anna, Calif.: Raylin Press.

Bodmer, W. F. 1983. The structure of HLA. *Transplant Proc.* 15:36–39.

Bodmer, W. F., J. R. Batchelor, J. G. Bodmer, H. Festenstein, and P. J. Morris., eds. 1978. *Histocompatibility testing 1977*, p. 205. Copenhagen: Munksgaard.

Boehm, C. D., C. E. Dowling, S. E. Antonarakis, G. R. Honig, and H. H. Kazazian. 1985. Evidence supporting a single origin of the β^C-globin gene in blacks. *Am. J. Hum. Genet.* 37:771–77.

Bolk, L. 1916. Problems of human dentition. *Am. J. Anat.* 19:9–148.

Bonney, G. E., M. Walker, K. Gbedenah, and F. I. D. Konotey-Ahulu. 1978. Multiple births and visible birth defects in 13,000 consecutive deliveries in one Ghanaian hospital. *Prog. Clin. Biol. Res.* 243:105–8.

Boreham, P. F., J. K. Lenahan, G. R. Port, and I. A. McGregor 1981. Haptoglobin polymorphism: Its relationship to malaria infections in the Gambia. *Trans. R. Soc. Trop. Med. Hyg.* 75:193–200.

Bornstein, P. E. and R. R. Peterson. 1966. Numerical variation of the presacral vertebral column in three population groups in North America. *Am. J. Phys. Anthropol.* 25:139–46.

Bosron, W. F., L. J. Magnes, and T.K. Li. 1981. Human liver dehydrogenase:

ADH Indianapolis results from genetic polymorphism at the ADH 2 gene locus. *Biochem. Genet.* 1983. 21:735–44.

Bowman, J. E. 1964a. Haptoglobin and transferrin differences in some Iranian populations. *Nature* 201:88.

————. 1964b. Comments on abnormal erythrocytes and malaria. *Am. J. Trop. Med. and Hyg.* (Suppl.) 13:159–161.

————. 1977. Genetic screening programs and public policy. *Phylon* 38:117–42.

————. ed. 1983. *Distribution and evolution of hemoglobin and globin loci.* New York: Elsevier.

Bowman, J. E., and R. W. Bloom. 1985. Genic interaction and falciparum malaria. Paper presented at the *International Symposium on Sickle Cell.* Lagos, Nigeria, November 3–7.

Bowman, J. E., P. E. Carson, and H. Frischer. 1969. The segregation in one family of three alleles at the glucose-6-phosphate dehydrogenase locus. *Hum. Hered.* 19:25–35.

Bowman, J. E., P. E. Carson, H. Frischer, and A. L. de Garay. 1966. Genetics of starch-gel electrophoretic variants of human 6-phosphogluconic dehydrogenase: Population and family studies in the United States and Mexico. *Nature* 210:811–13.

Bowman, J. E., P. E. Carson, H. Frischer, R. D. Powell, E. J. Colwell, L. J. Legters, A. J. Cottingham, S. C. Boone, and W. W. Hiser. 1971. Hemoglobin and red cell enzyme variation in some populations of the Republic of Vietnam with comments on the malarial hypothesis. *Am. J. Phys. Anthropol.* 34:313–24.

Bowman, J. E., and H. Frischer. 1964. Malaria, G-6-P.D. deficiency, and sickle cell trait. *Br. Med. J.* 1378.

Bowman, J. E., H. Frischer, F. Ajmar, P. E. Carson, and M. K. Gower. 1967. Population, family and biochemical investigation of human adenylate kinase polymorphism. *Nature* 214:1156–58.

Bowman, J. E., and E. Goldwasser. 1975. *Sickle Cell Fundamentals.* Chicago: University of Chicago Comprehensive Sickle Cell Center, with grant from National Heart, Lung, and Blood Institute, National Institutes of Health.

Bowman, J. E., and H. Rhonagy. 1967. Hemoglobin, glucose-6-phosphate dehydrogenase and adenylate kinase polymorphism in Moslems in Iran. *Am. J. Phys. Anthropol.* 27:119–24.

Bowman, J. E., and S. Maynard Smith. 1963. Theoretical evidence for an autosomal modifying gene pair in glucose-6-phosphate dehydrogenase deficient families. *Ann. Hum. Genet.* 26:213–18.

Bowman, J. E., and Walker, D. G. 1963. The origin of glucose-6-phosphate dehydrogenase deficiency in Iran: Theoretical considerations. *Proceedings, Second International Conference of Human Genetics* (Rome) 1:583–86.

Boyer, S. H., I. H. Porter, and R. Weilbacher. 1962. Electrophoretic heterogeneity of glucose-6-phosphate dehydrogenase and its relationship to enzyme deficiency in man. *Proc. Nat. Acad. Sci. (U.S.A.)* 48:1863–76.

Boyle, E. 1970. Biological patterns in hypertension by race, sex, body weight, and skin color. *JAMA* 213:1627–43.

Bradley, M. 1981. *The black discovery of America.* Toronto: Personal Library.

Brand, J. C., M. S. Gracy, R. M. Spargo, and S. P. Dutton. 1983. Lactose malabsorption in Australian Aborigines. *Am. J. Clin. Nutr.* 3:449–52.

Brewer, G. J., A. R. Tarlov, and A. S. Alving. 1962. The methemoglobin reduction test for primaquine sensitivity of erythrocytes. *JAMA* 180:386–88.

Brittenham, G. M. 1983. The geographic and ethnographic distribution of hemoglobinopathies in India. In J. E. Bowman, ed., *Distribution and evolution of hemoglobin and globin loci,* pp. 169–78. New York: Elsevier.

Brock, D. J., A. E. Bolton, and J. B. Scrimgeour. 1974. Prenatal diagnosis of spina bifida and anencephaly through maternal plasma-alpha-fetoprotein measurement. *Lancet* 1:767–69.

Brown, J. J., A. F. Lever, J. I. S. Robertson, P. F. Semple, R. F. Bing, A. M. Heagerty, J. D. Swales, H. Thurston, J. G. G. Lendingham, J. H. Laragh, L. Hansson, M. G. Nicholls, and A. E. Espiner. 1984. Salt and hypertension. *Lancet* 2:456.

Brown, K. H., L. Parry, M. Khatun, and G. Ahmed. 1979. Lactose malabsorption in Bangladeshi village children: Relation with age, history of recent diarrhea, nutritional status, and breast feeding. *Am. J. Clin. Nutr.* 32:1962–69.

Brown, W.M. 1980. Polymorphism in mitochondrial DNA of humans as revealed by restriction endonuclease analysis. *Proc. Natl. Acad. Sci. (U.S.A.)* 7:3605–9.

Browne, S. G. 1978. Tropical medicine—facing today's dilemmas. *Trans. R. Soc. Trop. Med. Hyg.* 72:1–5.

Bruce-Chwatt, L. J. 1979. Man against malaria: Conquest or defeat. *Trans. R. Soc. Trop. Med. Hyg.* 73:605–16.

Brues, A. M. 1974. Rethinking human pigmentation. *Am. J. Phys. Anthrop.* 43:387–92.

Buck v. Bell 1927. 274 U.S. 124.

Buckhout, R. 1971. *The war on people: A scenario for population control.* Collected papers of the Department of Psychology, California State College at Hayward.

Budowle, B., R. C. Go, B. O. Barger, and R. T. Acton. 1981. Properdin factor B polymorphism in black Americans. *J. Immunogenet.* 8:519–21.

Budowle, B., J. M. Rosman, R. C. Go, W. Louv, B. O. Barger, and R. T. Acton. 1983. Phenotypes of the fourth complement component (C4) in black Americans from the southeastern United States. *J. Immunogenet.* 10:199–203.

Bulmer, M. G. 1960. The twinning rates in Europe and Africa. *Ann. Hum. Genet.* 24:121–23.

Burgio, G. R., G. Flatz, C. Barbera, R. Patane, A. Boner, C. Cajozzo, and S. D. Flatz. 1984. Prevalence of primary adult lactose malabsorption and awareness of milk intolerance in Italy. *Am. J. Clin. Nutr.* 39:100–104.

Burkitt, D. B. Etiology and prevention of colorectal cancer. 1984. *Hosp. Pract.* 19:67–77.

Burns, E. B. 1980. *A history of Brazil.* New York: Columbia University Press.

Buschang, P. H., and R. M. Malina. 1980. Brachymesophalangia-V in five samples of children: A descriptive and methodological study. *Am. J. Phys. Anthropol.* 53:189–95.

Butler, M. G., D. D. Weaver, and F. J. Meaney. 1982. Prader-Willi syndrome: Are there population differences? *Clin. Genet.* 22:292–94.

Byard, P. J. 1981. Quantitative genetics of human skin color. *Yearbook of Phys. Anthropol.* 24:123–37.

Cahill, G. F. 1979. Current concepts of diabetic complications with emphasis on hereditary factors: A brief review. In C. F. Sing and M. H. Skolnick eds., *Genetic analysis of common diseases: Applications to predictive factors in coronary heart disease.*, pp. 113–129. New York: Alan R. Liss.

Cann, R. L. 1987. In search of Eve. A DNA trail leads to a single African woman 200,000 years old. *Sciences* 27:30–37.

Carpenter, J. C. 1979. A comparative study of metric and non-metric traits in a series of modern crania. *Am. J. Phys. Anthropol.* 45:337–44.

Carrasco, D. 1984. *Quetzalcoatl and the irony of empire: Myths and prophecies in the Aztec tradition.* Chicago: University of Chicago Press.

Carson, P. E., C. L. Flanagan, C. E. Ickes, A. S. Alving. 1956. Enzymatic deficiency in primiquine-sensitive erythrocytes. *Science* 124:484–85.

Carson, P. E., and H. Frischer. 1966. Glucose-6-phosphate dehydrogenase deficiency and related disorders of the pentose phosphate pathway. *Am J. Med.* 41:744–61.

Carter, C. O. 1966. Comments on genetic counseling. In *Proceedings of the Third International Congress of Human Genetics*, p. 75. Baltimore: Johns Hopkins University Press.

Cassel, J. C. 1971. Evans County cardiovascular and cerebrovascular epidemiologic study. *Arch. Int. Med.* 128:883–986.

Cavalli-Sforza, L. L., and W. F. Bodmer. 1971. *The genetics of human populations.* San Francisco: W. H. Freeman.

Chagula, W. K. 1960. The age of eruption of third permanent molars in male East Africans. *Am. J. Phys. Anthropol.* 18:77–82.

Chakraborty, R. 1986. Gene admixture in human populations. *Yearbook Phys. Anthropol.* 29:1–44.

Chakraborty, R., and P. E. Smouse. 1988. Recombination of haplotypes leads to biased estimates of admixture proportions in human populations. *Proc. Nat. Acad. Sci. (U.S.A.)* 85:3071–74.

Chakravarti, A., K. H. Buetow, S. E. Antonarakis, P. G. Waber, C. D. Boehm, and H. H. Kazazian. 1984a. Nonuniform recombination within the human β-globin gene cluster. *Am. J. Hum. Genet.* 36:1239–58.

Chakravarti, A., J. A. Phillips, K. H. Mellits, K. H. Buetow, P. H. Seeburg. 1984b. Patterns of polymorphism and linkage disequilibrium suggest independent origins of the human growth hormone gene cluster. *Proc. Nat Acad. Sci. (U.S.A.)* 81:6085–89.

Chamla, M.-C., 1962. La Réparation géographique des crêtes papillaires digitales dans le monde: Nouvel essai de synthèse. *L'Anthropologie* (Paris) 66:526–41.

Chang, J. C., and Y. W. Kan. 1982. A new sensitive prenatal test for sickle cell anemia. *N. Engl. J. Med.* 307:30–32.

Chavez, G. F., J. F. Cordero, and J. E. Becerra. 1988 Leading major congenital malformations among minority groups in the United States, 1981–1986. *MMWR* 37:17–24.

Chebloune, Y., J. Pagnier, G. Trabuchet, C. Faure, G. Verdier, D. Labie, and V. Nignon. 1988. Structural analysis of the 5′ flanking region of the β-globin gene in African sickle cell anemia patients: Further evidence for three origins of the sickle cell mutation in Africa. *Proc. Nat. Acad. Sci. (U.S.A.)* 85:4431–35.

Chehab, F. F., and Y. W. Kan. 1988. PCR based color complementation assay: A simple rapid nonradioactive method to detect gene deletions, chromosomal rearrangements, and etiologic agents. *Blood* (Abstr.) 72:57a.

Childs, B., M. M. Gordis, M. M. Kaback, and H. H. Kazazian. 1976. Tay Sachs screening: Social and psychological impact. *Am. J. Hum. Genet.* 28:537–49.

Childs, B., W. Zinkham, E. A. Browne, E. L. Kimbro, and J. V. Torbert. 1958. A genetic study of a defect in glutathione metabolism of the erythrocyte. *Johns Hopkins Med. J.* 102:21–37.

Christianson, R. E., B. J. van den Berg, L. Milkovich, and F. W. Oechsli. 1981. Incidence of congenital anomalies among white and black live births with long-term follow-up. *Am. J. Public Health* 71:1333–41.

Clarke, C. A. 1959. The relative fitness of human mutant genotypes. In D. F. Roberts and G. A. Harrison, eds., *Natural selection in human populations*, pp. 17–34. London: Pergamon Press.

Clavel, C., K. Mansinho, S. Chamaret, D. Guetard, V. Favier, J. Nina, M.-O. Santos-Ferreira, J.-L. Champalimaud, and L. Montagnier. 1987. Human immunodeficiency virus type 2 infection associated with AIDS in West Africa. *Nature* 316:1180–85.

Clemens, T. L., J. S. Adams, S. L. Henderson, and M. F. Holick. 1982. Increased skin pigment reduces the capacity of skin to synthesize vitamin D_3. *Lancet* 1:74–76.

Cockrum v. *Baumgartner* 1981. 425 N.E. 2d 968 (Ill. App., 1981).

Cockshoot, W. P. 1959. Carpal anomalies amongst Yorubas. *West Afr. Med. J.* 8:185–90.

———. 1963. Carpal fusions. *Am. J. Roentgen.* 89:1260–71.

Cohen, M. M., M. W. Shaw, J. W. Nac Clarer. 1966. Racial differences in the length of the human Y chromosome. *Cytogenet. Cell Genet.* 5:34–52.

Cohn, S. J., C. Abesamis, S. Yasumura, J. F. Aloia, I. Zanzi, and K. J. Ellis. 1977. Comparative skeletal mass and radial bone mineral content in black and white women. *Metabolism* 26:171–78.

Collins, F. S., C. D. Boehm, P. G. Waber, C. J. Stoeckert, S. M. Weissman, B. G. Forget, and H. G. Kazazian. 1984a. Concordance of a point mutation 5′ to the $^G\gamma$ globin gene with $^G\gamma\beta+$ hereditary persistence of fetal hemoglobin in the black population. *Blood* 64:1292–96.

Collins, F. S., J. L. Cole, W. K. Lockwood, and M. C. Iannuzzi. 1987. The deletion in both types of hereditary persistence of fetal hemoglobin is approximately 105 kilobases. *Blood* 70:1797–1803.

Collins, F. S., C. J. Stoeckert, G. R. Serjeant, B. G. Forget, and S. M. Weissman. 1984b. $^{G}\gamma\beta+-$ hereditary persistence of fetal hemoglobin: Cosmid cloning and identification of a specific mutation 5' to the $^{G}\gamma$ gene. *Proc. Nat. Acad. Sci. (U.S.A.)* 81:4894–98.

Committee on Nutrition, American Academy of Pediatrics. 1978. The practical significance of lactose intolerance in children. *Pediatrics* 62:240–45.

Community control of hereditary anemias: Memorandum from a WHO meeting. 1983. *Bull WHO.* 61:63–80.

Comstock, G. W. 1957. An epidemiologic study of blood pressure levels in a biracial community in the southern United States. *Am. J. Hyg.* 65:271–315.

Congdon, E. D., S. Rowhanavongse, and P. Varamisara. 1932. Human congenital auricular and juxta-auricular fossae, sinuses and scars (including the so-called aural and auricular fistulae) and the bearing of their anatomy upon the theories of their genesis. *Am. J. Anat.* 51:439–59.

Constans, J., P. Lefevre-Witier, P. Richard, and G. Jaeger. 1980. Gc (vitamin D binding protein) subtype polymorphism and variants distribution among Saharan, Middle East, and African populations. *Am. J. Phys. Anthropol.* 52:435–41.

Constans, J., P. Richard, and M. Viau. 1978. Relationship between Hp1S and Hp2 gene frequencies among human populations. *Hum. Hered.* 28:328–34.

Constans, J. and M. Viau. 1977. Group specific component: Evidence for two subtypes of the Gc1 gene. *Science* 198:1070–71.

Constans, J., M. Viau, C. Goraillard, and A. Clerc. 1981a. Haptoglobin polymorphism among Saharan and West African groups: Haptoglobin phenotype determination by radioimmunoelectrophoresis of Hp0 samples. *Am. J. Hum. Genet.* 33:606–12.

Constans, J., M. Viau, G. Jaeger, M. J. Palisson. 1981b. Gc, Tf, Hp subtype and alpha 1-antitrypsin polymorphisms in a pygmy Bi-Aka sample. *Hum. Hered.* 31:129–37.

Coogan, J. E. 1953. Eugenic sterilization holds jubilee. *Catholic World* 177:44–49.

Cook, G. C. 1979. Intestinal lactase status of adults in Papua New Guinea. *Ann. Hum. Biol.* 6:55–58.

Cook-Mozaffari, P. 1982. Symposium on tumors in the tropics: Carcinomas of the esophagus, bladder, cervix uteri and penis. *Trans. R. Soc. Trop. Med Hyg.* 76:157–63.

Cornell, J., M. M. Nelson, and P. Beighton. 1983. Neural tube defects in the Cape Town area, 1975–1980. *S. Afr. Med. J.* 64:83–89.

Corruccini, R. S. 1974. An examination of the meaning of cranial discrete traits for human skeletal biological studies. *Am. J. Phys. Anthropol.* 40:425–46.

Cowie, C. C., F. K. Port, R. A. Wolfe, P. J. Savage, P. P. Moll, and V. M. Hawthorne. 1989. Disparities in incidence of diabetic end-stage renal disease according to race and type of diabetes. *New. Engl. J. Med.* 321:1074–79.

Cowles, R. B. 1959. Some ecological factors bearing on the origin and evolution of pigment in the human skin. *Am. Naturalist* 93:283–93.

Craig-Holmes, A. P., and M. W. Shaw. 1971. Polymorphisms of human constitutive heterochromatin. *Science* 174:702–4.

Crandall, B. F., T. B. Lebherz, P. C. Schroth, and M. Matsumoto. 1983. Alpha-fetoprotein concentrations in maternal serum: relation to race and body weight. *Clin. Chem.* 29:531–33.

Crispin, J., K. G. Trigueiro, and F. F. R. Benevides. 1972. Third molar agenesis in a trihybrid Brazilian population. *Am. J. Phys. Anthropol.* 32:289–92.

Crombie, I. K. 1979. Racial differences in melanoma incidence. *Br. J. Cancer* 40:185–93.

Cruickshank, J. K., S. H. Jackson, D. G. Beavers, L. T. Bannan, M. Beavers, and V. L. Stewart. 1985 Similarity of blood pressure in blacks, whites, and Asians in England: The Birmingham factory study. *J. Hypertens.* 3:365–71.

Cruz-Coke, R., R. Etcheverry, and R. Nagel. 1964. Influence of migration on blood-pressure of Easter Islanders. *Lancet* 1:697–99.

Cummins, H., and C. Midlo. 1961. *Fingerprints, palms, and soles: An introduction to dermatoglyphics.* New York: Dover.

Curtin, P. D. 1975. *The Atlantic slave trade: A census.* Madison: University of Wisconsin Press.

Cutting, G. R., S. E. Antonarakis, K. E. Buetow, L. M. Kasch, B. J. Rosenstein, and H. H. Kazazian. 1989. Analysis of DNA polymorphism haplotypes linked to the cystic fibrosis locus in North American black and Caucasian families supports the existence of multiple mutations of the cystic fibrosis gene. *Am. J. Hum. Genet.* 44:307–18.

Czezei, A., G. Flatz, and S. D. Flatz. 1983. Prevalence of primary adult lactose malabsorption in Hungary. *Hum. Genet.* 64:398–401.

Dahl, R. A. 1982. *Dilemmas of pluralist democracy: Autonomy vs. control.* New Haven: Yale University Press.

Dahlberg, A. A. 1963. Analysis of the American Indian dentition. In D. R. Brothwell, ed., *Dental Anthropology*, 5:149–77. Oxford: Pergamon Press.

Dahr, W., G. Uhlenbruck, E. Janson, and R. Schmalish. 1977. Different N-terminal amino acids in the MN-glycoprotein from MM and NN erythrocytes. *Hum. Genet.* 35:335–43.

Daiger, S. P., N. S. Hoffman, R. S. Wildin, and T. S. Su. 1984. Multiple independent restriction site polymorphisms in human DNA detected with a cDNA probe to argininosuccinate synthetase (AS). *Am. J. Hum. Genet.* 36:736–49.

Daiger, S. P., D. P. Rummel, L. Wang, and L. L. Cavalli-Sforza. 1981. Detection of genetic variation with radioactive ligands. IV. X-linked polymorphic genetic variation of thyroxin-binding globin (TBG). *Am. J. Hum. Genet.* 33:640–48.

Daiger, S. P., M. S. Schanfield and, L. L. Cavalli-Sforza. 1975. Group-specific component (Gc) proteins bind vitamin D and 25-hydroxy vitamin D. *Proc. Nat. Acad. Sci. (U.S.A.)* 72:2076–80.

Dankmeijer, J. 1947. Finger prints of African Pygmies and Negroes. *Am. J. Phys. Anthropol.* 5:453–84.

Davee, M. A., T. Reed, and C. Plato. 1989. The effect of a pattern in palmar interdigital II on a-b ridge count in black and white Down syndrome cases and controls. *Am. J. Phys. Anthropol.* 78:439–47.

Davidson, B. 1980. *The African slave trade.* Boston: Little Brown.

Davidson, R. G., B. R. Migeon, M. Borden, and B. Childs. 1963. Dosage compensation in the regulation of erythrocyte glucose-6-phosphate dehydrogenase activity. *Bull. Johns Hopkins Hosp.* 112:318–22.

Davrinche, C., C. Rivat, L. Rivat-Peran, A. N. Helal, K. Boukef, M. P. Lefranc, and G. Lafranc. 1981. Genetic variants of human C3 and properdin factor B in a population from Tunisia. *Hum. Hered.* 31:299–303.

Dean, R. F. A., and P. R. M. Jones. 1959. Fusion of triquetral and lunate bones shown in serial radiographs. *Am. J. Phys. Anthropol.* 16:279–81.

de Ceulaer, K., D. R. Higgs, D. J. Weatherall, R. J. Hayes, B. E. Serjeant, and G. R. Serjeant. 1983. α-thalassemia reduces the hemolytic rate in homozygous sickle cell disease. *N. Engl. J. Med.* 309:189–90.

De Melo. M. J., E. Freitas, and F. M. Salzano. 1975. Eruption of permanent teeth in Brazilian whites and blacks. *Am. J. Phys. Anthropol.* 42:145–50.

de Moaar, E. W. J. 1979. The new policies of WHO in research for new tools to control six major tropical diseases. *Trans. R. Soc. Trop. Med. Hyg.* 73:147–49.

Demographic Yearbook (1983). 1985. New York: United Nations.

Denaro, M., H. Blanc, M. J. Johnson, K. H. Chen, E. Wilmsen, L. L. Cavalli-Sforza, and D. C. Wallace. 1981. Ethnic variations: the Hpa 1 endonuclease cleavage patterns of human mitochondrial DNA. *Proc. Nat. Acad. Sci. (U.S.A.)* 78:5768–72.

Deol, M. S. 1975. Racial differences in pigmentation and natural selection. *Ann. Hum. Genet.* 38:501–3.

Dern, R. J., E. Beutler, and A. S. Alving. 1954. The hemolytic effect of primaquine. II. The natural course of the hemolytic anemia and the mechanism of its self-limited character. *J. Lab. Clin. Med.* 44:171–75.

———. 1955. The hemolytic effect of primaquine. V. Primaquine sensitivity as a manifestation of a multiple drug sensitivity. *J. Lab. Clin. Med.* 45:36–39.

Destro-Bisol, G., and G. Spedini. 1989. Anthropological survey on red cell glutathione peroxidase (GP*1) polymorphism in central western Africa: A tentative hypothesis on the interaction between GPX1*2 and Hb beta *S allelic products. *Am. J. Phys. Anthropol.* 79:217–24.

Dhermy, D., P. Carnevale, I. Blot, and I. Zohoun. 1989. Hereditary elliptocytosis in Africa. *Lancet.* 1:225.

Diamond, J. M. 1987. AIDS: Infectious, genetic or both? *Nature* 328:199–200.

Dijkstra, B. K. 1977. Goldenhar's syndrome, oculoauricular malformation, in a Bantu girl. *ORL J. Otorhinotaryngol. Relat. Spec.* 39:101–6.

Dill, D. B., M. K. Yousef, A. Goldman, S. D. Hillyard, and T. P. Davis. 1983. Volume and composition of hand sweat of white and black women in desert walks. *Am. J. Phys. Anthropol.* 61:67–73.

Dobongnie, J. C., A. D. Newcomer, D. B. McGill, and S. F. Phillips. 1979.

Absorption of nutrients in lactase deficiency. *Dig. Dis. Sci.* 24:225–31.

Donnison, C. P. 1929. Blood pressure in the African native. *Lancet* 1:6–7.

Drennan, M. H. 1937. The torus mandibularis in the Bushman. *J. Anat.* 72:66–70.

Dunn, F. L. 1965. On the antiquity of malaria in the Western Hemisphere. *Hum. Biol.* 37:385–93.

Dunston, G. M., C. K. Hurley, R. J. Hartzman, and A. H. Johnson. 1987. Unique HLA-D region heterogeneity in American blacks. *Transplant. Proc.* 19:870–71.

Durham, W. H. 1983. Testing the malarial hypothesis. In J. E. Bowman, ed., *Distribution and evolution of hemoglobin and globin loci*, pp. 45–76. New York: Elsevier.

Dussalt, J. H., J. Letarte, H. Guyada, and C. Laberge. 1980. Thyroxin-binding globulin capacity and concentration evaluated from blood spots on filter-paper in a screening program for neonatal hypothyroidism. *Clin. Chem.* 26:463–65.

Eales, L.-J., K. Nye, and A. J. Pinching. 1987. Reply to letters to the editor. *Lancet* 1:1268–69.

Eales, L.-J., J. M. Parkin, S. M. Forster, K. E. Nye, J. N. Weber, J. R. W. Harris, and A. J. Pinching. 1987. Association of different allelic forms of group specific component with susceptibility to and clinical manifestation of human immunodeficiency virus infection. *Lancet* 1:999–1002.

Ebbesen, F., D. Edelsten, and J. Hertel. 1980a. Gut transit time and lactose malabsorption using photometry. I. A study using lactose-free human mature milk. *Acta. Paediatr. Scand.* 69:65–68.

———. 1980b. Gut transit time and lactose malabsorption during phototherapy. II. A study using raw milk from the mothers of infants. *Acta. Paediatr. Scand.* 69:69–71.

Eckerson, H. W., C. M. Wyte, and B. N. LaDu. 1983. The human serum paraoxonase/arylesterase polymorphism. *Am. J. Hum. Genet.* 35:1126–36.

Editorial. 1987. AIDS in Africa. *Lancet* 2: 192–94.

Editorial. 1988. AIDS in Sub-Saharan Africa. *Lancet* 1:1260–61.

Effertz, O. 1909. Malaria in tropical America and among Indians. *Janus* 14:246–61.

Eidelman, E., A. Chosack, and K. A. Rosernweig. 1973. Hypodontia: Prevalence amongst Jewish populations of different origin. *Am. J. Phys. Anthropol.* 39:129–34.

El-Hazmi, M. F. 1983. Abnormal hemoglobins and allied disorders in the Middle East in Saudi Arabia. In J. E. Bowman, ed., *Distribution and evolution of hemoglobin and globin loci*, pp. 239–49. New York: Elsevier.

Ellestad-Sayad, J. J., and C. Haworth. 1977. Disaccharide consumption and malabsorption in Canadian Indians. *Am. J. Clin. Nutr.* 30:698–703.

Embury, S. H., A. M. Dozy, A. M. Miller, J. Miller, J. R. Davis, K. M. Kleman, H. Preisler, E. Vichinsky, W. N. Larde. B. H. Lubin, Y. W. Kan, and W. C. Mentzer. 1982. Concurrent sickle cell anemia and α-thalassemia: Effect on severity of sickle cell anemia. *N. Engl. J. Med.* 306:270–74.

Embury, S. H., S. J. Scharf, B. A. Randall, K. Saiki, M. A. Gholson, M. Golbus,

N. Arnheim, and H. A. Ehrlich. 1987. Rapid prenatal diagnosis of sickle cell anemia by a new method of DNA analysis. *N. Engl. J. Med.* 316:656–61.

Emery, A. E. H. 1975. Genetic counseling. *Br. Med. J.* 3:219–29.

———. 1976. *Methodology in medical genetics: An introduction to statistical methods.* Edinburgh: Churchill Livingstone.

Emery, A. E. H., and D. L. Rimoin, eds. 1983. *Principles and practice of medical genetics*, vols. 1, 2. Edinburgh: Churchill Livingstone.

Epstein, C. J. 1973. Who should do genetic counseling and under what circumstances? In D. Bergsma, ed., *Contemporary genetic counseling. Orig. Art. Ser.* 9:39–48. White Plains, N.J.: National Foundation for Birth Defects.

Escobar, V., M. Melnick, and P. M. Conneally. 1976. The inheritance of bilateral rotation of maxillary central incisors. *Am. J. Phys. Anthropol.* 45:109–16.

Eveleth, P. B. 1978. Differences between populations in body shape of children and adolescents. *Am. J. Phys. Anthropol.* 49:373–82.

Excoffier, L., and A. Langaney. 1989. Origin and differentiation of human mitochondrial DNA. *Am. J. Hum. Genet.* 44:73–85.

Excoffier, L., B. Pellegrini, A. Sanchez-Mazas, C. Simon, and A. Langaney. 1987. Genetics and history of Sub-Saharan Africa. *Yearbook Phys. Anthropol.* 30:151–94.

Fabry, M. E., J. G. Mears, P. Patel, K. Schaefer-rego, L. D. Carmichael, G. Martinez, and R. L. Nagel. 1984. Dense cells in sickle cell anemia: The effects of gene interaction. *Blood* 64:10042–46.

Fagerhol, M. K., and M. Braend. 1965. Serum prealbumin polymorphism in man. *Science* 149:986–87.

Fagerhol, M. K., and C. B. Laurell. 1967. The polymorphism of prealbumins and alpha-1-antitrypsin in human sera. *Clin. Chim. Acta.* 16:199–203.

Falconer, D. S. 1967. The inheritance of liability to diseases with variable age of onset, with particular reference to diabetes mellitus. *Ann. Hum. Genet.* 31:1–20.

Falek, A., and S. Britten. 1974. Phases in coping: The hypothesis and its implications. *Soc. Biol.* 21:1–7.

Falkner, B. 1987. Is there a black hypertension? *Hypertension* 10:551–54.

Falusi, A. G., G. J. F. Esan, H. Ayyub, and D. R. Higgs. 1987. Alpha thalassemia in Nigeria: Its interaction with sickle cell disease. *Europ. J. Haematol.* 38:370–75.

Fang, T. C. 1950. *The inheritance of the a-b ridge count on the human palm, with a note on its relation to mongolism.* University of London, Ph.D diss.

Farrer, L. A., and F. J. Meany. 1985. An anthropometric assessment of Huntington's disease patients and families. *Am. J. Phys. Anthropol.* 67:185–94.

Feldman, J. G., S. C. Stein, R. J. Klein, S. Kohl, and G. Casey. 1982. The prevalence of neural tube defects among ethnic groups in Brooklyn, New York. *J. Chron. Dis.* 35:53–60.

Ferguson, A., D. M. MacDonald, and W. G. Brydon. 1984. Prevalence of lactase deficiency in British adults. *Gut* 2:163–67.

Fields, B. J. 1982. Ideology and race in American history. In J. Morgan

Kouser and J. M. McPherson, eds., *Race and reconstruction: Essays in honor of C. Vann Woodward*, pp. 143–77. Oxford: Oxford University Press.

Fildes, R. A., and H. Harris. 1966. Genetically determined variation of adenylate kinase in man. *Nature* 209:261–63.

Final Report of the Subcommittee on Definition and Prevalence of the 1984 Joint Committee. 1985. Hypertension prevalence and the status of awareness, treatment, and control in the United States. *Hypertension* 7:457–68.

Finkelstein, L., ed. 1949. *The Jews: Their history, culture, and religion*, vol. 1. New York: Harper & Row.

Fitch, N. 1979. Classification and identification of inherited brachydactylies. *J. Med. Genet.* 16:36–44.

Fitzpatrick, T. B., and W. C. Quevedo. 1966. Albinism. In J. B. Stanbury, J. B. Wyngaarden, and D. S. Frederickson. eds., *The Metabolic basis of inherited disease*, 3d ed., pp. 326–37. New York: McGraw-Hill.

Flatz, G., J. N. Howell, J. Doench, and S. D. Flatz. 1982. Distribution of physiological adult lactase phenotypes, lactose absorbers, and lactose malabsorbers in Germany. *Hum. Genet.* 62:152–57.

Flatz, G., and H. W. Rotthauwe. 1977. The human lactase polymorphism: Physiology and genetics of lactose absorption and malabsorption. In A. G. Steinberg, A. G. Bearn, A. G. Motulsky, and B. Childs, eds., *Prog. Med. Genet.* (n.s.) 2:205–49.

Flatz, G., C. Schildge, and H. Sekow. 1986. Distribution of adult lactase phenotypes in the Tuareg of Niger. *Am. J. Hum. Genet.* 38:515–20.

Fleming, A. F., J. Storey, L. Molineaux, E. A. Iroko, and E. D. E. Attai. 1979. Abnormal haemoglobins in the Sudan Savanna of Nigeria. I. Prevalence of haemoglobins and relationships between sickle cell trait, malaria and survival. *Ann. Trop. Med. Parasitol.* 73:161–72.

Fletcher, J. 1972. The brink: The parent-child bond in the genetic revolution. *Theol. Stud.* 33:457–85.

Flickinger, M. G., and K. M. Yarborough. 1976. Dermatoglyphics of Apache and Navajo Indians. *Am. J. Phys. Anthropol.* 45:117–22.

Flint, J., A. V. S. Hill, D. K. Bowden, S. J. Oppenheimer, P. R. Slii, S. W. Serjeantson, J. Bana-Koiri, K. Bhatia, M. P. Alpers, A. J. Boyce, D. J. Weatherall, and J. B. Clegg. 1986. High frequencies of α-thalassemia are the result of natural selection by falciparum malaria. *Nature* 321:744–50.

Fogh-Anderson, P. 1942. *Inheritance of harelip and cleft palate*. Copenhagen: Nyt Nordisk Forlag, A. Busch.

Ford, E. B. 1945. Polymorphism. *Biol. Rev.* 20:73–80.

Ford, E. B. 1964. *Ecological genetics*. New York: John Wiley.

Franco, L. J., M. P. Stern, M. Rosenthal, S. M. Haffner, H. P. Hazuda, and P. J. Comeaux. 1985. Prevalence, detection, and control of hypertension in a biethnic community: The San Antonio Heart Study. *Am. J. Epidemiol.* 121:684–96.

Franke, U. and M. A. Pellegrino. 1977. Assignment of the major histocompatibility complex to a region of the short arm of human chromosome 6. *Proc. Nat. Acad. Sci. (U.S.A.)* 74:1147–51.

Fraser, G. R., E. R. Giblett and A. G. Motulsky. 1966. Population genetic studies in the Congo. III. Blood groups (ABO, MNSs, Rh, Js). *Am. J. Hum. Genet.* 18:546–52.

Freiburghaus, A. V., J. Schmitz, M. Schindler, H. W. Rottauwe, P. Kutunen, K. Launiala, and B. Hadorn. 1976. Protein patterns of brush border fragments in congenital lactose malabsorption and in specific hypolactasia of the adult. *N. Engl. J. Med.* 294:1030–32.

Friedman, G. D., A. L. Klatsky, and A. B. Singelaub. 1982. Alcohol, tobacco, and hypertension. *Hypertension* 4:143–50.

Friedman, M. J. 1978. Erythrocytic mechanism of sickle cell resistance to malaria. *Proc. Nat. Acad. Sci. (U.S.A.)* 75:1994–97.

Friedman, M. J. 1979. Oxidant damage mediates variant red cell resistance to malaria. *Nature* 280:245–47.

Frischer, H., J. E. Bowman, P. E. Carson, K. H. Rieckmann, D. Willerson, Jr., and E. J. Colwell. 1973. Erythrocytic glutathione reductase, glucose-6-phosphate dehydrogenase, and 6-phosphogluconic dehydrogenase deficiencies in populations of the United States, South Vietnam, Iran, and Ethiopia. *J. Lab. Clin. Med.* 81:603–12.

Frischer, H., P. E. Carson, and J. E. Bowman. 1965. Methemoglobin reduction in intact G6PD deficent and HbM erythrocytes. *Proceedings of the Tenth Congress of the International Society of Blood Transfusion*, Stockholm, 1964, pp. 568–77.

Frischer, H., P. E. Carson, J. E. Bowman, and K. H. Rieckmann. 1973. Visual test for erythrocytic glucose-6-phospate dehydrogenase, 6-phosphogluconic dehydrogenase, and glutathione reductase deficiencies. *J. Lab. Clin. Med.* 81:613–24.

Froggatt, P. 1960. Albinism in Northern Ireland. *Ann. Hum. Genet.* 24:213–38.

Frohlich, E. D. 1987. Initial therapy for hypertension. *Hosp. Pract. [off.]* 22:89–96.

Frohlich, E. D., C. Grim, D. R. Labarthe, M. H. Maxwell, D. Perloff, and W. H. Weidman. 1988. Recommendations for human blood pressure determination by sphygmomanometers: Report of a special task force appointed by the Steering Committee of the American Heart Association. *Hypertension* 11:201A–22A.

Fulton, J. D., and P. T. Grant. 1956. Sulphur requirements of erythrocytic form of Plasmodium knowlesi. *Biochem. J.* 63:274–82.

Gabr, M., F. M. El-Beheiry, M. Soliman, M. El-Mahdi, M. El-Moughy, and N. El-Akkad. 1977. Lactose tolerance in normal Egyptian infants and children and in protein calorie malnutrition. *Gaz. Egypt Paediatr. Assoc.* 26:27–33.

Galanello, R., R. Ruggeri, E. Paglietti, M. Addis, M. A. Mellis, and A. Cao. 1983. A family with segregating triplicated alpha globin loci and beta thalassemia. *Blood* 62:1035–46.

Gans, H., H. L. Sharp, and B. H. Tan. 1969. Antiprotease deficiency and familial infantile liver cirrhosis. *Surg. Gyn. Obstet.* 129:289–99.

Garay, R. P., C. Nazaret, G. Dagher, E. Bertrand, and P. Meyer. 1981. A genetic approach to the geography of hypertension: Examination of Na$^+$–

K$^+$ cotransport in Ivory Coast Africans. *Clin. Exp. Hypertens.* 3:861–70.

Garn, S. M., S. L. Fels, and H. Israel. 1967. Brachymesophalangia of digit five in ten populations. *Am. J. Phys. Anthropol.* 27:205–10.

Garn, S. M., A. R. Frisancho, A. K. Poznanski, J. Schweitzer, and M. B. McCann. 1971. Analysis of triquetral-lunate fusion. *Am. J. Phys. Anthropol.* 34:431–34.

Garn, S. M., J. C. Gall, and J. M. Nagy. 1972. Brachymesophalangia-5 without cone-epiphysis mid-5 in Down's syndrome. *Am. J. Phys. Anthropol.* 36:253–56.

Garn, S. M., and A. B. Lewis. 1962. The relationship between third molar agenesis and reduction in tooth number. *Angle Orthodont.* 32:14–18.

Garn, S. M., and A. B. Lewis. 1963. Phylogenetic and intra-specific variations in tooth sequence polymorphism. In D. R. Brothwell, ed., *Dental anthropology*, 5:53–63. New York: Macmillan, Pergamon Press.

Garn, S. M., A. B. Lewis, and J. H. Vicinus. 1963. Third molar polymorphism and its significance to dental genetics. *J. Dent. Res.* 42:1344–63.

Garn, S. M., A. B. Lewis, and A. J. Walenga. 1968. Maximum-confidence values for human mesiodistal crown dimension of human teeth. *Arch. Oral Biol.* 13:841–44.

Garn, S. M., A. K. Pozanski, J. M. Nagy, and M. B. McCann. 1972. Independence of brachymesophalangia-5 from brachymesophalangia-5 with cone mid-5. *Am. J. Phys. Anthropol.* 36:295–98.

Gartler, S. M., R. M. Lisker, B. K. Campbell, R. Sparkes, and N. Gant. 1972. Evidence for two functional X-chromosomes in human oocytes. *Cell Differ.* 1:215–25.

Garza, C., and N. S. Scrimshaw. 1976. Relationship of latose intolerance to milk intolerance in young children. *Am. J. Clin. Nutr.* 29:192–96.

Gates, R. R. 1946. *Human Genetics*, 2:843–51. New York: Macmillan.

Gedde, H. W., H. G. Benkmann, D. P. Agarwal, U. Bienzle, R. Guggenmoos, F. Rosenkaimer, H. H. Hoppe, B. Brinkmann. 1979. Genetic Studies in the Cameroon: Red cell enzyme and serum protein polymorphisms. *Z. Morphol. Anthropol.* 70:33–40.

Geldmacher-von Mallinckrodt, M., T. L. Diepgen, C. Duhme, and G. Hommel. 1983. A study of the polymorphism and ethnic distribution differences of human serum paroxonase. *Am. J. Phys. Anthropol.* 62:235–41.

Gelpi, A. P. 1967. Glucose-6-phosphate dehydrogenase deficiency, the sickling trait, and malaria in Saudi Arab children. *J. Pediatr.* 71:138–46.

———. 1982. Sickle cell disease in the Middle East. In W. Fried, ed., *Comparative clinical aspects of sickle cell disease*, pp. 35–55. New York: Elsevier.

Gessain, M. 1957. Les dermatoglyphes digitaux des Noirs D'Afrique. *L'Anthropologie (Paris)* 61:239–67.

Gibbs, W. N., R. Gray, and M. Lowry. 1979. Glucose-6-phosphate dehydrogenase deficiency and neonatal jaundice in Jamaica. *Br. J. Haematol.* 43:263–74.

Giblett, E. R. 1969. *Genetic markers in human blood*, pp. 317–20. Philadelphia: F. A. Davis.

Giblett, E. R., and N. M. Scott. 1965. Red cell acid phosphatase: Racial distribution and report of a new phenotype. *Am. J. Hum. Genet.* 17:425–32.

Giblett, E. R., and A . Steinberg. 1960. The inheritance of serum haptoglobin types of American Negroes: Evidence for a third allele Hp2m. *Am. J. Hum. Genet.* 12:160–68.

Gilbert, B. M. 1976. Anterior femoral curvature: Its probable basis and utility as a criterion of assessment. *Am. J. Phys. Anthropol.* 45:601–4.

Gilles, H. M., K. A. Fletcher, R. G. Hendrickse, R. Lindner, S. Reddy, and N. Allan. 1967. Glucose-6-phosphate dehydrogenase deficiency, sickling, and malaria in African children in Southwestern Nigeria. *Lancet* 1:138–40.

Glanville, E. V. 1967. Perforation of the corono-olecranon septum: Humero-ulnar relationships in Netherlands and African populations. *Am. J. Phys. Anthropol.* 26:85–92.

———. 1969a. Nasal shape, prognathism, and adaptation in man. *Am. J. Phys. Anthropol.* 30:29–38.

———. 1969b. Digital ridge counts of Efe Pygmies *Am. J. Phys. Anthropol.* 31:427–28.

Golden, W. L., J. M. Hanchett, N. Breslin, and M. W. Steele. 1984. Prader-Willi syndrome in black females. *Clin. Genet.* 26:161–63.

Golenser, J., J. Miller, D. T. Spira, T., Navok, and M. Chevion. 1983. Inhibitory effect of a fava bean component on the in vitro development of Plasmodium falciparum in normal and glucose-6-phosphate dehydrogenase deficient erythrocytes. *Blood* 61:507–10.

Gonzalez-Redondo, J. M., T. A. Stoming, K. D. Lanclos, Y. C. Gu, A. Kutlar, F. Kutlar, T. Nakatsuji, B. Deng, I. S. Han, V. C. McKie, and T. H. J. Huisman. 1988. Clinical and genetic heterogeneity in black patients with homozygous β-thalassemia from the southeastern United States. *Blood* 72:1007–14.

Goodman, H. O. 1965. Genetic parameters of dentofacial development. *J. Dent. Res.* (suppl. to no. 1) 44:174–84.

Goodman, P. A., and P. C. Karn. 1983. Human parotid size polymorphism (Ps): Characterization of two allelic products, Ps1 and 2, by limited proteolysis. *Biochem. Genet.* 21:405–16.

Goodman, R. M., A. Feinstein, and M. Hertz. 1984. Stub thumbs in Israel revisited. *J. Med. Genet.* 21:460–62.

Gottlieb, M. S. 1980. Diabetes in offspring and siblings of juvenile and maturity-onset-type diabetes. *J. Chron. Dis.* 33:331–39.

Gould, S. J. 1978. Morton's ranking of races by cranial capacity. *Science* 200:503–9.

———. 1981. *The mismeasure of man.* New York: W. W. Norton.

Grace, H. H. 1981. Prenatal screening for neural tube defects in South Africa. An assessment. *S. Afr. Med. J.* 60:324–29.

Green, J. T., and F. H. Gray. 1939. High femoral neck fractures treated by multiple nail fixation. *Clin. Orthop.* 11:177–83.

Greenberg, F., L. M. James, and G. P. Oakley. 1983. Estimates of birth prevalence rates of spina bifida in the United States from computer-generated maps. *Am. J. Obstet. Gynecol.* 145:570–73.

Greenberg, J. H. 1963. *The languages of Africa.* The Hague: Mouton.

Greenwood, B. M. and H. C. Whittle. 1980. The pathogenesis of sleeping sickness. *Trans. R. Soc. Trop. Med. Hyg.* 74:716–25.

Grell, G. A. C. 1980. Race and hypertensive complications *Lancet* 2:744–45.

Griffin, M. L., C. M. Kavanaugh, and J. R. Sorenson. 1976. Genetic knowledge, client perspectives, and genetic counseling. *Soc Work Health Care* 2:2–10.

Grimaldi, S., L. Bartalena, C. Ramacciotti, and J. Robbins. 1983. Polymorphism of human thyroxine-binding globulin. *J. Clin. Endocrinol. Metab.* 57:186–92.

Gross, R. T., R. E. Hurwitz, and P. A. Marks. 1958. An hereditary enzymatic defect in erythrocyte metabolism: Glucose-6-phosphate dehydrogenase deficiency. *J. Clin. Invest.* 37:1176–84.

Guthrie, R. 1961. Blood screening for phenylketonuria. *JAMA* 178:863.

Gyepes, M., H. Z. Mellins, and I. Katz. 1962. The low incidence of fracture of the hip in the Negro. *JAMA* 181:1073–74.

Habib, Z. 1982. Thalassemia and haptoglobin polymorphism in Egypt. *Hereditas* 96:307–11.

———. 1983. Haptoglobin polymorphism in Egyptians. *Ann. Hum. Biol.* 10:385–87.

Hagema, M. J. 1978. Waardenburg's syndrome in Kenyan Africans. *Trop. Geogr. Med.* 30:45–55.

Hakomori, S. 1981. Blood group ABH and I: Antigens of human erythrocytes: Chemistry, polymorphisms and their developmental change. *Sem. Hematol.* 18:39–49.

Halberstein, R. A., and J. E. Davies. 1984. Biosocial aspects of high blood pressure in people of the Bahamas. *Hum. Biol.* 56:317–28.

Haldane, J. B. S. 1949. The rate of mutation of human genes. Proceedings of the Seventh International Congress of Genetics, *Hereditas* (suppl.) 35:267.

Hall, P. J., A. G. Levin, C. C. Entwistle, S. C. Knight, A. Wasunna and G. Brubaker. 1980. B15 heterogeneity in East African Blacks. *Tissue Antigens* 16:326–32.

Hamel, B.C., and A. Yohani. 1981. The Beckwith-Wiedemann syndrome in four African infants. *Ann. Trop. Paediatr.* 1:149–53.

Hanihara, K. 1963. Crown characteristics of the deciduous dentition of the Japanese-American hybrids. In D. R. Brothwell, ed., *Dent. Anthropol.* 5:105–24. New York: Pergamon Press.

Harburg, E., L. Gleibermann, P. Roeper, M. A. Schork, and W. J. Schull. 1978. Skin color, ethnicity, and blood pressure. I: Detroit blacks. *Am. J. Public Health* 68:1177–83.

Harper, H. A. S., A. K. Poznanski, and S. M. Garn. 1974. The carpal angle in American populations. *Invest. Radiol.* 9:217–21.

Harris, E. F. 1980. Sex differences in lingual marginal ridging on the human maxillary central incisor. *Am. J. Phys. Anthropol.* 52:541–48.

Harris, E. F., and T. A. Rathbun. 1989. Small tooth sizes in a nineteenth century South Carolina plantation slave series. *Am. J. Phys. Anthropol.* 78:411–20.

Harris, H. 1980. *The principles of human biochemical genetics*, 3d ed., p. 365. New York: Elsevier.

Harris, M. 1974. *Patterns of race in the Americas*. New York: W. W. Norton.

Harris v. McRae 1980. 48 *Law Week* 4941.

Harsha, D. W., R. R. Frerichs, and G. S. Berenson. 1978. Densitometry and anthropometry of black and white children. *Hum. Biol.* 50:261–80.

Harsha, D. W., A. W. Voors, and G. S. Berenson. 1980. Racial differences in subcutaneous fat patterns in children aged 7–15 years. *Am. J. Phys. Anthropol.* 53:333–37.

Hassanali, J. 1982. Incidence of Carabelli's trait in Kenyan Africans and Asians. *Am. J. Phys. Anthropol.* 59:317–19.

Haskell, P. T. 1977. Seventeenth seminar on tryponasomiasis. Problems of land use and tsetse control. *Trans. R. Soc. Trop. Med. Hyg.* 71:12–15.

Havlik, R. J., and M. Feinleib. 1982. Epidemiology and genetics of hypertension. *Hypertension* 4:121–27.

Havlik, R. J., H. B. Hubert, R. R. Fabsitz, and M. Feinleib. 1983. Weight and hypertension. *Ann. Int. Med.* 98:855–59.

Headings, V. E. 1976. Association between type of health profession and judgements about prevention of sickling disorders. *J. Med. Educ.* 51:682–88.

Health Information for International Travel 1988. U.S. Department of Health and Human Services, Public Health Service, Centers for Disease Control.

Health and Nutrition. 1981. U.S. Center for Health Statistics, p. 112.

Health United States. 1984. National Center for Health Statistics. U.S. Department of Health and Human Services publication (PHS) 85–1232 Public Health Service. Washington, D.C.: U.S. Government Printing Office.

Hempelmann, E., and R. J. M. Wilson. 1981. Detection of glucose-6-phosphate dehydrogenase in malarial parasites. *Mol. Biochem. Parasitol.* 2:197–204.

Hendrickse, R. G. 1976. Aspects of tropical pediatrics. *Trans. R. Soc. Trop. Med. Hyg.* 70:268–73.

Hertzberg, H. T. E. 1968. The conference on standardization of anthropometric techniques and terminology. *Am. J. Phys. Anthropol.* 28:1–16.

Hertzog, K. P. 1967. Shortened fifth medial phalanges. *Am. J. Phys. Anthropol.* 27:113–18.

Higginson, J. 1983. The face of cancer worldwide. *Hosp. Pract. [off.]* 18:145–57.

Higgs, D. R., B. E. Aldridge, J. Lamb, J. B. Clegg, D. J. Weatherall, R. J. Hayes, Y. Grandison, Y. L. Lowrie, K. P. Mason, B. E. Serjeant, and G. R. Serjeant. 1982. The interaction of alpha-thalassemia and homozygous sickle-cell disease. *N. Engl. J. Med.* 306:1441–46.

Higgs, D. R., M. A. Vickers, A. O. M. Wilkie, I.-M. Pretorius, A. P. Jarman, and D. J. Weatherall. 1989. A review of the molecular genetics of the human α-globin gene cluster. *Blood* 73:1081–1104.

Hijazi, S. S., A. Abulakan, Z. Ammarin, and G. Flatz. 1983. Distribution of adult lactase phenotypes in Bedouins and in urban and agricultural populations of Jordan. *Trop. Geogr. Med.* 35:157–61.

Hill, D. G. 1981. *The freedom seekers: Blacks in early Canada.* Agincourt: Book Society of Canada.

Hirono, A. and E. Beutler. 1988. Molecular cloning and nucleotide sequence of cDNA for human glucose-6-phosphate dehydrogenase variant A($-$). *Proc. Natl. Acad. Sci. (U.S.A.)* 85:3951–54.

Hirschfeld, J. 1959. Immunoelectrophoretic demonstration of qualitative differences in normal human sera and their relation to haptoglobins. *Acta Pathol. Microbiol. Scand.* 47:16068.

Hitzeroth, H. W., K. Bender, and R. Frank. 1981. South African Negroes: Isozyme polymorphisms (GPT, PGM1, PGM2, AcP, AK ADA) and tentative genetic distances. *Anthropol. Anz.* 39:20–35.

Hitzeroth, H. W., and K. Hummel. 1978. Serum protein polymorphisms Hp, Tf, Gm, Inv and Pt in Bantu speaking South African Negroids. *Anthropol. Anz.* 35:127–41.

Hitzeroth, H. W., H. Walter, and M. Hilling. 1977/8. Genetic markers and leprosy in South African Negroes. Part I. Serum protein polymorphisms. *S. Afr. Med. J.* 54:653–58.

———. 1979. Genetic markers and leprosy in South African Negroes: Part II. Erythrocyte enzyme polymorphisms. *S. Afr. Med. J.* 56:507–10.

Hobart, M. J. 1979. Genetic polymorphism of human plasminogen. *Ann. Hum. Genet.* 42:419–23.

Hodge, T. W., J. T. Demopolus, B. O. Berger, D. H. S. Bell, and R. T. Acton. 1988. Nucleotide sequence analysis of DQw3 genes in HLA-DR4 and DR5 American black insulin dependent diabetes. *Am. J. Hum. Genet.* (Abstr.) 43:A186.

Hoffman, M. J. 1974. Retinal pigmentation, visual acuity and brightness levels. *Am. J. Phys. Anthropol.* 43:417–24.

Hoffman, S. L., S. Masbar, P. R. Hussein, A. Soewarta, S. Harun, H. A. Marwotto, J. R. Campbell, L. Smrkovski, Purnomo, and I. Wiady. 1984. Absence of malaria mortality in villagers with chloroquine-resistant *Plasmodium falciparum* treated with chloroquine. *Trans. R. Soc. Trop. Med. Hyg.* 78:175–78.

Hollingsworth, M. J. 1965. Observations on the birth weights and survival of African babies: Single births. *Ann. Hum. Genet.* 28:291–300.

Hollingsworth, M. J., and C. Duncan. 1966. The birth weight and survival of Ghanaian twins. *Ann. Hum. Genet.* 30:13–24.

Holt, S. B. 1964. Finger-print patterns in mongolism. *Ann. Hum. Genet.* 27:279–82.

Holt, S. B. 1968. *The genetics of dermal ridges.* Springfield, Ill.: Charles C Thomas.

Honig, G. R., and J. G. Adams. 1986. *Human hemoglobin genetics.* New York: Springer-Verlag.

Hoosen, S., Y. K. Seedat, A. I. Bhigjee, and R. M. Neerahoo. 1985. A study of urinary sodium and potassium excretion rates among urban and rural Zulus and Indians. *J. Hypertens.* 3:351–58.

Hrdlička, A. 1920. Shovel-shaped teeth. *Am. J. Phys. Anthropol.* 30:53–56.

Hsia, Y. E. 1974. Choosing my children's genes: Genetic counseling. In M.

Lipkin, Jr., and P. T. Rowley, eds., *Genetic responsibility: On choosing our children's genes*, p. 43. New York: Plenum.

Hsu, L. Y., P. A. Benn, H. L. Tannenbaum, T. E. Perlis, and A. D. Carlson. 1987. Chromosomal polymorphisms of 1, 9, 16, and Y in 4 major ethnic groups: A large prenatal study. *Am. J. Med. Genet.* 26:95–101.

Huang, S., C. Wong, S. Antonarakis, T. Ro-lien, W. H. Y. Lo, and H. Kazazian. 1986. The same "TATA" box β-thalassemia mutation in Chinese and US blacks: Another example of independent origins of mutation. *Hum. Genet.* 74:162–64.

Hug, G., G. Chuck, and M. K. Fagerhol. 1981. Pi (P-Clifton): A new alpha-1-trypsin allele in an American Negro family. *J. Med. Genet.* 18:43–45.

Huisman, T. H. J. 1983. The occurrence of γ-chain variants and related anomalies in various populations of the world. In J. E. Bowman, ed., *Distribution and evolution of hemoglobin and globin loci*, pp. 119–42. New York: Elsevier.

Huisman, T. H., B. L. Abraham, H. F. Harris, M. E. Gravely, J. Henson, D. Williams, J. B. Wilson, A. Miller, S. Mayson, R. N. Wrightsone, E. Moss, B. Joseph, L. Walker, J. Brisco, and L. Brisco. 1980. Hemoglobinopathies observed in the population of the Southeastern United States (SE U.S.A.). *Hemoglobin* 4:449–67.

Huisman, T. H. J., C. Altay, B. Webber, A. L. Reese, M. E. Gravely, K. Okonjo, and J. B. Wilson. 1981. Quantitation of three types of chains of HbF by high pressure liquid chromatography: Application of this method to the HbF of patients with sickle cell anemia or the S-HPFH condition. *Blood* 57:75–82.

Hunt, J. C. 1983. Sodium intake and hypertension: A cause for concern. *Ann. Int. Med.* 98:724–28.

Hussein, L., S. D. Flatz, W. Kuhnan, and G. Flatz. 1982. Distribution of lactase phenotypes in Egypt. *Hum. Hered.* 32:94–99.

Hutchinson, J. 1986. Association between stress and blood pressure variation in a Caribbean population. *Am. J. Phys. Anthropol.* 71:69–79.

Hypertension Detection and Follow-up Program Cooperative Group. 1977. Blood pressure studies in 14 communities. *JAMA* 237:2385–91.

Ibraimov, A. I., and M. M. Mirrakhimov. 1983. Human chromosomal polymorphism. VI. Chromosomal Q polymorphism in Turkmen of the Kara-Kum desert of Central Asia. *Hum. Genet.* 63:380–83.

Ikemoto, S., K. Minqaguchi, K. Suzuki, and K. Komita. 1977. New genetic markers in human parotid saliva (Pm). *Science* 197:378–79.

Iregbulem, L. M. 1982. Incidence of cleft lip and palate in Nigeria. *Cleft Palate J.* 19:201–5.

Iscan, M. Y. 1983. Assessment of race from the pelvis. *Am. J. Phys. Anthropol.* 62:205–8.

Jackson, R. T., and M. C. Latham. 1979. Lactose malabsorption among Masai children of East Africa. *Am. J. Clin. Nutr.* 4:779–82.

Jacob, H. S., and J. H. Jandl. 1966. Effects of sulfhydryl inhibition on red blood cells. III. Glutathione in the regulation of the hexose monophosphate pathway. *J. Biol. Chem.* 241:4243–50.

Jaiyesimi, F., and A. U. Antia. 1981. Congenital heart disease in Nigeria: A ten-year experience at UCH, Ibadan. *Ann. Trop. Paediatr.* 1:77–85.

James, W. H. 1975. The declines in dizygotic twinning rates and in birth rates. *Ann. Hum. Biol.* 2:81–84.

Jamison, P. J., and S. L. Zegura. 1974. A univariate and multivariate examination of measurement error in anthropometry. *Am. J. Phys. Anthropol.* 40:197–204.

Jeffreys, A. J., V. Wilson, and S. L. Thein. 1985. Hypervariable minisatellite regions in human DNA. *Nature* 316:76–79.

Jenkins, T., and V. Corfield. 1972. The red cell polymorphism in Southern Africa: Population data and studies on the R, RA, and RB phenotypes. *Ann. Hum. Genet.* 35:379–91.

Jenkins, T., and P. V. Tobias. 1977. Nomenclature of population groups in southern Africa. *African Studies Q. J.* 36:49–55.

Jenkins, T., A. B. Lane, G. T. Nurse, and D. A. Hopkinson. 1979. Red cell adenosine deaminase (ADA) polymorphism in Southern Africa, with special reference to ADA deficiency among the Kung. *Ann. Hum. Genet.* 42:425–33.

Johnson, C. S., and A. J. Giorgio. 1981. Arterial blood pressure in adults with sickle cell disease. *Arch. Int. Med.* 141:891–93.

Johnson, R. C., R. E. Cole, and F. M. Ahern. 1981. Genetic interpretation of ethnic differences in lactose absorption and tolerance: A review. *Hum. Biol.* 53:1–13.

Johnson, R. C., R. E. Cole, F. M. Ahern, S. Y. Schwitters, E. H. Ahern, Y.-H. Huang, R. M. Johnson, and J. Y. Park. 1980. Reported lactose tolerance of members of various racial/ethnic groups in Hawaii and Asia. *Behav. Genet.* 10:377–85.

Johnson, J. D., F. J. Simoons, R. Hurwitz, A. Grange, F. R. Sinatra, P. Sunshine, W. V. Robertson, P. H. Bennet, and N. Kretchmer. 1978. Lactose malabsorption among adult Indians of the Great Basin and American Southwest. *Am. J. Clin. Nutr.* 3:381–87.

Johnston, F. E., P. V. V. Hamill, and S. Lemeshow. 1974. Skinfold thicknesses in a national probability sample of U.S. males and females aged 6 through 17 years. *Am. J. Phys. Anthropol.* 40:321–24.

Joshi, M. R. 1975. Carabelli's trait on maxillary deciduous molars and first permanent molars in Hindus. *Arch. Oral Biol.* 20:699–700.

Kahlon, D. P. S. 1976. Age variation in skin color: A study in Sikh immigrants in Britain. *Hum. Biol.* 48:419–28.

Kalman, F. J. 1965. Some aspects of genetic counseling. In J. V. Neel, M. W. Shaw, and W. T. Shaw, eds., *Genetics and the epidemiology of chronic diseases*, Public Health Service Publication 1163, pp. 385–95. Washington, D.C.: U.S. Department of Health, Education, and Welfare, Public Health Service.

Kalmus, H. 1962. Distance and sequence of the loci for protan and deutan defects and for glucose-6-phosphate dehydrogenase deficiency. *Nature* 194:214–15.

————. 1965. *Diagnosis and genetics of defective colour vision.* Oxford: Pergamon Press.

Kamboh, M. I., and C. Kirkwood. 1984. Genetic polymorphism of thryoxin-binding globulin (TBG) in the Pacific area. *Am. J. Hum. Genet.* 36:646–54.

Kamel, K., and N. Moafy. 1983. Some aspects of thalassemia and the world common hemoglobinopathies in the Middle East. In J. E. Bowman, ed., *Distribution and evolution of hemoglobin and globin loci,* pp. 209–20. New York: Elsevier.

Kan, Y. W., and A. M. Dozy. 1978. Polymorphism of DNA sequence adjacent to human-β-globin structural gene: Relationship to sickle mutation. *Proc. Natl. Acad. Sci. (U.S.A.)* 75:5631–35.

————. 1980. Evolution of the hemoglobin S and C genes in world populations. *Science* 209:388–91.

————. 1984. Human hemoglobins and the new genetics. In K. J. Isselbacher, ed., *Medicine, science, and society,* pp. 298–311. New York: John Wiley.

Kanki, P. J., F. Barin, S. M'Boup, J. S. Allan, J. L. Romet-Lemonne, R. Marlink, M. F. McLane, T.-H. Lee, B. Arbeille, F. Denis, and M. Essex. 1986. New human T-Lymphotropic retrovirus related to simian T-lymphotropic virus type III (STLV-IIIAGM). *Science* 232:238–43.

Kanno, H., I-Y Huang, Y. W. Kan, and A. Yoshida. 1989. Two structural genes on different chromosomes are required for encoding the major subunit of human red cell glucose-6-phosphate dehydrogenase. *Cell* 58:595–606.

Kaplan, N. M. 1983. Hypertension: Prevalence, risks, and effect of therapy. *Ann. Int. Med.* 98:705–9.

Karim, A. K., M. S. Elfellah, and D. A. Evans. 1981. Human acetylator polymorphism: Estimate of allele frequency in Libya and details of global distribution. *Med. Genet.* 18:325–30.

Karp, G. W., and H. E. Sutton. 1967. Some new phenotypes of human red cell acid phosphatase. *Am. J. Hum. Genet.* 35:379–91.

Kashgarian, M., and R.C. Rendtorff. 1969. Incidence of Down's Syndrome in American Negroes. *J. Pediatr.* 74:468–71.

Kass, E. H., B. Rosner, S. H. Zinner, H. S. Margolius, and Y.-H. Lee. 1977. Studies on the origin of human hypertension. *Postgrad. Med.* (Suppl. 2) 53:145–52.

Kaufman, P. de B. 1977. The number of vertebrae in the Southern African Negro, the American Negro and the Bushman (San). *Am. J. Phys. Anthropol.* 47:409–14.

Kazazian, H. H. 1985. The nature of mutation. *Hosp. Pract. [off.]* 20:55–69.

Kazazian, H., and C. D. Boehm. 1988. Molecular basis and prenatal diagnosis of β-thalassemia. *Blood* 72:1107–16.

Keene, H. J. 1965. The relationship of maternal age, parity, and birth weight to hypodontia in Naval recruits. *Am. J. Phys. Anthropol.* (abstr.) 23:330.

Keene, H. 1983. Criteria and classification of diabetes mellitus. In J. I. Mann, K. Pyorala, and A. Teuscher, eds., *Diabetes in epidemiological perspective,* pp. 174–77. New York: Churchill Livingstone.

Keil, J. E., H. A. Tyroler, S. H. Sandifer, and E. Boyle. 1977. Hypertension:

Effects of social class and racial admixture, the results of a cohort study in the black population of Charleston, SC. *Am. J. Public Health* 67:634–39.

Keller, D. F. 1971. *G-6-PD Deficiency*, pp. 24–25; 29–30. Cleveland: CRC Press.

Kellermeyer, R. W., A. R. Tarlov, G. J. Brewer, P. E. Carson, and A. S. Alving. 1962. Hemolytic effect of therapeutic drugs: Clinical considerations of the primaquine type hemolysis. *JAMA* 180:388–94.

Kendall, A. 1979. Human erythrocyte carbonic anhydrase polymorphism in Kenya. *Hum. Genet.* 52:259–61.

Kerr, G. E., P. Amante, M. Decker, and P. W. Callen. 1982. Ethnic patterns of salt purchase in Houston, Texas. *Am. J. Epidemiol.* 115:906–16.

Khaw, K.-T., and G. Rose. 1982. Population study of blood pressure and associated factors in St Lucia, West Indies. *Int. J. Epidemiol.* 11:372–77.

Khoury, M. J., J. D. Erickson, and L. M. James. 1983. Maternal factors in cleft lip with or without palate: Evidence from interracial crosses in the United States. *Teratology* 27:351–57.

Kibukamusoke, J. W. 1966. *The nephrotic syndrome in Uganda with special reference to the role of Plasmodium malaria*. M.D. Thesis, University of East Africa.

Kidson, C., and D. C. Gajdusek. 1962. Glucose-6-phosphate dehydrogenase deficiency in Micronesian peoples. *Aust. J. Sci.* 25:61–62.

Kidson, C., and J. G. Gorman. 1962. A challenge to the concept of selection by malaria in glucose-6-phosphate dehydrogenae deficiency. *Nature* 196:49–51.

Kieser, J. A. 1968. The incidence and expression of Carabelli's trait in two South African populations. *J. Dent. Assoc. S. Afr.* 33:5–9.

Kieser, J. A., and C. B. Preston. 1981. The dentition of the Lengua Indians. *Am. J. Phys. Anthropol.* 55:485–90.

King, R. A., C. Donnell, J. Cervenka, A. Okoro, and C. J. Witkop. 1980. Albinism in Nigeria with delineation of new recessive oculocutaneous type. *Clin. Genet.* 17:259–70.

Kirk, R. L., W. C. Parker, and A. G. Bearn. 1964. The distribution of the transferrin variants D1 and DCHI in various populations. *Acta. Genet.* 14:41–48.

Kirkman, H. N. 1962. Electrophoretic differences of human erythrocytic glucose-6-phosphate dehydrogenase. *Am. J. Dis. Child.* 140:566–67.

Klahr, S. 1989. The kidney in hypertension—villain and victim. *N. Engl. J. Med.* 320:731–33.

Klauda, P. T., W. E. Ollier, A. A. Hilali, and R. A. Bacchus. 1984. Properdin factor B (Bf) polymorphism in Saudi Arabs: High frequency of a "rare" allele BfSO.7. *Hum. Hered.* 34:269–72.

Kleihauer, E., and K. Betke. 1963. Elution procedure for demonstration of methaemoglobin in red cells of human blood smears. *Nature* 199:1196.

Knowles, W. J., M. L. Bologna, J. A. Chasis, S. L. Marchesi and V. T. Marchesi. 1984. Common structural polymorphisms in human erythrocyte spectrin. *J. Clin. Invest.* 3:973–79.

Konotey-Ahulu, F. I. D. 1970. Maintenance of high sickling rate in Africa—role of polygamy. *J. Trop. Med. Hyg.* 73:19–21.

———. 1971. Computer assisted analysis on 1,697 patients attending the Sickle Cell/Haemoglobinopathy Clinic of Korle Bu Teaching Hospital, Accra, Ghana: Clinical features: I. Sex, genotype, age, rheumatism and dactylitis frequencies. *Ghana Med. J.* 10:241–60.

———. 1972. History of sickle cell disease in West Africa: Geographical distribution and population dynamics of haemoglobins S and C with special reference to West Africa. *Ghana Med. J.* 11:397–412.

———. 1980. Male procreative superiority index (MPSI): The missing coefficient in African anthropogenetics. *Br. Med. J.* 281:1700–2.

———. 1982. Ethics of amniocentesis and selective abortion for sickle cell disease. *Lancet* 1:38–39.

———. 1987. AIDS in Africa: Misinformation and disinformation. *Lancet* 2:206–7.

Kraus, B. S. 1959. Occurrence of the Carabelli's trait in the south-west ethnic groups. *Am. J. Phys. Anthropol.* 17:117–24.

Kreiss, J. K., D. Koech, F. A. Plummer, K. K. Holmes, M. Lightfoote, P. Piot, A. R. Ronald, J. O. Ndinya-Achola, L. J. D'Costa, P. Roberts, E. N. Ngugi, and T. C. Quinn. 1986. AIDS virus infection in Nairobi prostitutes. *N. Engl. J. Med.* 314:414–18.

Krogman, W. M. 1967. The role of genetic factors in the human face, jaws and teeth: A review. *Eugen. Rev.* 59:165–92.

Kromberg, J. G. R., D. J. Castle, E. M. Zwang, and T. Jenkins. 1988. Albinism and skin cancer. *Am. J. Hum. Genet.* (abstr.) 43:A58.

Kromberg, J. G. R., E. Zwane, D. Castle, and T. Jenkins. 1987. Albinism in South African blacks. *Lancet* 2:388–89.

Kruatrachue, M., P. Charoenlarp, T. Chongsuphajaisiddhi, and C. Harinsuta. 1962. Erythrocyte glucose-6-phosphate dehydrogenase and malaria in Thailand. *Lancet* 2:1183–1186.

Kueppers, F., and M. J. Christopherson. 1978. Alpha-antitrypsin: Further heterogeneity revealed by isoelectric focussing. *Am. J. Hum. Genet.* 30:359–65.

Kueppers, F. and B. M. Harpel. 1979. Group-specific component (Gc) "subtypes" of Gc 1 by isoelectric focussing in U.S. blacks and whites. *Hum. Hered.* 29:242–49.

———. 1980. Transferrin C subtypes in U.S. blacks and whites. *Hum. Hered.* 30:376–82.

Kuhne, K. 1932. Die Verebung der menschlichen Wirbelsaule. *Z. Morph. Anthrop.* 30:1–221.

———. 1936 Die Zwillingswirbelsaule. *Z. Morphol. Anthropol.* 35:1–376.

Kulozik, A. E., B. C. Kar, G. R. Serjeant, B. E. Serjeant, and D. J. Weatherall. 1988. The molecular basis of thalassemia in India: Its interaction with the sickle cell gene. *Blood* 71:467–72.

Kulozik, A. E., J. S. Wainscoat, G. R. Serjeant, B. C. Kar, B. Al-Awamy, G. J. F. Essan, A. G. Falusi, S. K. Haque, A. M. Hilali, S. Kate., W. A. E. P. Ransinghe, and D. J. Weatherall. 1986. Geographical survey of βS-globin gene haplotypes: Evidence for an independent Asian origin of the sickle-cell mutation. *Am. J. Hum. Genet.* 39:239–44.

Kunikana, H., N. Ishikawa, and T. Nakayama. 1987. Detection of a novel HLA-DQ specificity. In M. Aizawa, ed., *Proceedings of the third Asia and Oceana histocompatibility workshop conference.*

Kustaloglu, O. A. 1962. Paramolar structures of the upper dentition. *J. Dent. Res.* 41:75–83.

Kwon, P. H., M. H. Rorick, and N. S. Scrimshaw. 1980. Comparative tolerance of adolescents of differing ethnic backgrounds to lactose-containing and lactose-free dairy drink. *Am. J. Clin. Nutr.* 33:22–26.

Ladas, S., J. Papanikos, and G. Arapakis. 1982. Lactose malabsorption in Greek adults: Correlation of small bowel transition time with the severity of lactose intolerance. *Gut* 23:968–73.

Laha, P. K., N. Saha, and R. A. Bayoumi. 1979. Red cell glyoxylase I polymorphism among the selected tribes of the Sudan. *Idengaku Zasshi* 24:259–64.

Landsman, S., H. Minkoff, S. Holman, S. McCalla, and O. Sijin. 1987. Serosurvey of human immunodeficiency virus infection in parturients. Implications for human immunodeficiency virus testing of women. *JAMA* 258:2701–3.

Landsteiner, K. 1900. Zur Kenntnis der antifermentativen, lytischen, und agglutinierenden Wirkunger des Blut serums und der Lymphe. *Zentralbl. Bakteriol.* 27:357–62.

Landsteiner, K., and A. S. Wiener. 1940. An agglutinable factor in human blood recognized by immune sera for rhesus blood. *Proc. Soc. Exp. Biol. Med.* 43:213.

Langford, H. G. 1983a. Dietary potassium and hypertension: Epidemiologic data. *Ann. Int. Med.* 98:770–72.

———. 1983b. Potassium in hypertension. *Postgrad. Med.* 73:227–33.

Lanier, R. R. 1939. The presacral vertebrae of American white and Negro males. *Am. J. Phys. Anthropol.* 25:341–420.

Laragh, J. H., and M. S. Pecker. 1983. Dietary sodium and hypertension: Some myths, hopes, and truths. *Ann. Int. Med.* 98:735–43.

Lasker, G. W. 1950. Genetic analysis of racial traits of the teeth. *Cold Spring Harbor Symp. Quant. Biol.* 15:191–203.

Laurell, C.-B., and S. Eriksson. 1963. The electrophoretic alpha-1-globulin pattern of serum in alpha-1-antitrypsin deficiency. *Scand. J. Clin. Lab. Invest.* 15:132–40.

Laurence, J. 1985. The immune system in AIDS. *Sci. Am.* 52:84–93.

Layde, P. M., J. D. Erickson, A. Falek, and B. J. McCarthy. 1980. Congenital malformation in twins. *Am. J. Hum. Genet.* 32:69–78.

Lebo, R. V., A. Chakravarti, K .H. Beutow, M. C. Cheung, H. Cann, B. Cordell, and H. Goodman. 1983. Recombination within and between the human insulin and beta-globin gene loci. *Proc. Nat. Acad. Sci. (U.S.A).* 80:4808–12.

Lees, F. C., and P. J. Byard. 1978. Skin colorimetry in Belize. 1. Conversion formulae. *Am. J. Phys. Anthropol.* 48:515–22.

Lees, F. C., P. J. Byard, and J. H. Relethford. 1979. New conversion formulae for light-skinned populations using Photovolt and E. E. L. Reflectometers. *Am. J. Phys. Anthropol.* 51:403–8.

Lestrange, M. de. 1953. Les Crêtes papillaires digitales de 1.491 Noirs d'Afrique occidentale. *Bull. Inst. franc. Afriq. noir* 15:1278–1315.

Levine, E. 1972. Carpal fusions in children of four South African populations. *Am. J. Phys. Anthropol.* 37:75–84.

Levine, P., and, R. E. Stetson. 1939. An unusual case of contragroup agglutination. *JAMA* 113:126–27.

Levitan, M., and A. Montague. 1977. *Textbook of human genetics*, 2d ed. New York: Oxford University Press.

Lewin, R. 1987. The unmasking of mitochondrial Eve. *Science* 238:24–26.

Lewis, R. A. 1967. Sickle cell anaemia in G6PD deficiency. *Lancet* 1:852–53.

Lewis, R. A., and M. Hathorn. 1965. Correlation of S hemoglobin with glucose-6-phosphate dehydrogenase deficiency and its significance. *Blood* 26:176–80.

Light, K. C., P. A. Obrist, A. Sherwood, S. A. James, and D. S. Strogatz. 1987. Effects of race and marginally elevated blood pressure on response to stress. *Hypertension* 10:555–63.

Linder, D., and S. M. Gartler. 1965. Glucose-6-phosphate dehydrogenase mosaicism: Utilization as a cell marker in different tissues of heterozygotes. *Science* 150:67–69.

Lisker, R., B. Gonzalez, and M. Daltabuit. 1975. Recessive inheritance of the adult type of intestinal lactase deficiency. *Am. J. Hum. Genet.* 27:662–64.

Lisker, R., R. Perez-Briceno, J. Granados, and V. Babinsky. 1988. Gene frequencies and admixture estimates in the State of Puebla, Mexico. *Am. J. Phys. Anthropol.* 76:331–35.

Livingstone, F. B. 1967. *Abnormal hemoglobins in human populations. A summary and interpretation.* Chicago: Aldine.

———. 1971. Malaria and human polymorphisms. *Ann. Rev. Genet.* 5:33–64.

———. 1983. The malaria hypothesis. In J. E. Bowman, ed., *Distribution and evolution of hemoglobin and globin loci*, pp. 15–44. New York: Elsevier.

Livingstone, F. B. 1989. Who gave whom hemoglobin S: The use of restriction site haplotype variation for the interpretation of the evolution of the β^S-globin gene. *Am. J. Hum. Biol.* 1:289–302.

Lonnerblad, L. 1935. Uber zwei seltene anomalien im carpus. (Verschmelzung von Os lunatum und Os triquetrum sowie von Os multangulum minus und Os capitatum). *Acta Radiol.* 16:682–90.

Loomis, W. F. 1967. Skin-pigment regulation of vitamin-D biosynthesis in man. *Science* 157:501–6.

Loukopoulos, D. 1985. Prenatal diagnosis of thalassemia and of the hemoglobinopathies: A review. *Hemoglobin* 9:435–59.

Lubahn, D. B., S. T. Lord, J. Bosco, J. Kirshtein, O. J. Jeffries, N. Parker, C. Levtzow, L. M. Silverman, and J. B. Graham. 1987. Population genetics of coagulant factor IX: Frequencies of two DNA polymorphisms in five ethnic groups. *Am. J. Hum. Genet.* 40:527–36.

Lubs, H. A., S. A. Patil, W. J. Kimberling, J. Brown, M. P. Cohen, P. Gerald, F. Hecht, N. Myrianthopoulos, and R. L. Summitt. 1977. Q and C banding polymorphisms in 7 and 8 year old children: Racial differences and clinical significance. In E. B. Hook and I. H. Porter, eds., *Population cytogenetics* pp. 133–59. New York: Academic Press.

Lubs, H. A. , and F. Ruddle. 1971. Chromosome polymorphism: Negro and white populations differ. *Nature* 233:134–36.

Ludmerer, K. M. 1972. The American Eugenics Movement. In *Genetics and American Society*, pp. 7–43. Baltimore: Johns Hopkins University Press.

Lundy, J. K. 1980. The mylohyoid bridge in the Khoisan of Southern Africa and its unsuitability as a Mongoloid genetic marker. *Am. J. Phys. Anthropol.* 53:43–48.

Luzzatto, L. 1973. Studies of polymorphic traits for the characterization of populations: African populations south of the Sahara. *Isr. J. Med. Sci.* 9:1181–94.

———. 1974. Genetic heterogeneity and pathophysiology of G-6-PD deficiency. *Br. J. Haematol.* 28:151–156.

Luzzatto, L., E. A. Usanga, and S. Reddy. 1969. Glucose-6-phosphate dehydrogenase deficient red cells: Resistance to infection by malaria parasites. *Science* 164:839–42.

Lyon, M. F. 1962. Sex chromatin and gene action in the mammalian X-chromosome. *Am. J. Hum. Genet.* 14:135–48.

McCarron, D. A. 1983. Calcium and magnesium nutrition in human hypertension. *Ann. Int. Med.* 98:800–805.

McCarron, D. A., C. D. Morris, H. H. Henry, and J. L. Stanton. 1984. Blood pressure and nutrient intake in the United States. *Science* 224:1392–98.

McConkey, E. H., B. J. Taylor, and D. Phan. 1979. Human heterozygosity: A new estimate. *Proc. Nat. Acad. Sci. (U.S.A.)* 76:6500–6504.

McCormick, M. C. 1985. The contribution of low birth weight to infant mortality and childhood morbidity. *N. Engl. J. Med.* 312:82–90.

McCracken, R. D. 1971. Lactase deficiency: An example of dietary evolution. *Curr. Anthropol.* 12:479–517.

McEvedy, C. 1980. *The Penguin Atlas of African History*. Middlesex, England: Penguin Books.

McGhee, R. B., and W. Trager. 1950. Cultivation of Plasmodium lophurae in vitro in chicken erythrocyte suspensions and effects of some constituents of culture medium upon its growth and multiplication. *J. Parasitol.* 36:123–27.

McIntosh, R., K. K. Merritt, M. R. Richards, M. H. Samuels, and M. T. Bellows. 1954. The incidence of congenital malformations: A study of 5,964 pregnancies. *Pediatrics* 14:505–22.

Mackay, D. H. 1952. Skeletal maturation in the hand: A study of development in East African children. *Trans. R. Soc. Trop. Med. Hyg.* 46:135–70.

McKenzie, W. H., and H. A. Lubs. 1975. Human Q and C chromosomal variations: Distribution and incidence. *Cytogenet. Cell Genet.* 14:97–115.

McKusick, V. A. 1986. *Mendelian inheritance in man*. 7th ed. Baltimore: Johns Hopkins University Press.

———. 1988. *Mendelian inheritance in man*. 8th ed. Baltimore: Johns Hopkins University Press.

————. 1989. Mapping and sequencing the human genome. *N. Engl. J. Med.* 320:910–15.

McLaren, M. J., A. S. Lachman, and J. B. Barlow. 1979. Prevalence of congenital heart disease in black schoolchildren of Soweto, Johannesburg. *Br. Heart J.* 41:554–58.

Maher v. Roe. 432 U.S. 464, 1973.

Malina, R. M., W. H. Mueller, and J. D. Holman. 1976. Parent-child correlations and heritability of stature in Philadelphia black and white children 6 to 12 years of age. *Hum. Biol.* 48:475–86.

Maniatis, T., E. F. Fritsch, J. Lauer, and R. Lawn. 1980. The molecular genetics of human hemoglobins. *Ann. Rev. Genet.* 14:145–178.

Mann, J. M. and J. Chin. 1988. AIDS: A global perspective (Editorial). *N. Engl. J. Med.* 319:302–3.

Marchesi, V. T., T. W. Tillach , R. L. Jackson, J. J. Sereat, and R. Scott. 1972. Chemical characterization and surface orientation of the major glycoprotein of the human erythrocyte membrane. *Proc. Nat. Acad. Sci. (U.S.A.)* 69:1445–49.

Marden, P. M., D. W. Smith, and M. J. McDonald. 1964. Congenital anomalies in the newborn infant, including minor variations. *J. Pediatr.* 64:357–71.

Marks, P. A., and R. T. Gross. 1959. Erythrocyte glucose-6-phosphate dehydrogenase deficiency: Evidence of difference between Negroes and Caucasians with respect to the genetically determined trait. *J. Clin. Invest.* 38:2253–62.

Marsh, W. L., R. Oyen, M. E. Nichols, and H. Charles. 1974. Studies of NNSSU antigen activity of leukocytes and platelets. *Transfusion* 14:462–66.

Martin, S. K. 1980. Modified G-6-PD malaria hypothesis. *Lancet* 1:51.

Martin, S. K., L. H. Miller, D. Alling, V. C. Okoye, G. J. F. Esan, B. O. Osunkoya, and M. Deane. 1979. Severe malaria and glucose-6-phosphate dehydrogenase deficiency: A reappraisal of the malaria/G-6-P.D. hypothesis. *Lancet* 1:524–26.

Marx, J. L. 1986. The slow, insidious natures of the HTLV'S. *Science* 231:450–51.

Materson, B. J. 1985. Black/white differences in response to antihypertensive therapy. *J. Natl. Med. Assoc.* 77:9–13.

Mauff, G., F. D. Gauchel, and H. W. Hitzeroth. 1976. Polymorphism of properdin factor B in South African Negroid, Indian and Coloured populations. *Hum. Genet.* 33:319–22.

Mbanefo, C. C., E. A. Bababunmi, A. Mahgoub, T. P. Sloan, J. R. Idle, R. L. Smith. 1980. A study of the debrisoquine hydroxylation polymorphism in a Nigerian population. *Xenobiotica* 10:811–18.

Mears, J. G., H. M. Lachman, R. Cabannes, K. P. E. Amegnizin, D. Labie, and R. L. Nagel. 1981. Sickle cell gene: Its origin and diffusion from West Africa. *J. Clin. Invest.* 68:606–10.

Mears, J. G., H. M. Lachman, D. Labie, and R. L. Nagel. 1983. Alpha thalassemia is related to prolonged survival in sickle cell anemia. *Blood* 62:286–90.

Melnick, M., and N. C. Myrianthopoulos. 1979. *External ear malformations:*

Epidemiology, genetics, and natural history, Orig. Art. Ser. 15:69–82. White Plains, N.J.: National Foundation for Birth Defects.

Meredith, H. V. 1979. Relationship of lower limb height to sitting height in black populations of Africa and the United States. *Am. J. Phys. Anthropol.* 51:63–66.

Metneki, J., A. Czeizel, S. D. Flatz, and G. Flatz. 1984. A study of lactose absorption capacity in twins. *Hum. Genet.* 67:296–300.

Metras, D., H. Turquin, A. O. Coulibaly, and K. Ouattara. 1979. Congenital cardiopathies in a tropical environment: Study of 259 cases seen at Abidjan from 1969–1976. *Arch. Mal. Coeur* 72:305–10.

Miall, W. E., E. H. Kass, J. Ling, and K. l. Stuart. 1962. Factors influencing arterial pressure in the general population in Jamaica. *Br. Med. J.* 2:497–506.

Miller, B. A., N. Olivieri, M. Salameh, M. Ahmed, G. Antognetti, T. H. J. Huisman, D. G. Nathan, and S. Orkin. 1987. Molecular analysis, of the high-hemoglobin-F phenotype in Saudi Arabian sickle cell anemia. *N. Engl. J. Med.* 316:244–50.

Miller, B. A., M. Salameh, J. Waistcoat, G. Antognetti, S. Orkin, D. Weatherall, and D. G. Nathan. 1986. High fetal hemoglobin production in sickle cell anemia in the Eastern Province of Saudi Arabia is genetically determined. *Blood* 67:1404–10.

Miller, J., J. Golenser, D. T. Spira, and N. S. Kosower. 1984. Plasmodium falciparum: Thiol status and growth in normal and glucose-6-phosphate dehydrogenase deficient human erythrocytes. *Exp. Parasitol.* 57:239–47.

Mittal, K. K. 1976. The HLA polymorphism and susceptibility to disease. *Vox Sang.* 73:161–73

———. 1979. Human histocompatibility antigens. *J. Sci. Ind. Res.* 38:37–46.

———. 1984. Immunobiology of the major histocompatibility complex: Association of HLA antigens with disease. *Acta Anthropogenet.* 8:245–68.

Mohrenweiser, H. W., and R. S. Decker. 1982. Identification of several electrophoretic variants of human ceruloplasmin including Cp Michigan, a new polymorphism. *Hum. Hered.* 32:369–73.

Money, J. 1975. Counseling in genetics and applied behavior genetics. In K. W. Schale, V. E. Anderson, G. E. McClearn, and J. Money, eds., *Developmental human behavior genetics.* pp. 151–78. Lexington, Mass: Lexington Books, D.C. Heath.

Morrison, E. Y. St. A. 1982. Diabetes mellitus—a third syndrome. *Bull. Del. Health Care Diabet. Dev. Countries.* 3:14–15.

———. 1983. Diabetes mellitus in Jamaica. *West Ind. Med. J.* 32:199–200.

Morton, W. E. 1975. Hypertension and color blindness in young men. *Arch. Int. Med.* 135:653–56.

Moss, M. L., P. S. Chase, and R. I. Howes. 1967. Comparative odontometry of the permanent post-canine dentition of American whites and Negroes. *Am. J. Phys. Anthropol.* 27:125–42.

Motulsky, A. G. 1964. Hereditary red cell traits and malaria. *Am. J. Trop. Med. Hyg.* (Suppl.) 13:147:58.

Motulsky, A. G. 1988. Invited editorial: Normal and abnormal color-vision genes. *Am. J. Hum. Genet.* 42:405–7.

Mourant, A. E., A. C. Kopec, and K. Domaniewska-Sobczak. 1976. *The distribution of the human blood groups and other polymorphisms.* 2d ed. Oxford: Oxford University Press.

Mourant, A. E., D. Tills, and K. Domaniewska-Sobczak. 1976. Sunshine and the geographical distribution of the alleles of the Gc system of plasma proteins. *Hum. Genet.* 33:307–14.

Mueller, W. H., and R. M. Malina. 1977. Differential contribution of stature phenotypes to assortative mating in parents of Philadelphia black and white school children. *Am. J. Phys. Anthropol.* 45:269–76.

Munn v. Illinois 1876. 97 U.S. 124.

Murphy, E. A. 1973. Probabilities in genetic counseling. In D. Bergsma, ed. *Contemporary genetic counseling, Orig. ser.* 9. White Plains, N.J.: National Foundation for Birth Defects.

Murray, F. G. 1934. Pigmentation, sunlight, and disease. *Am. Anthropol.* 36:438–45.

Murray, R. F., Jr., R. Bolden, V. E. Headings, B. A. Quinton, and R. B. Surana. 1974. Information transfer in genetic counseling for sickle cell trait. *Am. J. Hum. Genet.* 26:63A.

Myrdal, G. 1962. *An American dilemma: The Negro problem and modern democracy.* New York: Harper & Row.

Myrianthopoulos, N. C. 1970. An epidemiological survey of twins in a large prospectively studied population. *Am. J. Hum. Genet.* 22:611–29.

Myrianthopoulos, N. C., and C. S. Chung. 1974. *Congenital malformations in singletons: Epidemiologic survey.* In D. Bergsma, ed., *Contemporary genetic counseling,* pp. 1–20, *Orig. ser.* 10, no. 11. White Plains N.J.: National Foundation for Birth Defects.

———. 1975. *Factors affecting risks of Congenital Malformations I. Epidemiologic Analysis. II. Effect of Maternal Diabetes Birth Defects.* In D. Bergsma, ed., *Contemporary genetic counseling, Orig. ser.* 11, no. 10. White Plains, N.J.: National Foundation for Birth Defects.

Nagel, R. L. 1984. The origin of the hemoglobin S gene: Clinical, genetic, and anthropological consequences. *Einstein Q. J. Biol. Med.* 2:53–62.

———. 1985. Hematologically and genetically distinct forms of sickle cell anemia in Africa. The Senegal and the Benin type. *N. Engl. J. Med.* 312:880–84.

Nagel, R. L., K. S. K. Rao, O. Dunda-Belkhodja, M. M. Connolly, M. E. Fabry, A. Georges, R. Krishnamoorthy, and D. Labie. 1987. The hematologic characteristics of sickle cell anemia bearing the Bantu haplotype: The relationship between $^G\gamma$ and Hb F level. *Blood* 69:1026–30.

Nagylaki, T. 1977. *Selection in 1- and 2-locus systems,* pp. 166–209. Berlin: Springer.

Nasrallah, S. M. 1979. Lactose intolerance in the Lebanese population and in "Mediterranean lymphoma." *Am. J. Clin. Nutr.* 32:1994–96.

Nathans, J., T. P. Piantanida, R. L. Eddy, T. B. Shows, and D. S. Hogness.

1986. Molecular genetics of human color vision: The genes encoding blue, green, and red pigments. *Science* 232:193–202.

Nathans, D., and H. O. Smith. 1985. Restriction endonucleases in the analysis and restructuring of DNA molecules. *Ann. Rev. Biochem.* 44:23–293.

National Diabetes Data Group. 1979. Classification and diagnosis of diabetes mellitus and other categories of glucose intolerance. *Diabetes* 28:1039–57.

Naylor, A. F., and N. C. Myrianthopoulos. 1967. The relation of ethnic and selected socio-economic factors to human birth-weight. *Ann. Hum. Genet.* 31:71–83.

Neer, R. M. 1974. The evolutionary significance of vitamin D, skin pigment, and ultraviolet light. *Am. J. Phys. Anthropol.* 43:409–16.

Nei, M. 1975. Gene diversity within populations and gene diversity in subdivided populations. *In Molecular population genetics and evolution*, pp.132–54. New York: Elsevier.

Nelson, M. M., and J. O. Forfar. 1969. Congenital abnormalities at birth: Their association in the same patient. *Dev. Med. Child Neurol.* 11:3–10.

Nerl, C., and G. T. O'Neill. 1982. Factor B polymorphism in North American blacks: A study of a new variant Bf F135. *Hum. Genet.* 61:357–59.

Newcomer, A. D., H. Gordon, P. J. Thomas, and D. B. McGill. 1977. Family studies of lactase deficiency in the American Indian. *Gastroenterol.* 73:985–88.

Nicoloff, J. T., J. T. Dowling, and D. D. Patton. 1964. Inheritance of decreased thyroxine-binding by the thyroxine-binding globulin. *J. Clin. Endocrinol.* 24:294–98.

Noguchi, C. T., and A. N. Schecter. 1981. The intracellular polymerization of sickle hemoglobin and its relevance to sickle cell disease. *Blood* 58:1057–68.

Nomenclature for factors of the HLA system. 1984. *Tissue Antigens* 24:73–80.

Nose, O., Y. Ida, H. Kai, T. Harda, M. Ogawa, and H. Yabuuchi. 1979. Breath hydrogen test for detecting lactose markers in infants and children: Prevalence of lactose malabsorption in children and adults. *Arch. Dis. Child.* 6:436–40.

Nylander, P. P. S. 1971a. Ethnic differences in twinning rates in Nigeria. *J. Biosoc. Sci.* 3:151–57.

———. 1971b. The incidence of triplets and higher multiple births in some rural and some urban populations in Western Nigeria. *Ann. Hum. Genet.* 34:409–16

———. 1978. Causes of high twinning frequencies in Nigeria. In W. E. Nance, ed., *Twin research. Part B. Biology and epidemiology. Progress in clinical and biological research*, 24B:35–43. New York: Alan R. Liss.

Nylander, P. P. S., and G. Corney. 1977. Placentation and zygosity of twins in Northern Nigeria. *Ann. Hum. Genet.* 40:323–29.

Nzilambi, N., K. M. De Cock, D. N. Forthal, H. Francis, R. W. Ryder, I. Malebe, J. Getchell, M. Laga, P. Piot, and J. McCormick. 1988. The prevalence of infection with human immunodeficiency virus over a 10-year period in rural Zaire. *N. Engl. J. Med.* 318:276–79.

Odenheimer, D. J., S. A. Sarnaik, C. F. Whitten, D. L. Rucknagel, and C. F.

Sing. 1987. The relationship between fetal hemoglobin and disease severity in children with sickle cell anemia. *Am. J. Med. Genet.* 27:525–35.

O'Donnell, F. E., R. Green, J. A. Fleischman, and G. W. Hambrick. 1978a. X-linked ocular albinism in blacks. Ocular albinism cum pigmento. *Arch. Ophthalmol.* 96:1189–92.

O'Donnell, F. E., R. A. King, W. R. Green, and C. J. Witkop. 1978b. Autosomal recessively inherited ocular albinism. *Arch. Ophthalmol.* 96:1621–25.

Ojikutu, R. O., G. T. Nurse, and T. Jenkins. 1977. Red cell enzyme polymorphisms in Yoruba. *Hum. Hered.* 27:444–53.

O'Keefe, S. J., and J. K. Adam. 1983. Primary lactose intolerance in Zulu adults. *S. Afr. Med. J.* 63:778–80.

Okoro, A. N. 1975. Albinism in Nigeria. A clinical and social study. *Br. J. Dermatol.* 92:485–92.

Old, J. M., M. Petrou, B. Modell, and D. J. Weatherall. 1984. Feasibility of antenatal diagnosis of β thalassemia by DNA polymorphisms in Asian Indian, and Cypriot populations. *Br. J. Haematol.* 57:255–63.

Olowe, S. A., and O. Ransome-Kuti. 1980. The risk of jaundice in glucose-6-phosphate dehydrogenase deficient babies exposed to menthol. *Acta Paediatr. Scand.* 69:341–45.

Omoto, K. 1979. Carbonic anhydrase-I polymorphism in a Philippine aboriginal population. *Am. J. Hum. Genet.* 31:747–50.

Oppenheimer, S. J., D. R. Higgs, D. J. Weatherall, J. Barker, and R. A. Spark. 1984. Alpha thalassemia in Papua New Guinea. *Lancet* 1:424–26.

O'Rahilly, R. 1953. Epitriquetrum and lunatotriquetrum. *Acta Radiol.* 39:401–11.

———. 1957. Developmental deviations in the carpus and tarsus. *Clin. Orthop.* 10:9–18.

Orkin, S. H., S. E. Antonarakis, and H. H. Kazazian. 1983. Polymorphism and molecular pathology of the human beta-globin gene. *Prog. Hematol.* 13:49–73.

Orkin, S. H., and H. H. Kazazian. 1984. The mutation and polymorphism of the human β-globin gene and its surrounding DNA. *Ann. Rev. Genet.* 18:131–71.

Orkin, S. H., P. F. R. Little, H. H. Kazazian, and C. D. Boehm. 1982. Improved detection of the sickle mutation by DNA analysis. *N. Engl. J. Med.* 307:32–36.

Oski, F. A., and P. M. Growney. 1965. A simple micromethod for the detection of erythrocyte glusoce-6-phosphate dehydrogenase deficiency. *J. Pediatr.* 66:90–93.

Ostrander, C. R., R. S. Cohen, A. O. Hopper, S. M. Shahin, J. A. Kerner, J. D. Johnson, and D. K. Stevenson. 1983. Breath hydrogen analysis: A review of the methodologies and clinical applications. *J. Pediatr. Gastroenterol. Nutr.* 2:525–33.

Oviasu, V. O. 1978. Arterial blood pressure and hypertension in a rural Nigerian community. *Afr. J. Med. Sci.* 7:137–43.

Oviasu, V. O., and F. E. Okupa. 1980. Relation between hypertension and occupational factors in rural and urban Africans. *Bull. WHO* 58:485–89.

Page, C. 1987. In limbo: A woman without a race. *Chicago Tribune*, sec. 4, p. 3, May 24.

Page, I. H. 1979. Two cheers for hypertension. *JAMA* 242:2559–61.

Page, H. S. and A. S. Asire. 1985. *Cancer rates and risks*. National Institutes of Health Publication 85-961, U.S. Department of Health and Human Services, Public Health Service, National Institutes of Health.

Pagnier, J., D. Labie, H. M. Lachman, O. Dunda-Belkhodja, L. Kaptue-Noche, I. Zohoun, R. L. Nagel, and J. G. Mears. 1983. Human globin gene polymorphism in West and Equatorial Africa. In J. E. Bowman, ed., *Distribution and evolution of hemoglobin and globin loci*, pp. 145–58. New York: Elsevier.

Pagnier, J., J. G. Mears, O. Dunda-Belkhodja, K. E. Schaefer-Rezo, C. Beldjord, R. L. Nagel, and D. Labie. 1984a. Evidence for the multicentric origin of the sickle cell hemoglobin gene in Africa. *Proc. Nat. Acad. Sci. (U.S.A.)* 81:1771–73.

Pagnier, J., O. Dunda-Belkhodja, I. Zohoun, J. Teyssier, H. Baya, G. Jaeger, R. L. Nagel, and D. Labie. 1984b. α thalassemia among sickle cell anemia patients in various African populations. *Hum. Genet.* 68:318–23.

Paige, D. M., and T. M. Bayless. 1981. *Lactose digestion*. Baltimore: Johns Hopkins University Press.

Paige, D. M., T. M. Bayless, and W. S. Dellinger. 1975a. Relationship of milk consumption to blood glucose rise in lactose intolerant individuals. *Am. J. Clin. Nutr.* 28:677–80.

Paige, D. M., T. M. Bayless, S. S. Huang, and R. Wexler. 1975b. Lactose hydrolyzed milk. *Am. J. Clin. Nutr.* 28:818–22.

Paige, D. M., T. M. Bayless, E. D. Mellitis, and L. Davis. 1977. Lactose malabsorption in preschool black children. *Am. J. Clin. Nutr.* 30:1018–22.

Parker, B. F. 1950. The incidence of mongoloid imbecility in newborn infants: A ten year study covering 27,931 live births. *J. Pediatr.* 36:493–94.

Pasvol, G., and R. J. M. Wilson. 1982. The interaction of malaria parasites with red blood cells. *Br. Med. Bull.* 38:133–40.

Pasvol, G., D. J. Weatherall, and R. J. M. Wilson. 1978. Cellular mechanism for the protective effect of haemoglobin S against P. falciparum malaria. *Nature* 274:701–3.

Patil, S. R., and H. A. Lubs. 1977. Classification of qh regions in human chromosomes 1, 9, and 16 by C-banding. *Hum. Genet.* 38:35–38.

Pearn, J. H. 1973. Patients' subjective interpretation of risks offered in genetic counseling. *J. Med. Genet.* 10:129–34.

Penrose, L. S. 1963. Fingerprints, palms, and chromosomes. *Nature* 197:933–38.

Penrose, L. S., and D. Loesch. 1970. Topological classification of palmar dermatoglyphics. *J. Ment. Defic. Res.* 14:111–28.

Penrose, L. S., and P. T. OHara. 1973. The development of the epidermal ridges. *J. Med. Genet.* 10:201–8.

Perrine, R. P., A. P. Gelpi, and S. P. Perrine. 1983. Population genetics of hemoglobinopathies and thalassemia in Eastern Saudi Arabia. In J. E.

Bowman, ed., *Distribution and evolution of hemoglobin and globin loci*, pp. 221–37. New York: Elsevier.

Piomelli, S., L. M. Corash, D. D. Davenport, J. Miraglia, and E. L. Amorosi. 1968. In vivo liability of glucose-6-phosphate in Gd and Gd Mediterranean deficiency. *J. Clin. Invest.* 47:940–48.

Plato, C. C. 1970. Polymorphism of the C line of palmar dermatoglyphics with a new classification of the C line terminations. *Am. J. Phys. Anthropol.* 33:413–20.

Plato, C. C., H. A. Brown, and D. C. Gajdusek. 1975. The dermatoglyphics of the Elema people from the Gulf District of Papua New Guinea. *Am. J. Phys. Anthropol.* 42:241–250.

Plato, C. C., and W. Wertlecki. 1972. A method for subclassifying the interdigital patterns: A comparative study of the palmar configurations. *Am. J. Phys. Anthropol.* 37:97–110.

Pobee, J. O. M., E. B. Larbi, D. W. Belcher, F. K. Wurapa, and S. R. A. Dodu. 1977. Blood pressure distribution in a rural Ghanaian population. *Trans. R. Soc. Trop. Med. Hyg.* 71:66–72.

Pol, R. 1921. "Brachydactylie"-"Klinodaktylie"-Hyperphylangie und ihre Grundlagen. *Virchows Arch. [A]* 229:388–530.

Pollitzer, W. S. 1958. The Negroes of Charleston (S.C.): A study of hemoglobin types, serology, and morphology. *Am. J. Phys. Anthropol.* 16:241–63.

Pompe van Meerdervoort, H. F. 1976. Congenital musculoskeletal malformation in South African blacks: a study of incidence. *S. Afr. Med. J.* 50:1853–55.

Ponzone, A., G. F. Voglino, and A. Tognolo. 1975. Milk casein polymorphism in the Kikuyu population. *Ann. Hum. Genet.* 18:203–5.

Porter, G. A. 1983. Chronology of the sodium hypothesis and hypertension. *Ann. Int. Med.* 98:720–23.

Porter, I. H., S. H. Boyer, E. J. Watson-Williams, A. Adam, A. Szeinberg, and M. Siniscalco. 1964. Variation of glucose-6-phosphate dehydrogenase in different populations. *Lancet* 1:895–99.

Porter, I. H., J. Schulze, and V. A. McKusick. 1962. Genetical linkage between the loci for glucose-6-phosphate dehydrogenase deficiency and colour-blindness in American Negroes. *Ann. Hum. Genet.* 26:107–22.

Post, P. W., F. Daniels, and R. T. Binford. 1975. Cold injury and the evolution of "white" skin. *Hum. Biol.* 47:65–80.

Post, P. W., A. N. Krauss, S. Waldman, and P. A. M. Auld. 1976. Skin reflectance of newborn infants from 25 to 44 weeks gestational age. *Hum. Biol.* 48:541–57.

Post, P. W., and D. C. Rao. 1977. Genetic and environmental determinants of skin color. *Am. J. Phys. Anthropol.* 47:399–402.

Post, R. H. 1969. Tear duct size differences of age, sex, and race. *Am. J. Phys. Anthropol.* 30:85–88.

Powars, D. R., W. A. Schroeder, J. N. Weiss, L. S. Chan, and S. P. Azen. 1980. Lack of influence of fetal hemoglobin levels or erythrocyte indices on the severity of sickle cell anemia. *J. Clin. Invest.* 65:732–40.

Powell, G. F., M. A. Rasco, and R. M. Maniscalco. 1974. A prolidase deficiency in a man with iminopeptiduria. *Metabolism* 23:505–13.

Powell, R. D., and G. J. Brewer. 1965. Glucose-6-phosphate dehydrogenase deficiency and falciparum malaria. *Am. J. Trop. Med. Hyg.* 14:358–62.

Poznanski, A. K., and J. F. Holt. 1971. The carpals in congenital malformation syndromes. *Am. J. Roentgenol.* 112:443–59.

Pronk, J. C., W. J. Jansen, A. v.d. Pronk, C. F. Pol, R. R. Frants, and A. W. Eriksson. 1984. Salivary protein polymorphism in Kenya: Evidence for a new AMY 1 allele. *Hum. Hered.* 34:212–16.

Pyeritz, R. E., J. E. Tumpson, and B. A. Bernhardt. 1987. The economics of clinical genetics services. I. Preview. *Am. J. Hum. Genet.* 41:549–58.

Qazi, Q. H., H. C. Mapa, and J. Woods. 1977. Dermatoglyphics of American blacks. *Am. J. Phys. Anthropol.* 47:483–88.

Quevedo, W. C., T. B. Fitzpatrick, M. A. Pathak, and K. Jimbo. 1974. Role of light in human skin color variation. *Am. J. Phys. Anthropol.* 43:393–408.

Race, R. R., and R. Sanger. 1975. *Blood groups in man*, 6th ed. p. 206. Oxford: Blackwell Scientific Publications.

Ramot, B., S. Fisher, A. Szeinberg, A. Adam, C. Sheba, and D. Ganni. 1959. A study of subjects with erythrocyte glucose-6-phosphate dehydrogenase deficiency: Investigation of leukocyte enzymes. *J. Clin. Invest.* 38:2234–37.

Ramsay, M., R. Bernstein, E. Zwane, D. C. Page, and T. Jenkins. 1988. XX true hermaphroditism in Southern African blacks: An enigma of primary sexual differentiation. *Am. J. Hum. Genet.* 43:4–13.

Ramsay, M., and T. Jenkins. 1984. Alpha-thalassemia in Africa: The oldest malaria protective trait? *Lancet* 2:410.

———. 1987. Globin gene-associated restriction-fragment-length polymorphisms in Southern African peoples. *Am. J. Hem. Genet.* 41:1132–44.

Ramsay, M. R., and T. Jenkins. 1988. Alpha-globin gene cluster haplotypes in the Kalahari San and Southern African Bantu-speaking blacks. *Am. J. Hum. Genet.* 43:527–33.

Ransome-Kuti, O., N. Kretchmer, J. D. Johnson, and J. T. Gribble. 1975. A genetic study of lactose digestion in Nigerian families. *Gastroenterology* 68:431–36.

Rao, D. C., N. E. Morton, and S. Yee. 1974. Analysis of family resemblance. II. A linear model for family correlation. *Am. J. Hum. Genet.* 26:331–59.

Raum, D., D. Marcus, and C. A. Alper. 1980. Genetic polymorphism of human plasminogen. *Am. J. Hum. Genet.* 32:681–89.

Raum, D., M. A. Spence, D. Balavitch, S. Tideman, A. D. Merritt, R. T. Taggert, B. H., Petersen, N. K. Day, and C. A. Alper. 1979. Genetic control of the eighth component of complement. *J. Clin. Invest.* 64:858–65.

Ree, G. H. 1976. Arterial pressures in a West African (Gambian) population. *J. Trop. Med. Hyg.* 76:65–70.

Reed, T. E. 1969. Caucasian genes in American Negroes. *Science* 165:762–68.

Reed, W. E. 1981. Racial differences in blood pressure levels of adolescents. *Am. J. Public Health* 71:1165–67.

Reich, R. B. 1982. Ideologies of survival. *New Republic* 187:32–37.

Reitnauer, P. J., R. C. P. Go, and R. T. Acton. 1982. Evidence for genetic admixture as a determinant in the occurrence of insulin-independent diabetes mellitus in U.S. blacks. *Diabetes* 31:532–37.

Reitnauer, P. J., J. M. Roseman, B. O. Barger, C. C. Murphy, K. A. Kirk, and R. T. Acton. 1981. HLA association with insulin-dependent diabetes mellitus in a sample of the American black population. *Tissue Antigens* 1:286–93.

Report of a WHO working group on the community control of hereditary anemias. 1982. *WHO Rep.*

Ride, L. (1935). 1951. Cited in Malformations of the auricle and the external auditory meatus: A critical review. *Arch. Otolaryngol.* 54:115–39.

Rinaldi, E., L. Albini, C. Costagliola, G. De Rosa, G. Aurricchio, B. De Vizia, and S. A. Aurricchio. 1984. High frequency of lactose absorbers among adults with high prevalence of primary adult lactose malabsorption. *Lancet* 1:355–57.

Ringelhann, B. 1972. A simple laboratory procedure for the recognition of A-(African type) G6PD deficiency in acute haemolytic crisis. *Clin. Chim. Acta* 36:272–74.

Ritchie, J. C., and J. R. Idle. 1982. Population studies of polymorphism in drug oxidation and its relevance to carcinogensis. *IARC Sci. Publ.* 32:381–94.

Rivas, M. L., A. D. Merritt, and L. Oliner. 1971. *Genetic variants of thyroxine-binding globulin (TBG). Birth defects, Orig. ser.* 7 (6):34–41. White Plains, N.J.: National Association for Birth Defects.

Roberts, D. F., H. Lehmann, and A. E. Boyo. 1960. Abnormal hemoglobins in Bornu. *Am. J. Phys. Anthropol.* 18:5–11.

Robinson, D., J. Day, and A. Bailey. 1980. Blood pressure in urban and tribal Africa. *Lancet* 2:424.

Roe v. Wade 1973. 410 U.S. 116.

Roper, A. 1976. Hip dysplasia in the African Bantu. *J. Bone Joint Surg. (Br.)* 58B:155–58.

Rosado, J. L., and N. W. Solomons. 1983. Sensitivity and specificity of the hydrogen breath-analysis test for detecting malabsorption of physiological doses of lactose. *Clin. Chem.* 29:545–48.

Rosatelli, C., A. M. Falchi, M. T. Scalas, T. Tuveri, M. Furbetta, and A. Cao. 1984. Hematological phenotype of the double heterozygous state for alpha and beta thalassemia. *Hemoglobin* 8:25–35.

Rosen-Bronson, S., T. F. Tang, D. D. Eckels, F. M. Robbins, G. Dunston, R. J. Hartzman, and A. H. Johnson. 1987. DR3 heterogeneity recognized by HTC and alloproliferative T cell clones. *Transplant. Proc.* 19:842–44.

Rosenkranz, W., B. Hadorn, W. Muller, P. Heinz-Erian, C. Henson, and G. Flatz. 1982. Distribution of human lactase phenotypes in the population of Austria. *Hum. Genet.* 62:158–61.

Rostand, S. G., G. Brown, K. A. Kirk, E. A. Rutsky, and H. Dustan. 1989. Renal insufficiency in treated essential hypertension. *N. Engl. J. Med.* 320:684–88.

Roth, E. F., M. Friedman, Ueda, Y., Tellez, I., Trager, W., and R. L. Nagel.

1978. Sickling rates of human AS red cells infected in vitro with Plasmodium falciparum malaria. *Science* 202:650–52.

Rotter, J. I., and D. L. Rimoin. 1987. The genetics of diabetes. *Hosp. Pract. [off.]* 22:79–88.

Rowley, P. T., E. J. Benz, and A. N. Nienhuis. 1986. Molecular genetics for the hematologist. *Curr. Hematol. Oncol.* 4:1–38.

Roux, P., C. D. Karabus, and P. S. Hartley. 1982. The effect of glucose-6-phospate dehydrogenase deficiency on the severity of neonatal jaundice in Cape Town. *S. Afr. Med. J.* 61:781–82.

Rucknagel, D. L., and C. J. Bruzdzinski. 1983. The α-globin loci in human populations. In J. E. Bowman, ed., *Distribution and evolution of hemoglobin and globin loci*, pp. 79–92. New York: Elsevier.

Ruiz, L., and D. Penaloza. 1977. Altitude and hypertension. *Mayo Clin. Proc.* 52:442–45.

Sadre, M., and K. Karbasi. 1979. Lactose tolerance in Iran. *Am. J. Clin. Nutr.* 32:1948–54.

Saha, N. 1981. Erythrocyte glutathione reductase polymorphism in a Sudanese population. *Hum. Hered.* 31:32–34.

Saha, N., K. A. Gumaa, A. P. Samuel, and H. El-Naeim. 1979. Placental alkaline phosphatase in a Sudanese population: Polymorphism and enzyme activity. *Hum. Biol.* 51:335–39.

Saha, N., and N. Patgunarajah. 1981. Phenotypic and quantitative relationship of red cell acid phosphatase with hemoglobin, haptoglobin, and G6PD phenotypes. *J. Med. Genet.* 18:271–75.

Sahi, T. 1978. Intestinal lactase polymorphism and dairy foods. *Hum. Genet.* (suppl.) 1:115–23.

Sahi, T., K. Launiala, and H. Lartinen. 1983. Hypolactasemia in a fixed cohort of young Finnish adults. A follow-up study. *Scand. J. Gastroenterol.* 18:865–70.

Saiki, R., S. Scharf, F. Falcona, K. B. Mullis, G. T. Horn, H. A. Erlich, and N. Arnheim. 1985. Enzymatic amplification of β-globin genomic sequences and restriction site analysis for diagnosis of sickle cell anemia. *Science* 230:1350–54.

Salman, C., J. P. Cartron, and P. Rouger. 1984. The Kell system. In *The human blood groups*, pp. 240–48. New York: Masson.

Sansone, G., A. Quartino Rasore, and G. Veneziano. 196 . Two red cell populations in the human female heterozygous for G-6-PD deficiency. *Lancet* 1:329.

Santachiara-Benerecetti, A. S. 1980. Population genetics of red cell enzymes in Pygmies: A conclusive account. *Am. J. Hum. Genet.* 32:934–54.

Santachiara-Benerecetti, A. S., G. N. Ranzani, and G. Antonini. 1977. Studies on African pygmies. V. Red cell acid phosphatase polymorphism in Babinga Pygmies: High frequency of ACPR allele. *Am. J. Hum. Genet.* 29:635–38.

Santachiara-Benerecetti, A. S., G. N. Ranzani, G. Antonini, and M. Beretta. 1982. Subtyping of phosphoglucomutase locus 1 (PGM1) polymorphism in some populations of Rwandas: Description of variant phenotypes, "haplotype" frequencies, and linkage disequilibrium data. *Am. J. Hum. Genet.* 34:337–48.

Santos, M. C. N., and E. S. Azevedo. 1981. Generalized joint hypermobility and black admixture in school children of Bahia, Brazil. *Am. J. Phys. Anthropol.* 55:43–46.

Sass-Kuhn, S. P., R. Mogbel, J. A. Mackay, O. Cromwell, and A. B. Kay. 1984. Human granulocyte/pollen binding protein: Recognition and identification as transferrin. *J. Clin. Invest.* 73:202–10.

Satori, E., F. Panizon, and F. Zacchello. 1966. Bimodal distribution of erythrocytes in heterozygotes for strong Mediterranean glucose-6-phosphate dehydrogenase deficiency. *J. Med. Genet.* 3:42–46.

Savilahti, E., K. Launiala, and P. Kuitunen. 1983. Congenital lactase deficiency: A clinical study on 16 patients. 1983. *Arch. Dis. Child.* 58:246–52.

Schachter, H., M. A. Michaels, C. A. Tilley, M. C. Crookson, and J. Crookston. 1973. Qualitative differences in the N. acetyl-D-galactosminyl tranferases produced by human A1 and A2 genes. *Proc. Nat. Acad. Sci. (U.S.A).* 70:220–24.

Schneckloth, R. E., A. C. Corcoran, and K. L. Moore. 1962. Arterial blood pressure and hypertensive disease in a West Indian Negro population. *Am. Heart J.* 63:607–28.

Schreffler, D. C., G. J. Brewer, J. C. Gall, and M. S. Honeyman. 1967. Electrophoretic variation in human serum ceruloplasmin: A new genetic polymorphism. *Biochem. Genet.* 1:101–16.

Schroeder, W. A., T. H. J. Husiman, J. R. Shelton, E. F. Kleihauer, A. M. Dozy, and A. M. Robberson. 1968. Evidence for multiple structural genes for the γ-chain of fetal hemoglobin. *Proc. Nat. Acad. Sci. (U.S.A.)* 60:537–44.

Schutte, J. E. 1980. Growth differences between lower and middle income black male adolescents. *Hum. Biol.* 52:193–204.

Schwartz, E., and S. Surrey. 1986. Molecular diagnosis of the hemoglobinopathies. *Hosp. Pract. [off.]* 21:163–78.

Scotch, N. A. 1963. Sociocultural factors in the epidemiology of Zulu hypertension. *Am. J. Public Health* 53:1205–13.

Scott, G. R. 1977. Classification, sex dimorphism, association, and population variation of the canine distal accessory ridge. *Hum. Biol.* 49:453–69.

———. 1980. Population variation of Carabelli's trait. *Hum. Biol.* 52:63–68.

Scott-Emuakpor, A. B., and E. D. Madueke. 1976. The study of genetic variation in Nigeria. II. The genetics of polydactyly. *Hum. Hered.* 26:198–202.

Scozzari, R., A. Torroni, O. Semino, G. Sirugo, A. Brega, and A. S. Santachiara-Benerecetti. 1988. Genetic studies on the Senegal population. I. Mitochondrial polymorphisms. *Am. J. Hum. Genet.* 43:534–44.

Scriver, C. R., M. Bardaris, L. Cartier, C. L. Clow, G. A, Lancaster, and J. Ostrowsky. 1984. β-thalassemia disease prevention: Genetic medicine applied. *Am. J. Hum. Genet.* 36:1024–38.

Seedat, Y. K. 1982. Hypertension and ischemic heart disease in Indian people living in South Africa and in India. *S. Afr. Med. J.* 61:965–67.

Seedat, Y. K., and J. Reddy. 1976. The clinical pattern of hypertension in the

South African black population: a study of 1000 patients. *S. Afr. J. Med. Sci.* 5:1–7.

Seedat, Y. K., M. A. Seedat, and D. B. Hackland. 1982. Biosocial factors and hypertension in urban Zulus. *S. Afr. J. Med. Sci.* 61:999–1002.

Seedat, Y. K., M. A. Seedat, and M. T. Veale. 1980. The prevalence of hypertension in urban whites. *S. Afr. Med. J.* 57:1025–30.

Seftel, H. C. 1978. The rarity of coronary heart disease in South African blacks. *S. Afr. Med. J.* 54:99–105.

Segal, I., P. P. Gagjee, A. R. Europe, and A. M. Noormohamed. 1983. Lactase deficiency in the South African black population. *Am. J. Clin. Nutr.* 38:901–5.

Selkirk, T.E. 1935. Fistula auris congenita. *Am J. Dis. Child.* 49:431–47.

Sellars, S. and P. Beighton. 1983. The Waardenburg syndrome in deaf children in southern Africa. *S. Afr. Med. J.* 63:725–28.

Seltzer, W. K., T. C. Abshire, J. K. Wolford, P. A. Lane, J. S. Roloff, and J. H. Githens. 1988. Molecular genetic analysis of five American black families with sickle cell disease and non-deletional hereditary persistence of fetal hemoglobin (ndHPFH), presenting with unusually high HbF levels. *Blood* (Abstr.) 72:73a.

Selvin, S. 1971. The number of pregnancies prior to the birth of twins. *Ann. Hum. Genet.* 34:427–29.

Selwyn, P. A. 1986. AIDS: What is now known. II. Epidemiology. *Hosp. Pract. [off.]* 21:127–64.

Semino, O., A. Torroni, R. Scozzari, A. Brega., G. De Benedictis, and A. S. Santachiara Benerectti. 1989. Mitochondrial DNA polymorphism in Italy. III. Population data from Sicily: A possible quantitation of African ancestry. *Ann. Hum. Genet.* 53:193–202.

Senewiratne, B., S. H. Thambipillai, and H. Perera. 1977. Intestinal lactase deficiency in Ceylon (Sri Lanka). *Gastroenterology* 72:1257–59.

Sepehrnia, B., M. I. Kamboh, L. L. Adams-Campbell, M. Nwankwo, and R. E. Ferrell. 1988. Genetic studies of human apolipoproteins. VII. Population distribution of polymorphisms of apolipoproteins A-I, A-II, A-IV, C-II, E, and H in Nigeria. *Am. J. Hum. Genet.* 43:847–53.

Sertima, I. V. 1976. *They came before Columbus: The African presence in ancient America*. New York: Random House.

———. 1983. They came before Columbus: The African presence in ancient America, an update. *Dollars Sense* 8:36–58.

Seva-Pereira, A., and B. Beiguelman. 1982. Primary lactose malabsorption in healthy Brazilian adult caucasoid, negroid, and mongoloid subjects. *Arq. Gastroenterol.* 19:133–38.

Sever, P. S., D. Gordon, W. S. Peart, and P. Beighton. 1980. Blood pressure and its correlates in urban and tribal Africa. *Lancet* 2:60–64.

Shahidi, N. T., and L. K. Diamond. 1959. Enzyme deficiency in erythrocytes in congenital nonspherocytic hemolytic anemia. *Pediatrics* 24:245–53.

Shaper, A. G., and G. A. Saxton. 1969. Blood pressure and body build in a rural community in Uganda. *East Afr. Med. J.* 46:228–45.

Shaper, A. G., D. H. Wright, and J. Kyobe. 1969. Blood pressure and body build in three nomadic tribes of Northern Kenya. *East Afr. Med. J.* 46:273–78.

Shapiro, L. R., R. O. Petterson, P. L. Wilmot, D. Warburton, P. A. Benn, and L. Y. F. Hisu. 1984. Pericentric inversion of the Y chromosome and prenatal diagnosis. *Prenat. Diagn.* 4:463–65.

Shapiro, M. M. J. 1949. The anatomy and morphology of the tubercle of Carabelli. *J. Dent. Assoc. S. Afr.* 4:355–62.

Shaw, M. W. 1984. Conditional prospective rights of the fetus. *J. Leg. Med.* 5:63–115.

Sheba, C. 1963. Environmental vs. ethnic factors determining the frequency of G-6-PD deficiency. In E. Goldschmidt, ed., *The genetics of migrant and isolate populations*, pp. 100–106. Baltimore: Williams & Wilkins.

Sheba, C., and A. Adam. 1962. A survey of some genetical characters in Ethiopian tribes. *Am. J. Phys. Anthropol.* 20:167.

Shokeir, M. H., K. D. C. Schreffler, and J. C. Gall. 1970. Two new ceruloplasmin variants in Negroes: data on three populations. *Biochem. Genet.* 4:517–28.

Shore, M. F. 1975. Psychological issues in counseling the genetically handicapped. In C. Birch and P. Abrecht, eds., *Genetics and the quality of life*, pp. 160–72. New York: Pergamon Press.

Sidel, V. W. 1985. Destruction before detonation: The impact of the arms race on health and health care. *Lancet* 2:1287–89.

Siegel, B. 1979. A racial comparison of cleft patients in a clinic population: Associated anomalies and recurrence rates. *Cleft Palate* 16:193–97.

Sigerist, H. E. 1965. *Civilization and disease.* Chicago: University of Chicago Press.

Simkiss, M., and A. Lowe. 1976. Congenital abnormalities in the African newborn. *Arch. Dis. Child.* 36:404–6.

Simoons, F. J. 1981. Geographic patterns of primary adult lactose malabsorption: A further interpretation of evidence for the Old World. In D. M. Page and T. M. Bayless, eds., *Lactose digestion*, pp. 23–48. Baltimore: Johns Hopkins University Press.

Siniscalco, M., L. Bernini, G. Filippi, B. Latte, P. M. Khan, S. Piomelli, and M. Rattazgi. 1966. Population genetics of haemoglobin variants, thalassemia and glucose-6-phosphate dehydrogenase deficiency, with particular emphasis to the malaria hypothesis. *Bull. WHO* 34:379–93.

Singer, R., and K. Kimura. 1981. Body height, weight, and skeletal maturation in Hottentot (Khoikhoi) children. *Am. J. Phys. Anthropol.* 54:401–13.

Singer, R., K. Kimura, and S. Gajisin. 1980. Brachymesophalangia V in Hottentot and "Cape Coloured" children in Namibia (South West Africa) and South Africa. *Am. J. Phys. Anthropol.* 52:533–39.

Skirving, A. P., and W. J. Scadden. 1979. The African neonatal hip and its immunity from congenital dislocation. *J. Bone Joint Surg.* 61-B:339–41.

Slaughter, C. A., D. A. Hopkinson, and H. Harris. 1975. Aconitase polymorphism in man. *Ann. Hum. Genet.* 39:193–202.

Smitham, J. H. 1948. Some observations on certain congenital abnormalities of the hand in African natives. *Br. J. Radiol.* 21:513–18.

Smith, M., D. A. Hopkinson, and H. Harris. 1971. Developmental changes and polymorphism in human alcohol dehydrogenase. *Ann. Hum. Genet.* 34:251–57.

―――. 1972. Alcohol dehydrogenase isozymes in adult human stomach and liver: Evidence for activity of the ADH 3 locus. *Ann. Hum. Genet.* 35:243–53.

―――. 1973. Studies on the subunit structure and molecular size of the human alcohol dehydrogenase isoenzymes determined by the different loci, ADH 1, ADH 2, and ADH 3. *Am. J. Hum. Genet.* 36:401–14.

Smithies, O. 1955. Zone electrophoresis in starch gels: Group variations in the serum proteins of normal human adults. *Biochem. J.* 61:629–41.

―――. 1957. Variations in human serum β-globulins. *Nature* 180:1482–84.

―――. 1958. Third allele at the serum β-globulin locus in humans. *Nature* 181:1203–6.

Smithies, O., G. E. Connell, and G. H. Dixon. 1962. Inheritance of haptoglobin subtypes. *Am. J. Hum. Genet.* 14:14–21.

―――. 1966. Gene action in the human haptoglobins. I. Dissociation into constituent polypetide chains. *J. Mol. Biol.* 21:213–24.

Smithies, O., and N. F. Walker. 1955. Genetic control of some serum proteins in normal humans. *Nature* 176:1265–69.

Snell, G. D., J. Dausset, and S. Nathanson. 1976. *Histocompatibility*. New York: Academic Press.

Snook, C. R., J. N. Mahmoud, and W. P. Chang. 1976. Lactose tolerance in adult Jordanian Arabs. *Trop. Geogr. Med.* 24:333–35.

Socha, J., J. Ksiazk, G. Flatz, and S. D. Flatz. 1984. Prevalence of primary adult lactase malabsorption in Poland. *Ann. Hum. Biol.* 11:311–16.

Sorsby, A. 1958. Noah—An albino. *Br. Med. J.* 2:1587–89.

Sorensson, J. R. 1973. Counselors: Self-portrait. *Genet. Counsel.* 1:31–35.

Sowers, M. F., and E. Winterfeldt. 1975. Lactose intolerance among Mexican Americans. *Am. J. Clin. Nutr.* 28:704–5.

Spedini, G., E. Capucci, N. Crosti, M. E. Danubio, and S. Romagnoli. 1982. Erthrocyte glyoxalase I and superoxide dismutase polymorphisms in the Mugu and some other populations of the Central African Republic. *Hum. Hered.* 32:253–58.

Spedini, G., E. Capucci, O. Richards, M. Fuciarelli, L. Graccaia, M. L. Aebischer, E. Mannella, and O. Loreti. 1981. Some genetic erythrocyte polymorphisms in the Mugu and other populations of the Central African Republic with an analysis of genetic distances. *Anthropol Anz.* 39:10–19.

Spedini, G., M. Fuciarelli, and O. Richards. 1980. The AcP polymorphism frequencies in the Mugu and Sango of Central Africa. (Correlation between the Pr allele frequencies and some climatic factors in Africa). *Ann. Hum. Biol.* 7:125–28.

Spencer, N., D. A. Hopkinson, and H. Harris. 1968. ADA polymorphism in man. *Ann. Hum. Genet.* 32:9–14.

Spielman, R. S., H. Harris, W. J. Mellman, and H. Gershowitz. 1978. Dissection of a continuous distribution: Red cell galactokinase activity in blacks. *Am. J. Hum. Genet.* 30:237–48.

Spranger, J., K. Benirsche, J. G. Hall, W. Lenz, R. B. Lowry, J. M. Opitz, L. Pinsky, H. G. Schwarzacher, and D. W. Smith. 1982. Errors of morphogenesis: Concepts and terms. *J. Pediatr.* 100:1160–65.

Spurgeon, J. H., and H. V. Meredith. 1979. Body size and form of black and white male youths: South Carolina youths compared with youths measured at earlier times and other places. *Hum. Biol.* 51:187–200.

Spurgeon, J. H., E. M. Meredith, and H. V. Meredith. 1978. Body size and form of children of predominately black ancestry living in West and Central Africa, North and South America, and the West Indies. *Ann. Hum. Biol.* 5:229–46.

Stamler, R., J. Stamler, W. F. Riedlinger, G. Algera, and R. H. Roberts. 1978. Weight and blood pressure: Findings in hypertension screening of 1 million Americans. *JAMA* 240:1607–10.

Stanbury, J. B., J. B. Wyngaarden, and D. S. Frederickson eds. 1975. *The metabolic basis of inherited disease.* 4th ed. New York: McGraw Hill.

Stannus, H. S. 1914. Congenital anomalies in a native African race. *Biometrika* 10:1–24.

Statistical Abstract of the United States. 1984. 105th ed. Washington, D.C.: Bureau of the Census.

Stavig, G. R., Igra, A., and Leonard, A. R. 1984. Hypertension among Asians and Pacific Islanders in California. *Am. J. Epidemiol.* 119:677–91.

Steinberg, F. S., J. J. Cereghino, and C. C. Plato. The dermatoglyphics of American Negroes. 1975. *Am. J. Phys. Anthropol.* 42:183–94.

Steinberg, M. H., and S. H. Embury. 1986. α-Thalassemia in blacks: Genetic and clinical aspects and interactions with the sickle hemoglobin gene. *Blood* 68:985–90.

Steinberg, M. H., W. Rosenstock, M. B. Coleman, J. G. Adams, O. Platica, M. Cedano, R. F. Rieder, J. T. Wilson, D. Milner, and S. West. 1984. Effects of thalassemia and microcytosis on the hematologic and vaso-oclusive severity of sickle cell anemia. *Blood* 63:1353–60.

Stern, C. 1953. Model estimates of the frequency of white and near-white segregants in the American Negro. *Acta Genet. (Basel)* 4:281–98.

———. 1970. Model estimates of the number of gene pairs involved in pigmentation variability of the Negro-American. *Hum. Hered.* 20:165–68.

Stevens, M. C. G., G. H. Maude, M. Beckford, Y. Grandison, Mason, B. Taylor, B. E. Serjeant, D. R. Higgs, H. Teal, D. J. Weatherall, and G. R. Serjeant. 1986. α thalassemia and the hematology of homozygous sickle cell disease in childhood. *N. Engl. J. Med.* 67:411–14.

Stevenson, A. C. 1961. Frequency of congenital and hereditary disease, with special reference to mutation. *Br. Med. Bull.* 17:254–59.

Stewart, T. D. 1962. Anterior femoral curvature: Its utility for race identification. *Hum. Biol.* 34:49–62.

Suk, V. 1919. Eruption and decay of permanent teeth in whites and Negroes, with comparative remarks on other races. *Am. J. Phys. Anthropol.* 2:351–68.

Sunderland, E., and E. Coope. 1973. The tribes of South and Central Ghana: A dermatoglyphic investigation. *Man* 8:228–65.

Svejgaard, A., P. Platz, and L. P. Ryder. 1983. HLA and disease—a survey. *Immunol. Rev.* 70:193–218.

Swan, D. A. 1964. Juan Comas on "Scientific racism again?": A scientific analysis. *Mankind Monogr.* 6:24–36.

Szabo, G., A. B. Gerald, M. A. Pathak, and T. B. Fitzpatrick. 1969. Racial differences in the fate of melanosomes in human epidermis. *Nature* 222:1081–82.

Szeinberg, A., Y. Asher, and C. Sheba. 1958. Studies on glutathione stability in erythrocytes of cases with past history of favism or sulfa-induced hemolysis. *Blood* 13:348–58.

Takahashi, N., T. L. Ortel, and F. W. Putnam. 1984. Single-chain structure of human ceruloplasmin: The complete amino acid sequence of the whole molecule. *Proc. Nat. Acad. Sci. (U.S.A.)* 81:390–94

Takenori, T., I.-Y. Huang, T. Ikuta, and A. Yoshida. 1986. Human glucose-6-phosphate dehydrogenase: Primary structure and cDNA cloning. *Proc. Natl. Acad. Sci. (U.S.A.)* 83:4157–61.

Takizawa, T. and A. Yoshida. 1987. Molecular abnormality of the common glucose-6-phospate dehydrogenase variant A(+), and restriction-fragment-length polymorphism *Am. J. Hum. Genet.* (Abstr.) 41:A241.

Taleb, N., J. L. Loiselet, F. Guorra, and H. Sebir. 1964. Sur la déficiencesiens glucose-6-phosphate-dehydrogenase dans les populations authochones du Liban. *C R Acad. Sci. (Paris)* 258:5449–51.

Tandon, R. K., J. K. Joshi, D. S. Singh, M. Narendranathan, and V. Balakrishan. 1981. Lactose intolerance in North and South Indians. *Am. J. Clin. Nutr.* 34:943–46.

Tarlov, A. R., G. J. Brewer, P. E. Carson, and A. S. Alving. 1962. Primaquine sensitivity: Glucose-6-phosphate dehydrogenase deficiency, an inborn error of metabolism of medical and biological significance. *Arch. Int. Med.* 109:209–34.

Tentori, L., and M. Marinucci. 1983. Hemoglobinopathies and thalassemias in Italy and Northern Africa. In J. E. Bowman, ed., *Distribution and evolution of hemoglobin and globin loci*, pp. 299–313. New York: Elsevier.

Thomson, G. E. 1980. Hypertension: Implications of comparisons among blacks and whites. *Urban Health* 9:31–41.

Tills, D., A. C. Kopec, A. Warlow, N. A. Barnicot, A. E. Mourant, A. Marin, F. J. Bennett, and J. C. Woodburn. 1982. Blood group, protein, and red cell enzyme polymorphisms of the Hadza of Tanzania. *Hum. Genet.* 61:525–29.

Tipler, T. D., D. S. Dunn, and T. Jenkins. 1982. Phosphoglucomutase first locus polymorphism as revealed by isoelectric focussing in South Africa. *Hum. Hered.* 32:80–93.

Tobian, L. 1978. Hypertension and obesity. *N. Engl. J. Med.* 298:46–47.

Tobias, P. V. 1970. Brain-size, grey matter and race—fact or fiction. *Amer. J. Phys. Anthropol.* 32:3–26.

Topping, M. O., and W. M. Watkins. 1975. Isoelectric points of the human blood group A1 and A2 and B gene-associated glycosyl transferases in ovarian cyst fluids and serum. *Biochem. Biophys. Res. Commun.* 64:89–96.

Trinh-Dinh-Khoi, I., D. Glaise, A. Le Treut, R. Fauchet, Y. Godin, and J. Y. Le

Gall. 1979. Genetic polymorphism of alpha-L-fucosidase in Brittany (France). *Hum. Genet.* 51:293–96.

Troller, H. A., and S. A. James. 1978. Blood pressure and skin color. *Am. J. Public Health* 68:1170–72.

Truswell, A. S., B. M. Kennelly, J. D. L. Hansen, and R. B. Lee. 1972. Blood pressures of !Kung Bushmen in Northern Botswana. *S. Afr. Med. J.* 84:5–12.

Turner, C. G., and H. Hanihara. 1977. Additional features of Ainu dentition. V. Peopling of the Pacific. *Am. J. Phys. Anthropol.* 46:13–24.

Turner, C. G., and D. R. Swindler. 1978. The dentition of New Britain West Nakanai Melanesians. *Am. J. Phys. Anthropol.* 49:361–72.

Usanga, E. A., and L. Luzzatto. 1985. Adaptation of Plasmodium falciparum to glucose-6-phosphate dehydrogenase-deficient host red cells by production of parasite-coded enzyme. *Nature* 313:793–95.

Vance, J. M., M. A. Pericak-Vance, R. C. Elston, P. M. Conneally, K. K. NambooDoro, R. S. Wappner, and P. L. Yu. 1980. Evidence of genetic variation for alpha-N-acetyl-D-glucosaminidase in black and white populations: A new polymorphism. *Am. J. Med. Genet.* 7:131–40.

van Loghem, E., L. Salimonu, A. I. Williams, B. O. Osunkoya, A. M. Boyd, G. de Lange, and L. E. Nijenhuis. 1978. Immunoglobulin allotypes in African populations. I. Gm-Am haplotypes in a Nigerian population. *J. Immunogenet.* 5:143–47.

Vecchi, F. 1981. Geographical variation of digital dermatoglyphics in Africa. *Am. J. Phys. Anthropol.* 54:565–80.

Veiga, R. V., and R. E. Taylor. 1986. Beta blockers, hypertension, and blacks—is the answer really in? *J. Nat. Med. Assoc.* 78:851–56.

Vergnes, H., A. Sevin, J. Sevin, and G. Jaeger. 1979. Population genetic studies of the Aka pygmies (Central Africa): A survey of red cell and serum enzymes. *Hum. Genet.* 48:343–55.

Verma, R. S., J. Rodriguez, and H. Dosik. 1979. Human chromosome heteromorphisms in American blacks. II. Higher incidence of pericentric inversions of secondary constriction regions (h). *Am. J. Med. Genet.* 8:17–25.

Verma, R. S., and H. Dosik. 1981 Human chromosomal heteromorphism in American blacks. V. Racial differences in size variation of the short arm of acrocentric chromosomes. *Experientia* 37:241–43.

Vogel, F., and A. G. Motulsky. 1979. *Human genetics: Problems and approaches*, p. 203. New York: Springer-Verlag.

Von Wartburg, J. P., J. Papenberg, and H. Aebi. 1965. An atypical human alcohol dehydrogenase. *Can. J. Biochem.* 43:889–98.

Von Wuthenau, A. 1975. *Unexpected faces in ancient America: 1500 B.C.–A.D. 1500*. New York: Crown Publishers.

Vrydagh-Laourex, S. 1979. Digital and palmar dermatoglyphics in a sample of Moroccans. *Hum. Biol.* 51:537–49.

Vulliamy, T. J., M. D'Urso, G. Battistuzzi, M. Estrada, N. S. Foulkes, G. Martini, V. Calabro, V. Poggi, R. Giordano, M. Town, L. Luzzatto, and M. G. Persico. 1988. Diverse point mutations in the human glucose-6-phosphate dehydrogenase gene cause enzyme deficiency and mild or severe hemolytic anemia. 1988. *Proc. Nat. Acad. Sci. (U.S.A.)* 85:5171–75.

Wagner, D. K., and L. L. Cavalli-Sforza. 1975. Ethnic variation in genetic disease: Possible roles of hitchhiking and epistasis. *Am. J. Hum. Genet.* 27:348–64.

Wainscoat, J. S., J. I. Bell, S. L. Thein, D. R. Higgs, G. R. Serjeant, T. E. A. Peto, and D. J. Weatherall. 1983. Multiple origins of the sickle mutation: Evidence from β^S globin gene cluster polymorphisms. *Mol. Biol. Med.* 1:191–97.

Walaas, E., R. A. Lovstad, and O. Walaas. 1967. Interaction of dimethyl-p-phenylenediamine with ceruloplasmin. *Arch. Biochem. Biophys.* 121:480–85.

Walensky, N. A. 1965. A study of anterior femoral curvature in man. *Anat. Rec.* 151:559–70.

Walensky, N. A., and M. P. O'Brien. 1968. Anatomical factors relative to the racial selectivity of femoral neck fracture. *Am. J. Phys. Anthropol.* 28:93–96.

Walker, D. G., and J. E. Bowman. 1960. In vitro effect of Vicia faba extracts upon reduced glutathione of erythrocytes. *Proc. Soc. Exp. Biol. Med.* 103:476–77.

Walsh, J. 1986. River blindness: A gamble pays off. *Science* 232:922–25.

Wang, A.-C., and H. E. Sutton. 1965. Human transferrins C and D1: Chemical difference in a peptide. *Science* 149:435–37.

Wang, A,.-C., H. E. Sutton, and P. N. Howard. 1967. Human transferrins C and DCHI: An amino acid difference. *Biochem. Genet.* 1:55–60.

Wang, A.-C., H. E. Sutton, and A. Riggs. 1966. A chemical difference between transferrins B2 and C. *Am. J. Hum. Genet.* 18:454–58

Wang, Y. G., Y. S. Yan, J. J. Xu, R. F. Du, S. D. Flatz, W. Kuhnau, and G. Flatz. 1984. Prevalence of primary adult lactose malabsorption in three populations of Northern China. *Hum. Genet.* 67:103–6.

Warkany, J. 1971. *Congenital malformations.* Chicago: Year Book Medical Publishers.

Wasi, P. 1983. Hemoglobinopathies in Southeast Asia. In J. E. Bowman, ed., *Distribution and evolution of hemoglobin and globin loci,* pp. 179–208. New York: Elsevier.

Watkins, W. M. 1980. Biochemistry and genetics of the ABO, Lewis and P blood group systems. In H. Harris and K. Hirschhorn, eds., *Advances in human genetics,* 10:1–136. New York: Plenum Press.

Weatherall, D. J. 1983. The diagnostic features of the different features of thalassemias. In D. J. Weatherall, ed., *The thalassemias,* pp. 1–26. New York: Churchill Livingstone.

Weatherall, D. J. 1984. DNA in medicine: Implications for medical practice and human biology. *Lancet* 2:1440–1444.

Weatherall, D. J., J. Mold, S. L. Thein, J. S. Wainscoat, and J. B. Clegg. 1985. Prenatal diagnosis of the common hemoglobin disorders. *J. Med. Genet.* 22:422–30.

Welch, S. G., C. A. Swindlehurst, I. A. McGregor, and K. Williams. 1979. Serum protein polymorphisms in a village community from the Gambia, West Africa (Hp, Tf and Gc). *Hum. Genet.* 48:81–84.

Welch, S. G. , I. A. McGregor, and K. Williams. 1980. Alpha 1-antitrypsin (Pi) phenotypes in a village population from the Gambia, West Africa: Evidence of a new variant occurring at a polymorphic frequency. *Hum. Genet.* 54:119–24.

Wetherington, R. K. 1960. A note on the fusion of the lunate and triquetral centers. *Am. J. Phys. Anthropol.* 18:251–53.

Wijsman, E., and W. A. Neves. 1986. The use of nonmetric traits in estimating human population admixture: A test case with Brazilian, blacks, whites, and mulattos. *Am. J. Phys. Anthropol.* 70:395–405.

Willkox, M. 1984. α-Thalassemia and the malaria hypothesis. *Lancet* 2:980.

Williams, A. W. 1969. Blood pressure differences in Kikuyu and Samburu communities in Kenya. *E. Afr. Med. J.* 46:262–72.

Williams, B. 1972. *Morality: An introduction to ethics* New York: Harper & Row.

Williams, C. 1976. *The destruction of black civilization: Great issues of a race from 4500 B.C. to 2000 A.D.* Chicago: Third World Press.

Williams, E. 1984. *From Columbus to Castro: The history of the Caribbean.* New York: Vintage Books.

Wilson, T. 1961. Malaria and glucose-6-phosphate dehydrogenase. *Br. Med. J.* 2:245–46; 895–96.

Wilson, T. W. 1986. History of salt supplies in West Africa and blood pressures today. *Lancet* 1:784–85.

Wilson, W. J. 1980. *The declining significance of race.* 2d. ed. Chicago: University of Chicago Press.

Winter, W. E., N. K. Maclaren, W. J. Riley, D. W. Clarke, M. S. Kappy, and R. P. Spillar. 1987. Maturity-onset diabetes of youth in black Americans. *N. Engl. J. Med.* 316:285–91.

Witkop, C. J. 1973. Albinism. In H. Harris and K. Hirschhorn, eds., *Advances in Human Genetics.* 2:61–142. New York: Plenum Press.

Woolf, C. M. 1964. Albinism among Indians in Arizona and New Mexico. *Am. J. Hum. Genet.* 17:23–35.

———. 1971. Congenital cleft lip: A genetic study of 496 propositi. *J. Med. Genet.* 8:65–83.

Woolhouse, N. M., B. Andoh, A. Mahgoub, T. P. Sloan, J. R. Idle, R. L. Smith. 1979. Debrisoquin hydroxylation polymorphism among Ghanaians and Caucasians. *Clin. Pharmacol. Ther.* 26:584–91.

Woolridge, E. Q. and R. F. Murray, Jr. 1988. The health orientation scale: A measure of feelings about sickle cell trait. *Soc. Biol.* 35:123–26.

World malaria situation 1986–1987. 1989. *Week. Epidemiol. Rec.* 64(32–33): 241–54.

Woteki, C. E., E. Weser, and E. A. Young. 1977. Lactose malabsorption in Mexican American adults. *Am. J. Clin. Nutr.* 30:470–75.

Yin, S. J., W. F. Bosron, T. K. Li, K. Ohnmishi, K. Okuda, H. Ishii, and M. Tschiya. 1984. Polymorphism of human liver alcohol dehydrogenase identification of ADH 2–1 and ADH 2–2 phenotypes in Japanese by isoelectric focussing. *Biochem. Genet.* 22:169–80.

Yoshida, A. 1967. A single amino acid substitution (asparagine to aspartic acid) between normal B+) and the common Negro variant (A+) of human glucose-6-phospate dehydrogenase. *Proc. Nat. Acad. Sci. (U.S.A.)* 57:835–40.

———. 1978. Tabulation of human glucose-6-phosphate dehydrogenase variants. In V. A. McKusick, *Mendelian inheritance in Man*, 5th ed., pp. 732–45. Baltimore: Johns Hopkins University Press,

Yoshida, A., and E. Beutler. 1978. Human glucose-6-phosphate variants: Supplementary tabulation. *Ann. Hum. Genet.* 41:347–55.

Yoshida, A., and E. Beutler. 1982. Tabulation of human glucose-6-phosphate dehydrogenase variants. In V. A. Mckusick, *Mendelian inheritance in man*, 6th ed., pp. 1110–29. Baltimore: Johns Hopkins University Press.

Yoshida, A., E. Beutler, and A. G. Motulsky. 1971. Table of human glucose-6-phosphate dehydrogenase variants. *Bull. WHO* 45:243–53.

Yoshida, A., and E. F. Roth. 1987. Glucose-6-phosphate dehydrogenase of malaria parasite Plasmodium falciparum. *Blood* 69:1528–30.

Yoshida, A., T. Takizawa, and J. T. Prchal. 1988. RFLP of the X chromosome-linked glucose-6-phosphate dehydrogenase locus in blacks. *Am. J. Hum. Genet.* 42:872–76.

Ziai, M., G. H. Amirhakimi, J. G. Reingold, M. Tabatabaee, M. E. Gettner, and J. E. Bowman. 1967. Malaria prophylaxis and treatment in G-6-PD deficiency: An observation on the toxicity of primaquine and chloroquine. *Clin. Pediatr.* 6:242–43.

Zuckerman, A. J. 1977. Symposium on liver carcinoma Hepatocallular carcinoma and hepatitis B. *Trans. R. Soc. Trop. Med. Hyg.* 71:459–61.

Zuckerman, A. J. 1982. Primary hepatocellular carcinoma and hepatitis B virus. *Trans. R. Soc. Trop. Med. Hyg.* 76:711–18.

INDEX

Abetalipoproteinemia, infrequency in blacks, 397

Abortion, in U.S.: *Harris v. McRae,* and restriction of abortion services, 368; Hyde amendment, 368; *Maher v. Roe,* and denial of state aid for, 367–68; and possible effect on siblings, 372; *Roe v. Wade,* 368

Acatalassemia, infrequency in blacks, 397

Acetyl transferase: population differences, 140; population frequencies, 403

Acid phosphatase, alleles, phenotypes, and population studies, 132–33, 402

Aconitase: in Nigerians, 402; and population studies, 132

Adenylate kinase, variants and population differences, 133–34, 402. *See also* Admixture estimates

Admixture estimates: adenylate kinase, 134, 167–68; cystic fibrosis and (*see* Cystic fibrosis), DNA haplotypes and, 168–69; galactokinase and, 168; glucose-6-phosphate dehydrogenase deficiency and, 168; hemoglobin S and, 168; problems of calculation of, 165–70; and São Paulo mulattos, 29

Afghanistan, lactose malabsorption in various populations in, 255

Africa: eras of history of, 1; language families in, 3; migrations within, 1; peoples of, 1–3

Africans, pre-Columbian Africans in Latin America and Caribbean, 11–15

AIDS: genetic component, 336–42; incidence of in Africa, 340–41; Kaposi's sarcoma and, 338; nomenclature, 337; opportunistic infections and population differences, 337; patterns, region-al, of, 336; problems in simian and, 340; sources of, on various continents, 336; transmission of, 336–37, 338–39

Ainu: brachymesophalangia V in, 57; Carabelli's trait in, 51; high frequency of torus palatinus in, 50; mylohyoid bridge in, 54

Albinism, ocular autosomal recessive, clinical characteristics of in whites, 91

Albinism, oculocutaneous: African names for, 84; biochemical and clinical forms, 85–90; brown, tyrosinase negative and positive, yellow mutant, 86–87; classification of, 84; consanguinuity and, 88; definition of, 83; inheritance of, 87; morphologic patterns and population frequencies of, 85–90; precancerous lesions and skin cancer in, 91–94. *See also* Nigeria

Albinism, X-linked: clinical differences between blacks and whites, 90–91

Alcohol dehydrogenase: frequency in blacks, 403; population studies, 137–38

Algeria, *Hpa*I β-globin gene polymorphism in, 209

Alpha-N-acetyl-D-glucosaminidase deficiency (Sanfillippo syndrome type B), in U.S. blacks and whites, 140, 403

Alpha-1-antitrypsin, 147; deficiency, 397; population studies, 148; variants and prevalence of some alleles of, 404

Amerindians: albinism of, in New Mexico and Arizona, 89–90; blood group *A* allele in, 125–26; brachymeso-phalangia V in Blackfeet, Peru, and Pima, 57; canine distal accessory ridge in, 46–47; Carabelli's trait in Lengua and Pima, 51; ceruloplasmin in, 147;

James E. Bowman, M.D., received both his undergraduate and medical degrees from Howard University. He has served on advisory committees to the Food and Drug Administration, the National Institutes of Health, the Centers for Disease Control, and the Office of Technology Assessment of the U.S. Congress, and has served on a number of international appointments. He is a fellow of the Hastings Center and a member of the American Society of Human Genetics Committee on Cystic Fibrosis. Previously, he has published more than eighty scientific papers as well as edited the book *Distribution and Evolution of Hemoglobin and Globin Loci.* The subjects of his writings have included the areas of hematological population genetic studies, genetic variations among diverse peoples, and ethical, legal, and public policy issues in human genetics programs.

Robert F. Murray, Jr., M.D., M.S., is a graduate of Union College, Schenectedy, NY, received his master's degree in genetics from the University of Washington, and his medical degree from the University of Rochester School of Medicine. He has served on National Institutes of Health advisory committees as well as on the Bioethics Advisory Committee to the Secretary of the U.S. Department of Health, Education, and Welfare, is a fellow and director of the Hastings Center, and a fellow of the Institute of Medicine of the National Academy of Sciences. Previously, he co-edited *Genetic, Metabolic and Developmental Aspects of Mental Retardation* (with P.L. Rosser).

Designed by Nighthawk Design

Composed by The Composing Room of Michigan, Inc.
in Baskerville text and Optima display.
Printed by the Maple Press Company
on 60-lb. Glatfelter Hi-Brite Offset
and bound in Holliston Roxite B.